WATER: ECONOMICS, MANAGEMENT AND DEMAND

T0136296

Other titles from E & FN Spon

Clay's Handbook of Environmental Health
17th Edition
W.H. Bassett

The Coliform Index and Waterborne Disease
C. Gleeson and N. Gray

Ecological Effects of Wastewater
2nd Edition
E.B. Welch

Hydraulics in Civil and Environmental Engineering
2nd Edition
A. Chadwick and J. Morfett

Hydraulic Structures
2nd Edition
P. Novak, A. Moffat, C. Nalluri and R. Naryanan

International River Water Quality
G. Best, T. Bogacka and E. Neimircyz

Water Policy
P. Howsam and R. Carter

Water Pollution Control
R. Helmer and I. Hespanhol

Water Quality Assessments
2nd Edition
D. Chapman

Water Quality Monitoring
J. Bartram and R. Ballance

For more information about these and other titles please contact:
The Marketing Department, E & FN Spon, 2-6 Boundary Row, London, SE1 8HN. Tel: 0171 865 0066

WATER: ECONOMICS, MANAGEMENT AND DEMAND

Edited by

Melvyn Kay
Silsoe College, Cranfield University, UK

Tom Franks
University of Bradford, UK

and

Laurence Smith
Wye College, University of London, UK

CRC Press
Taylor & Francis Group
Boca Raton London New York

CRC Press is an imprint of the
Taylor & Francis Group, an **informa** business

CRC Press
Taylor & Francis Group
6000 Broken Sound Parkway NW, Suite 300
Boca Raton, FL 33487-2742

First issued in paperback 2019

ISBN-13: 978-0-419-21840-1 (hbk)
ISBN-13: 978-0-367-44783-0 (pbk)

A catalogue record for this book is available from the British Library

Publisher's note
This book has been produced from camera ready copy supplied by the individual contributors in order to make the book available for the conference.

Visit the Taylor & Francis Web site at
http://www.taylorandfrancis.com

and the CRC Press Web site at
http://www.crcpress.com

CONTENTS

PREFACE

Water is often regarded as a naturally occurring, freely available commodity. However, its supply, especially in an unpolluted state, is often limited compared to the demands placed upon it. Hence, water is a constraint on economic activity and decisions are required with respect to its allocation between competing uses. Furthermore, considerable resources are committed to delivering required standards of water services in terms of adequate quantities and qualities. In addition, water related environmental qualities, such as those associated with wetland regimes, are recognised as having a value to society. These characteristics of water imply that it can be regarded as an economic commodity - an economic good - whose use can be guided by economic principles.

Irrigation is a major player in the demand for water and already accounts for between 70-80% of the total world consumption. It is concerned with water management for crop production in conditions of water deficit which implies a high value for extra water. Drainage too is a key player but is concerned with managing a water surplus which implies negative values for water, whereby the removal of excess water by field drainage and flood defence generates extra benefit.

This is consistent with the Dublin Statement made in January 1992 and developed in the Agenda Item 21 of the United Nations Conference of the Environment held in Rio de Janeiro in 1992 which addressed the need for the integrated development, management and use of water resources. Furthermore the European Ministers of Environments' conference held in Lucerne in 1993 emphasised the need for careful management and protection of water resources.

The papers selected for presentation at this ICID conference focus primarily on the role of irrigation and drainage in the debate on water use as an economic good. They highlight experiences in both developed and developing countries in six key areas:

- the value of water for irrigation
- the value of drainage and flood control
- the social and environmental value of water
- paying for services
- management systems and
- policy - legal and institutional issues.

Melvyn Kay, Tom Franks and Laurence Smith
June 1997

SECTION A

THE VALUE OF WATER FOR IRRIGATION

WATER AS AN ECONOMIC GOOD: A SOLUTION, OR A PROBLEM?

C. J. PERRY
International Irrigation Management Institute, Colombo, Sri Lanka
M. ROCK and D. SECKLER
Winrock International Institute for Agricultural Development, Arlington, USA

Abstract

There is wide interest in, and support for the idea of treating water as an economic good. However, the role of water—as a basic need, a merit good, and a social, economic, financial and environmental resource—make the selection of an appropriate set of prices exceptionally difficult. Further, the application of price-based instruments, once an appropriate value system has been agreed, is particularly difficult in the case of water, because the flow of water through a basin is complex, and provides wide scope for externalities, market failure, and high transaction costs. While judiciously applied market tools can be expected to have benefits, in many cases the necessary and sufficient conditions, especially defined and enforced water rights, are not yet in place. In the absence of such preconditions, interventions that introduce market forces may have unpredictable, and possible very negative effects.
Keywords: Water management, water as an economic good, economic analysis

1   Introduction

The now famous proclamation that water should be treated "as an economic good" originated in the Dublin Conference [1]. The proclamation was a compromise between those, mainly economists, who wanted to treat water in the same way as other private goods, subject to allocation through competitive market pricing, and those who wanted to treat water as a basic human need that should be largely exempted from competitive market pricing and allocation.

This distinction is further complicated by the distinction in economics between the "economic" value of a good and its "financial" value. The two values rarely correspond, and as will be argued, for water the divergences are exceptionally complex and important. Thus it does not follow from the declaration that water is an economic good that it should be allocated by competitive market prices that reflect only financial, and not necessarily economic, social, or environmental, values.

Water: Economics, Management and Demand. Edited by Melvyn Kay, Tom Franks and Laurence Smith. Published in 1997 by E & FN Spon. ISBN 0 419 21840 8.

The question is not whether or not water is an economic good—as shown below, it certainly is an economic good in most cases, like almost everything else we have to worry about. Rather the question is whether it is an essentially *private* good that can reasonably be left to free market forces, or a *public* good that requires some degree of extra-market management to effectively and efficiently serve social objectives.

Water is far too important to its users to be the basis for socio-economic experiments. Much is already known about the nature of successful policies and procedures for allocating water: understanding and incorporating the implications of this knowledge will avoid some potentially enormous financial, economic, and social costs.

2   Is water an economic good?

Robbins' [2] defined economics as "the science which studies human behaviour as a relationship between ends and scarce means which have alternative uses." Water meets these requirements: it serves a multiplicity of ends (drinking and bathing, irrigation and industrial uses, power generation; recreation and environmental use, waste disposal), and   thus   satisfies   the   condition   of   "alternative uses." Increasingly too, water is scarce in the sense that it cannot fully satisfy all its alternative uses simultaneously.

In sum, there can be no doubt that water is an economic good—a fact with which every irrigator who has fought for water, stolen it, or had it stolen from him, every women who has walked or hours to fetch drinking water from a well, and every irrigation official who been offered a bribe to divert water, is fully conversant, in practice, if not in theory.

The important question is not whether or not water is an economic good, but what *kind of economic good* it is.

In the following pages we address the question: What does it mean to treat water in irrigated agriculture as an economic good? We focus on irrigated agriculture for several reasons. First, it is the largest consumer of water virtually everywhere that irrigation is practiced. Second, policies governing water use in irrigated agriculture are fraught with disagreements over both values (should it be treated as a purely private good, a public good, or a basic human need) and facts (what is the most cost-effective allocation policy and method if water is treated as a purely private good, or a public good, and in what instances should it be treated in one way rather than the other way).

The argument proceeds in three steps. Section 3 examines the economic analysis of different values involved in treating water as a private and public good. Section 4 then examines the facts involved in this debate in terms of the causes of both market failure and public failure. Section 5 summarises the present state of knowledge, and Section 6 suggests a practical plan of action.

3 Values versus Prices

The proponents of water as a private good contend that water is just like any other good, that its production allocation should be determined by the overriding value of *consumer's sovereignty*—"The value of water to a user is the maximum amount the user would be willing to pay for the use of the resource." [3]. Using this clear and simple criterion to allocate water totally ignores a number of alternative and arguably equally valid value sets: If water is a basic need, its allocation should first meet that need for each member of society; if water is a merit good, its allocation should be such as to ensure that its merit impact is achieved; if water serves environmental purposes, its allocation should reflect national (or even international) preferences for environmental protection and improvement; if certain uses of water today have significant implications for future use, allocation mechanisms should reflect such issues; if individual incentives for water use result in collective actions that damage society's interest, then constraints must be placed on individual behavior.

Where market-determined prices are allowed to dominate the allocation of a resource, market (price) equilibrium results in a balance between supply and demand that maximizes consumer welfare; divergence from this price causes 'deadweight' losses (Figure 1). With the price held at P^1, below the equilibrium price, demand rises from X^* to X^1, and the increase in costs thus incurred exceeds the increase in benefits to society, measured at financial prices.

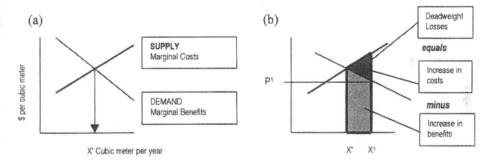

Fig. 1. Equilibrium Supply and Demand, and "Deadweight losses"

Of course, if the good in question is basic food, and the price is held below the market-determined price, we can translate the "deadweight loss" into numbers of poor people whose needs have been more fully met—a gain rather than a loss in the view of many. But these arguments are well known and documented; our point more generally is that is no simple set of prices, especially in the case of water, is likely to provide the basis for the introduction of economic incentives to ensure optimum water allocation and use.

However, if, like good economists, we assume away the difficulties of defining the price set, there can be still be important disagreements on potential impacts. There are several well-known conditions under which economic instruments and markets fail to function effectively. We now review some of the main options being considered to move away from public sector allocation of water towards privatization.

4   Facts: Public failure and market failure

Given the generally dismal record of the public sector in the field of water resources management, one must sympathize with anyone who is interested in alternative institutional arrangements for water management. Many have understandably turned to economic instruments, market systems, and prices as an alternative on the persuasive grounds that the market works so spectacularly well in hosts of other areas of economic activity. But even if one accepts the value of consumer's sovereignty without qualification, water is unfortunately a field beset with the classic problems of market failure. In this section, we first examine the rationale for privatization, the approaches that can be followed, and their appropriateness. We then turn to the question of market failure—a critical problem in the debate about the role of economic instruments in the management of water resources.

4.1   Privatizing irrigation systems

In the public sector truly, "the road to hell is paved with good intentions." While there are some notable exceptions to the rule, the public sector has generally performed miserably in all forms of water management—whether in irrigation, or in domestic and industrial water supplies, or in protecting resources and environmental quality. Some of the well-known and interrelated ills of the public sector can be listed readily:

- Rent seeking, either economically, in the form of direct bribes and corruption, or socio-politically in the form of empire building, high costs, and excessive supplies [4 and 5]
- The divorce of incentives from performance—indeed, sometimes almost an inverse relationship.
- Capture of public agencies and funds by politically powerful interests and their clients.
- Administrative operations, "by the book," rather than management in terms of objectives and results.

The problems with public sector management and allocation of water have created the movement toward privatizing irrigation systems [6]. Privatization can take several forms, from turnover of operation and maintenance to farmer associations, to volumetric or quasi-volumetric pricing at the farm level, to development of water markets and tradable water rights[1] that would allow water to flow to the highest-value uses.

Which of these privatization options is best? A few aspects are clear.

An important size problem—the small size of most farms in most irrigation command areas in developing countries—virtually rules out volumetric metering at the farm level. In most instances, the incremental costs of the infrastructure, , and administration required for volumetric metering at the farm level will exceed marginal benefits—especially where systems have not been designed and constructed to meet these objectives. Quasi-pricing schemes (such as a crop charge linked to evapotranspiration) can, to some degree, overcome these problems. An additional problem in existing irrigation systems is that the value of water rights is capitalized in the value of agricultural land. When water is volumetrically priced by metering or quasi-priced, full marginal pricing amounts to the expropriation of those rights (with consequent capital losses in irrigated land). Farmers who have purchased land on the basis of the accepted, if not legally specified, water entitlement will strongly oppose this. It should also be noted that the levels of prices required to induce a significant demand response are high, indeed extremely high in relation to the level of charges required simply to recover O&M costs (and that level is politically infeasible in many countries).

In principle, turnover is a desirable option. It can reduce an onerous financial burden of financially strapped public irrigation agencies. It can provide a locally negotiated basis to link services to user charges, contributing to an improvement in the quality of services. And it can simplify irrigation users' involvement in investment decisions: Two drawbacks characterize turnover. First, there is little evidence that turnover increases productivity [7]. Second, there are serious questions regarding the long-term sustainability of turned-over systems, because farmers may not pay their dues and the dues do not include adequate provision for replacement of major facilities [16]. And often the withdrawal of the government agency leaves important gaps in oversight areas—overexploitation of groundwater; pollution of canals, drains, and aquifers; dam safety; competition among urban and agricultural demands; and drought planning.

This leaves water markets. At the local level of farmers trading small amounts of water on the watercourse and buying and selling water from tube wells, water markets are already thriving in most irrigation systems and should be encouraged. Water prices and markets can also serve a valuable function in larger transfers, whether within or among sectors under the appropriate conditions, noted before, and with suitable regulation, as discussed in the last section.

4.2   Water resource management and market failure
In the following discussion, we shall examine three of the major causes of market failure in irrigation and water resource management, "externalities," "transaction costs," and "property rights."
Irrigation water and externalities
External effects may be defined as uncompensated costs or benefits incurred by one party by virtue of the activities of another party. There are few, if any, economic activities that have as high an incidence of external effects, both costs and benefits, as water.

One of the most important, yet least appreciated, facts about water is that in a basin, a substantial amount of it is recycled. When water is diverted from a stream (or pumped from groundwater) for use in agriculture or other activities, some of it is consumed—for example, through evaporation—and some, returns via drains or percolation to the stream or aquifer, thus becoming available for a further cycle of diversion at another time, another place, and at another quality. Eventually, as demand increases, the fresh water available in a stream is fully utilized—all outflows are of sufficiently poor quality, or to places where the cost of recovery is too high, to be justified for potential uses. Once this happens a water basin is "closed" and no usable water supply is left.

The description set out above is highly simplified but indicates that if we wish to privatize the allocation of water, the first change that needs to be made is to base accounting not diverted water, but consumed water, with additional adjustments for time, location, and quality.

The importance of accounting for water in terms of consumption is well understood in the western US. There, there is substantial resistance to transfers of water out of irrigation districts because it is realized that once trading starts all of the secondary effects of recharging aquifers and recycling are disturbed.

Because of this problem, some states in the US now assign property rights on the basis of consumptive use rather than water diversions, for purposes of sale and transfer [9]. In some instances, state laws expand this by requiring irrigation districts to document reductions in consumption at the local level before water can be traded. While this is a step in the right direction, it does not completely solve the problem of externalities. If farmers sell the consumptive use right, they will no longer irrigate. In this situation, losses in canals will tend to form a higher proportion of deliveries; the fixed operating costs of the system must be borne by fewer users; aquifer recharge is reduced, and other farmers, who depended on return flows will have to reduce their irrigation accordingly. Keeping these accounts in order presents a major challenge.

Privatization in water and transactions costs

The second efficiency problem associated with treating irrigation water as an economic good relates to transaction costs. In many irrigation projects, the irrigation infrastructure—both the physical and infrastructure—required to allow delivery of water to serve market purposes—that is, to price water at its marginal cost—is entirely absent. Perry [10] recently calculated the cost of introducing the infrastructure required to give farm-level measurements of water deliveries in Egypt. By his estimate, they were such as to more than offset the benefits that would flow from being able to set prices closer to marginal cost. For the most part, in the real world, water is allocated first to municipal and domestic use, second to industrial and commercial use, and third to agriculture. (Environmental allocations are also growing in volume and priority.) This sequence of priorities is generally consistent with social and economic objectives that many would share; the question then is what would the incremental benefit be of fully liberating the allocation process (with agreed values structures!), and what would the infrastructural and transaction costs be.

Property rights

The third point that needs to be considered is that effective water markets and water pricing are utterly dependent on secure and effective property rights in water. Yet many would agree that the single greatest problem in water resource management in the developing world is that property rights in water are very insecure and ineffective. Or said another way, a central problem of irrigation performance is the ability of some farmers to steal water from other farmers.

Since property rights in water are not, in this first instance, secure or effective, it is difficult to see how privatization will contribute to more efficient allocations of water use unless substantial efforts (and costs) are made in advance to establish and protect property rights.

5   Summing Up

How might these considerations of water as an economic good affect water policies, particularly those in irrigated agriculture? There are two answers to this question. First, it is important to recognize that much of the discussion of irrigated agriculture is taking place under conditions of conflict—where there is intense disagreement over values, and facts are unknown. This is what makes working in water so interesting and exciting. Second, we propose a conceptually simple solution to this problem: To designers, constructors, and managers (including farmers) of irrigation systems we say: consider water as a basic human need; be sure to internalize externalities; take account of transaction costs and pay close attention to cost-effectiveness and economic efficiency. Doing these things and doing them well lead to different policy recommendations in different places.

Finally, it is important to take advantage of opportunities to enhance efficiency by not letting the best be the enemy of the good. This might require recognizing, for example, that user fees for operation and maintenance costs and public sector management may be more efficient than turnover, even though turnover is attractive on other grounds. Or it might require recognizing the physical limits of delivery systems, at least in the short run. This may mean, for example, that gains in economic efficiency from water trades in supply-based irrigation systems that ration water allocations might be limited to those between farmers within given watercourses.

If all of these things are done and done well, we stand a better chance of meeting our objectives. But if they are not, or if the current fad or ideology (of getting prices right in water) replaces a search for more understanding, we may find ourselves no better off, or worse off, a decade from now.

6   Conclusion: Toward improved water management

In our arguments above, we have concluded that the case for privatization and the introduction of economic forces into the allocation of water is a priori neither good nor bad. Those who argue the merits of such initiatives on the basis of economic dogma are probably wrong; those who reject the potentially beneficial role of markets in the allocation of water among competing users will also often be wrong.

But our position can be stated more positively, and to do this we define a necessary and sequential set of preconditions for the beneficial introduction of market forces into the allocation of water, namely that:

- the entitlement of all users under all levels of resource availability are defined
- infrastructure is in place to deliver the defined entitlements
- measurement standards are acceptable to the delivering agency and users
- effective recourse is available to those who do not receive their entitlements
- effective recourse is available to third parties affected by changes in use
- reallocations of water can be measured and delivered, and third-party impacts (in quality, time, quantity, and place) can be identified
- users must be legally obligated to pay their user fees through effective legal and police procedures
- large-scale transfers of water with and between sectors must be subject to approval and relevant charges by regulatory agencies.

With these sequentially interdependent preconditions in place, we believe that the privatization of water allocation would have significant benefits; in their absence impacts are uncertain. We also believe the implications of running experiments with peoples' livelihoods, especially where water is involved, to be unacceptable.

Our hypothesis is that development and the efficient use of water will be better served by the widespread, indeed universal introduction of the necessary underpinnings and prerequisites to good water management (assigned water rights, delivery of a defined service). With these extraordinarily difficult steps in place, further pursuit of market forces in the allocation of water will be useful. Privatizing water, in the sense of giving farmers a greater role in both the financing and management of irrigation, is a promising development. Its major benefits are likely to be more in the long-run than in the short-run, by "inducing technological and institutional innovations [11] in irrigation management. A common observation of scientists working with farmer-managed systems is that *they really are interested* in how to improve their systems! For some of those scientists this is a unique and refreshing experience.

7   References

1.   International Conference on Water and the Environment, (1992) The Dublin Statement and Report of the Conference, WMO, Geneva, Switzerland.

2.   Robbins, L. (1935) *An Essay on the Nature and Significance of Economic Science.* London: MacMillan.

3.   Briscoe, J. (1996) *Water as an Economic Good: The Idea and What It Means in Practice.* Paper presented to World Congress of ICID, Cairo, Egypt.

4.   Wade, R. (1982) *The System of Administrative and Political Corruption: Canal Irrigation in South India.* Jn of Development Studies. Vol.18. No.3. 287-325.

5.   Repetto, R. (1986) *Skimming the Water: Rent-Seeking and the Performance of Public Irrigation Systems.* Washington, D. C., USA: World Resources Institute.

6.   Seckler, D. (1993) *Privatizing Irrigation Systems.* Discussion paper No. 12. Arlington, Va.: Center for Economic Policy Studies, Winrock International.

7.   Vermillion, D. (1996) *Impacts of Irrigation Management Turnover: A Review of the Evidence to Date.* Draft Paper. Colombo, Sri Lanka: International Irrigation Management Institute.

8.   Svendsen, M., and Vermillion, D. (1994) *Irrigation Management Transfer in the Columbia Basin: Lessons and International Implications.* Colombo, Sri Lanka: International Irrigation Management Institute. (Research paper No. 12).

9.   Rosegrant, M. and Binswanger, H. (1994) *Markets in Tradable Water Rights: Potential for Efficiency Gains in Developing Country Water Allocation.* World Development. Vol. 22. No. 22. 1613-1625.

10.   Perry, C. J. (1995) *Alternative Approaches to Cost Sharing for Water Service to Agriculture in Egypt.* Research Report 2. Colombo, Sri Lanka: International Irrigation Management Institute

11.   Hayami, Y. and V. Ruttan. (1985) *Agricultural Development: An International Perspective.* Baltimore: Johns Hopkins University Press

[1]We differentiate deliberately, though there are areas of overlap, between water markets and tradable water rights. The latter require both the formal definition of entitlements, and the specification of the conditions under which the entitlement may be traded. Water markets, on the other hand, have developed widely, for example in areas where private tube wells provide competitive services to numerous potential buyers, in the absence of the formal definition of water rights, tradable or otherwise.

EXPOST ECONOMIC ANALYSIS OF IRRIGATION IN THE BAS-RHÔNE LANGUEDOC REGION

Expost analysis of an irrigation scheme

I. CARRIERE & F. Le LANDAIS
BRL*ingénierie*, Nîmes, France

Abstract

The Bas-Rhône Languedoc region was primarily a vine-growing area until 1955, when the first French regional development programme introduced irrigation. Forty years later, what conclusions can be drawn from an analysis of the qualitative and quantitative effects brought about as a result of this development programme, both on agriculture and on other induced activities?

Keywords: Agricultural trends, expost economic analysis, irrigation, Mediterranean region.

1   Introduction

The second world war left the economy of the Bas-Rhône Languedoc region in the South of France suffering with agriculture as its dominant activity. The main trend was single crop-farming and 62% of the surface area was planted with vineyards. Vine-growing was hit by a harsh economic crisis, which soon spread to agriculture in the entire region because of the lack of diversification.

In 1955, Philippe Lamour founded a regional development company: BRL. The French government charged it with the implementation of its first overall regional development programme. The main orientation of this development programme was hydroagricultural development. Today, 130,000 hectares have been equipped for irrigation and are run by BRL.

A first expost economic analysis on the induced effects of the oldest irrigation scheme (32,000 hectares) was carried out in 1980 [1]. The effects method was applied using the data from national accounts with a 90-branch input-output table. The results showed that the effects of the scheme were considerable with rapid convergence.

The objective of the present study [2] is to provide an overall economic appraisal of the effects of irrigation on the development of the region in qualitative and quantitative terms. The equipped area dealt with is that supplied by Philippe Lamour

Water: Economics, Management and Demand. Edited by Melvyn Kay, Tom Franks and Laurence Smith. Published in 1997 by E & FN Spon. ISBN 0 419 21840 8.

Canal, altogether 66,500 hectares. Using the effects method adapted to regional conditions, this study is an attempt to assess how much direct and induced prosperity –measured in terms of value added, employment and fiscal resources– is injected into the entire regional economy.

The mechanism propagating the effects of irrigation comprises several stages: the introduction of irrigation modifies activity and agricultural production levels leading to variations in value added, in agricultural employment and in fiscal resources. These are the direct effects of irrigation. At a different level, the use of the additional agricultural commodities produced is a source of indirect, downstream effects, whilst the additional intermediate consumption (inputs) demand is a source of indirect, upstream effects. All of these together constitute the overall induced effects.

The first part of the study deals with the analysis of the direct effects of irrigation. This paper reports on the first results of this analysis. In a second stage of study, we shall proceed to examine the induced effects. There are 89 communes either partially or totally located in the area involved in this study.

2   Evolution of agriculture since 1955

Table 1 shows the evolution of crops throughout the period studied. These data are taken from official agricultural censuses.

In 1955, the surface area under development was 98,610 hectares and there were 17,640 farms, which means that the average area per farm was 5.64 hectares.

The irrigation infrastructure on the scheme was built from 1959 onwards. By 1970, 50,000 hectares were equipped for irrigation. Later on, another 16,000 hectares were also equipped.

Except for the vine, which is a special case, most of the evolution took place between 1955 and 1970.

Since 1955, a 10% reduction in cropped land can be observed (approximately 10,000 hectares). This reduction is only really significant after the 1979 census. The number of farms was divided by 2.4 during the same period. Simultaneously, the average cropped land per farm more than doubled (12.26 hectares in 1988 against 5.64 hectares in 1955).

The main trends were:

- A reduction of more than one third in the area used as vineyards as a result of up-rooting policy and the possibilities of diversification afforded by irrigation.
- A constant increase in fallow land since 1970 due to European agricultural policy.
- A rise in industrial crops between 1979 and 1988, mainly sunflower, in relation with European incentive.
- A strong increase in the proportion of orchards (10%) and a transition as early as 1970 from extensive orchards (mainly olives, cherries and apples) to intensive, irrigated orchards producing peaches, apricots and apples.
- Fourfold multiplication of land used to grow market-garden crops since 1955, with notable development of specific regional crops such as asparagus, melons, tomatoes and lettuce.

Table 1. Evolution of agricultural land (%)

Crop categories	1955	1970	1979	1988
Vineyard	62%	59%	52%	40%
Fruit trees	5%	12%	10%	10%
Vegetables, potatoes and flowers	2%	4%	5%	7%
Grassland	5%	4%	3%	3%
Cereals	13%	15%	15%	16%
Industrial crops, legumes, annual fodder	*	1%	1%	7%
Fallow land and domestic gardens	13%	5%	6%	8%
Loss in cropland	-	0,1%	7%	10%

*included in previous category

3   Reference situation

To measure the effects caused by the introduction of irrigation, it is necessary to compare the present situation with a reference situation to reflect how agriculture would have evolved if irrigation infrastructure had not been built. This reference is based on observations of a neighbouring area (89 communes) where no community irrigation facilities were built. The comparative evolution of the area with irrigation and that without irrigation is shown in table 2.

The share of land occupied by the different types of crops was analogous in 1955 in both zones. In the area without community irrigation, cropped land diminished three times more than in the zone with irrigation.

In the zone with community irrigation facilities, the number of farms dropped more than in the other zone.

3.1   Comparison per type of crop

Vineyards: the reduction in surface area is greater in the zone with irrigation despite the vast areas of AOC vines (quality controlled appellation), which were on the rise during this period. The existence of various possibilities of diversifying crops (or conversion to a different type of farming) thanks to irrigation is most certainly the explanation of this trend.

- Fruit trees: the percentage of land used to grow fruit trees virtually doubled with irrigation, whilst it shrunk by 2% in the zone without community irrigation. Also the varieties used in each case differ to a great extent.
- Vegetables: the proportion of land used for market-garden crops was multiplied by four with irrigation, whilst there is virtually no evolution without irrigation.
- Large crops: the land used for cereals, industrial crops and grassland underwent a strong rise in the zone with irrigation, but fell by almost half in the area with no community irrigation equipment.

Table 2. Comparison of the evolution of crops with and without irrigation (ZI = zone with irrigation; ZNI = zone which is not irrigated)

Crop categories	1955		1970		1979		1988	
	ZI	ZNI	ZI	ZNI	ZI	ZNI	ZI	ZNI
Vineyard	62%	62%	59%	57%	52%	52%	40%	45%
Fruit trees	5%	7%	12%	8%	10%	6%	10%	6%
Vegetables, potatoes and flowers	2%	2%	4%	2%	5%	2%	7%	3%
Grassland	5%	5%	4%	2%	3%	2%	3%	1%
Cereals	13%	11%	15%	8%	15%	7%	16%	6%
Industrial crops, legumes, annual fodder	*	*	1%	0%	1%	1%	7%	2%
Fallow land and domestic gardens	13%	12%	5%	5%	6%	6%	8%	7%
Loss in cropland	-	-	0,1%	17%	7%	25%	10%	30%

*included in previous category

Table 3. Surface area in 1988 for reference situation and net surfaces

Crop categories	Surface area (hectares)			% additional surface
	Situation in 1988	Reference Situation	Difference in surface	
Vineyards	39 515	46 308	-6 793	-15%
Fruit trees	10 001	5 665	4 336	77%
Vegetables	6 616	2 616	4 000	153%
Grassland	3 442	1 284	2 158	168%
Cereals	15 431	9 234	6 197	67%
Industrial crops	6 431	2 267	4 164	184%
Fallow land	7 406	7 361	45	1%
Total cropland	88 842	74 735	14 107	19%

The introduction of irrigation infrastructure plays a part in maintaining the extent of farm land and diversifying the crops grown. Two stages can be identified:

* The equipment stage (1955-1970), with a considerable spread of fruit trees and market-garden crops on land not occupied by vineyards. Diversification began in parallel with vine-growing.
* After 1970, a relative stagnation of the agricultural surface area thanks to the possibility of using irrigation to convert areas where vines had been pulled out. The existence of equipment made it easier to apply up-rooting policy and to sustain agricultural activity on the irrigation scheme.

According to this information, we were able to reconstitute the would be evolution in the study area had the community irrigation infrastructure not been built. The figures obtained are given in table 3.

The introduction of irrigation facilities enables the preservation of 14,107 hectares, i.e. 19% of the additional cropland. Irrigation changes the structure of crop shares and allows greater diversification, despite the still dominant proportion of vineyards (see Fig. 1).

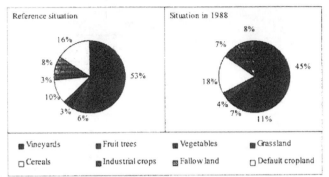

Fig. 1.   Comparison of the composition of cropped land in 1988 with the reference situation

4   Calculation of direct effects

4.1   Production increase in volume

The increase in the volume of commodities (sold) is determined by multiplying the various surface areas of crops in the additional cropland by the corresponding yield per hectare.

The following overall results are obtained:

- 28,392 tonnes of cereals, i.e. 67% surplus.
- 10,216 tonnes of industrial crops, i.e. 202% surplus.
- 3,275 tonnes of various crops (fodder and legumes) i.e. 65% surplus.
- 97,611 tonnes of vegetables, i.e. 195% surplus (+6.840 million lettuces, i.e. 60% surplus).
- 124,275 tonnes of fruit, i.e. 122% surplus.
- a reduction in wine production by 479,590 hectolitres, i.e. -15%, and a reduction by 450km in vine nurseries, i.e. -14.5%.

The use of the extra volume of agricultural commodities produced gives rise to induced downstream effects. These will be studied at a later stage.

4.2   Gross product before tax

The gross product before tax corresponds to the total volume of production sold multiplied by the average factor sales price plus the amount of compensation incentive granted for large crops (cereals, industrial crops).

The extra gross product before tax for the scheme is FFr504.9m, i.e. 41% more gross product before tax than in the reference situation (see Table 4 for results per crop category).

The sharing of the gross product before tax is substantially affected by irrigation. There is a 30pts fall for vineyards, whilst fruit trees and market-garden crops mark a respective progression of 16pts and 9pts (see fig. 2).

Table 4.   Gross product before tax

| Crop categories | Situations | | | Additional % |
	Situation in 1988	Reference Situation	Difference	
Cereals	151 878	89 045	62 833	71%
Industrial crops	64 408	23 493	40 915	174%
Miscellaneous	9 340	5 770	3 570	62%
Vegetables	415 841	186 417	229 424	123%
Fruit trees	354 507	66 811	287 696	431%
Vineyards	731 109	850 635	-119 526	-14%
Total	1 727 083	1 222 171	504 912	41%

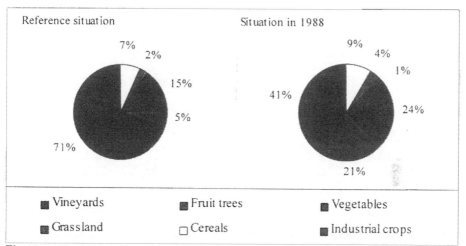

Fig. 2.   Structure of gross product before tax in 1988 and in reference situation

4.3   Production factor balance for additional agricultural commodities

The production factor balance for each crop provides enlightenment as to:

- quantified intermediate consumption (inputs) used to cultivate one hectare of a given crop and the unit price of each product before tax. There are 377 different products falling into nine main categories (miscellaneous products, fertilizers, chemicals, seeds and young plants, irrigation water, equipment...);
- the number of hours of skilled or unskilled labour required to cultivate one hectare of the given crop;
- the number of hours of tractor use required to cultivate one hectare of the given crop (4 types of tractor);
- the total amount of tax per hectare (11 types of tax).

The total production factor balance for the additional production in the irrigated area is obtained as follows:

- term by term total of the balance of all the production factor accounts multiplied by the surface area concerned;
- application of V.A.T. at the rate of 5.5% of commodity value.

This gives the total production factor balance for the additional agricultural production from the scheme as shown in Table 5. The total additional intermediate consumption amounts to FFr190.8m. The production of these intermediate products is a source of induced upstream effects to be studied at a later stage.

Table 5. Additional production factor balance for agricultural commodities from the irrigated area

	Amount (FFr10^3)		Amount (FFr10^3)
Intermediate consumption before tax		Gross product before tax	
Miscellaneous products	10 466	Value of production	471 474
Fertilizers	28 329	Compensation incentives	33 437
Chemicals	10 919		
Seeds and young plants	31 531		
Irrigation water	28 857		
Specific equipment	22 840		
Tractors	9 047		
Servicing, fuel, lubricants, etc.	26 913		
	21 964		
Sub-total	190 866	Sub-total	504 911
		VAT	25 931
Value added incl. VAT	339 976		
Grand total	530 842	Grand total	530 842

4.4   Value added

4.4.1   Amount of additional value added

Irrigation earns an annual additional agricultural value added amounting to FFr340 million, i.e. 46.3% more than if no development programme had been conducted. The following figure illustrates the contribution of each type of crop to this supplement.

4.4.2 Distribution of value added

The value added falls under various headings: salaries, social contributions, taxes, financial costs, insurance, structural and operating costs and farm income. Each heading corresponds to a particular agent in the economy. Value added is said to be divided between the different agents.

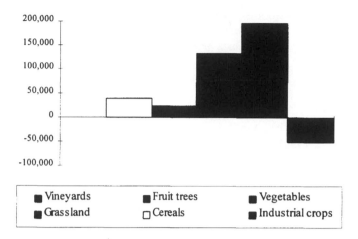

Fig.3. Contribution of the various types of crops to the additional value added (FFr million)

1. Salaries and social contributions

The total amount of salaries and social contributions (family labour and farmer included) is FFr229.5 million, i.e. 40.7% more than in the reference situation. The sharing of this amount between agricultural households and the State has not yet been determined.

2. Fiscal resources.

Value added tax - The additional V.A.T. received by the State is FFr25.9 million per annum, i.e. 33.5% more than in the reference situation.

Tax on large crops - Tax levied on large crops scores an annual FFr1.1 million, i.e. 112% more than in the reference situation.

Tax on non built-up land property - This tax is based on cadastral revenue of agricultural land. It is collected by the local authorities, Chambers of Agriculture and the State. We plan to determine the exact amounts involved.

4.5   Agricultural employment

Using the production factor balance for the various crops and referring to Agricultural Labour Units (one ALU is the equivalent of 2,200 hours of work per annum) as full-time employment for one person, the irrigation scheme created a total of 1,973 full-time employment units (including 242 ALUs for skilled labourers), i.e. 41.2% more ALUs than in the reference situation.

5   Conclusions

This preliminary work is yet to be completed before the end of the year 1997 with the calculation of upstream and downstream regional value added, fiscal resources and employment at regional scale as induced by irrigation.

Decision-makers –mainly at regional level– will thus be able to use these criteria in comparison with investments to assess the interest of irrigation for the regional community at a time when strategic choices are difficult.

The interest of this work is principally the quality of the data used and the fact that they are a true reflection of regional economic reality and not a series of imperfect regional statistics improper for such practice.

6   References

1.   Le Landais, F. (1980) *The induced effects of irrigation on a Bas-Rhône Languedoc scheme.*

2.   Carriere, I. (1996) *Expost analysis of irrigation in the Bas-Rhône Languedoc Region,* interim report.

TO BE OR NOT TO BE? SOUTH AFRICAN IRRIGATION AT THE CROSSROADS

A.H. CONLEY
Department of Water Affairs and Forestry, Pretoria, South Africa

Abstract

South Africa has become a water-stressed country which, after providing for basic human needs, environmental protection and international sharing, is replacing riparian water rights with a system of dynamic, competitive water allocations of limited duration to achieve optimum economic growth and social equity. Under growing competition from other water users, a new policy for the irrigation sector, the largest user of water, is being developed through public consultation.
Keywords: South Africa, water scarcity, law, irrigation, economic optimisation, social equity, environmental protection.

1   Introduction

South Africa (SA) has developed to the stage where its meagre water resources will in future have to be used for the most worthwhile purposes only. In a radical departure from riparian water rights, SA's future agricultural use of water will have to recognise not only the need to treat water as an economic good, but also satisfy the requirements of social equity and ecological sustainability.

2   South Africa's difficult water situation

The greater part of SA is semi-arid and is noted for its hydrological extremes. The country's average annual rainfall of 500 mm is only about 60 per cent of the world average. Sixty-five per cent of the land receives an average annual rainfall of less that 500 mm, which is usually seen as the minimum for successful rain-fed agriculture. Over most of the country, the average potential evaporation, which ranges from about 1 100 mm in the east to more than 3 000 mm in the west, is well in excess of the average rainfall. Surface runoff is SA's main water source notwithstanding that on average only about nine per cent of the rain reaches the rivers. The economically

Water: Economics, Management and Demand. Edited by Melvyn Kay, Tom Franks and Laurence Smith. Published in 1997 by E & FN Spon. ISBN 0 419 21840 8.

usable portion of the runoff is estimated at 30 000 million cubic metres per year. It is poorly distributed relative to areas showing economic growth and even more variable than the rainfall, particularly in the west. In general, the region is characterised by long droughts interrupted by severe floods.

As SA is underlain by hard rock formations, only about 5 400 million cubic metres of groundwater may be obtainable per year. In many areas it is saline.

SA's major dams have a combined capacity of 50 per cent of the average runoff and intercept almost all of the rivers in the interior part of the country. During droughts the water may have to be stored for very long periods (even exceeding ten years) under high evaporation, which increases its salinity. In addition, as all effluent is required to be returned to the rivers, increasing urbanization, industrialization and irrigation is causing the quality of many sources of water to decline, mainly through salination, eutrophication and pollution by trace-metals, micro-pollutants and bacterial contamination. Therefore, downstream users are vulnerable to diminished quality as well as to reductions in flow. Relative to growing demand, water is scarce in many areas of the country and is expected to become scarce almost everywhere. Ultimately, SA's water situation may worsen through global warming.

As the unit costs of water supply accelerate, the water that can be developed at costs that are affordable to many classes of user, particularly for irrigation, is limited. If the present rate of growth in use is not checked, SA could reach the limits of its water availability by the year 2030. Accordingly determining the optimum use of water in the public interest, on an informed basis from many possibly best solutions, has become a major national aim which requires a sound approach to demand management with an appropriate emphasis on the value of water as an economic good. It has therefore become essential to take account of the full economic, social and environmental cost of using water, and not the financial cost only.

3   Momentous changes to South Africa's water law

After 1652, when Dutch settlers established a vegetable garden at the southern tip of Africa to supply their ships trading with the East, the Dutch continued to respect the *dominus fluminus* principle derived from Roman-Dutch law, according to which the right to use public water could be exercised only on State authority. This principle accords with African customary law. Having annexed the settlement in 1806, Britain introduced riparian water rights by the mid 19th-century (when agriculture was still modest) according to which landowners along a river had an exclusive right to use its water. Groundwater was seen as private water.

Early 20th-century water legislation entrenched the rights of riparian farmers and authorised the State to promote irrigation. However as towns and industries grew, the static riparian approach clashed with the dynamic water needs of diversifying development. Accordingly, in 1956 a new Water Act[1] was promulgated to facilitate an equitable distribution of water for industrial and other competing uses. While acknowledging riparian use, it provided for the re-introduction of State control over water that exceeded existing users' rights. Nevertheless, as its allocation mechanisms were still derived from a poor initial appreciation of SA's climatic conditions, over time the law became less and less suitable for equitable and lasting solutions to the growing intricacy of SA's problems.

During the 20th-century most water schemes were built to support SA's economically and politically influential entrepreneurial sector, which showed modest popula-tion growth, while the underdeveloped sector, which was largely rural, grew immensely, but had fewer services. After democratic elections in 1994, the new government has accordingly undertaken to renew SA's water law to comply with the needs of the country's new Constitution[2] and to achieve equity and ongoing optimisation in the use of what has become a scarce fundamental resource. This implies the resumption of full State control on behalf of 43 million citizens and the demise of riparianism which favours certain land owners, particularly the fewer than half of the 60 000 commercial farmers for whom irrigation is a major activity.

After two years of public consultation[3], in November 1996 SA's Cabinet approved a set of principles and objectives for new water legislation[4,5]. Amongst others, they determine that the objective of managing the quantity, quality and reliability of the nation's water is to achieve optimum, long term, environmentally sustainable social and economic benefit from its use as an indivisible national asset. To be able to exercise its duty as a public trust, the national government, as custodian, has ultimate authority over water resource management, the equitable allocation and usage of water and its transfer between catchments, and international water matters. There shall be no ownership of water, but only a right for basic human needs and environmental sustainability (called the Reserve) and non-perpetual authorisations for all other uses. The location of the water resource in relation to land shall not confer preferential rights to usage. An authorisation to use water shall be timely, clear, secure and predictable in respect of assurance of availability, extent and duration. The purpose for which it may be used shall not arbitrarily be restricted and the conditions for authorisation shall consider the user's investment in developing infrastructure to use the water.

The adoption of the principles was followed in May 1997 by a government White Paper on a national water policy for SA[6], according to which all water occurring anywhere in the water cycle will be treated as part of the common resource. The new system is to be phased in, beginning in water managementareas which are under stress. Except for the reserve and international water-sharing priorities, it will use dynamic, competitive limited term allocations and other administrative mechanisms, including water pricing, to bring supply and demand into balance. To terminate the riparian system of allocation, transitional arrangements are to be put in place to ensure an orderly, efficient and gradual shift in water use allocations as and when necessary. Future allocations will be granted for a reasonable period, which may be renewable, and it will be possible to transfer or possibly trade these rights with Ministerial consent.

In addition to charging users the full costs of providing and operating waterworks, all water use will be subject to a catchment management charge to cover actual costs (such as for removing invasive vegetation), as well as a resource conservation charge where beneficial uses compete or significantly impair other uses. Waste disposal into watercourses will also be subject to appropriate management and resource conservation charges.

All user-groups will need to continually re-evaluate their impact on the country's water, and will have to meet the real cost to society and the environment of their activities. In the case of irrigation, a small percentage reallocation, by achieving a better yield per unit of water, could make a proportionately larger contribution to the needs of smaller user sectors, some of which can multiply the effect by recycling.

All major water user sectors will have to develop a water use, conservation and protection policy. Since water ignores national and international political boundaries, it will be managed according to regional or catchment management areas of which the boundaries will coincide with natural river catchments, groups of catchments, sub-catchments or areas with linked supply systems having common socio-economic interests. Provision will be made to phase in the establishment of nationally authorised management agencies in these areas.

No claim to an existing right will be recognised if it limits the water which is available for basic human needs. Also, as allocations for irrigation already exceed the available water in many places, intervention may be required to protect the resource base where the environmental reserve cannot be met. Where future allocations to redress past racial discrimination result in the reduction of existing valid allocations, the reallocations will be protected by constitutional provisions for corrective action which recognise the government's right to establish such legislative programmes.

Information in a register of allocations will enable the calibre of resource development, monitoring, assessment and catchment management to be improved. While this procedure can function as an administrative system, it is compatible with the creation of a market in water use allocations, should that become desirable.

4   Growing competition for water and its future role as an economic good

The consideration of water as an economic good requires the value of the water itself to be taken into account, in addition to the financing of waterworks. The largest portion of SA's runoff is consumed by irrigation, although the contribution of agriculture to the gross national product has declined during this century to about a third of its earlier value, while that of manufacturing has increased six-fold. Mining, manufacturing and power generation together use about a quarter of the water consumed under irrigation, and add far more value and create more jobs per kilolitre of water than agriculture or forestry. The relative qualities and assurances of supply to user sectors is also a vital consideration. For instance, power generation would warrant a higher assurance than agriculture. Nevertheless, irrigation is important and many people will continue to depend on the agricultural economy. The new government's land reform programme also needs equitable access to water if resettled land is to be productive. In addition, the macro-economic Growth, Employment and Redistribution Strategy adopted by SA in 1996 emphasises that its economy cannot continue to grow satisfactorily only through the continued exploitation of depletable natural resources such as minerals. The strategy aims to boost growth by lowering protective barriers in a number of industrial sectors, and to promote small and medium industries as well as greater integration with the other countries of the Southern African Development Community.

Despite its economic worth, much of the water used in SA is not paid for directly by landowners, such as riparian abstractions, groundwater and rain-fed cultivation. In addition, many users of the State's water infrastructure do not even meet the operation and maintenance costs of waterworks, a White Paper of 1984[7] having determined that while the full financial costs of water schemes should be paid by industrial and domestic users, irrigation farmers need only pay what was affordable, which has never been quantified. Accordingly, in 1996 the new government determined that charges for all major users should be raised progressively to meet the full financial costs of making water available and to reflect its value to society as a scarce resource, with only the quantity reserved for basic human needs being available without charge. While the concept of realistic pricing holds promise for a better allocation of water, it has to be introduced in a manner that will not penalise communities whose opportunities were limited during the apartheid era. Examples of burdens placed on society by water uses which are not paid for arise from rainfall interception and waste disposal. For instance, afforestation and dry-land farming reduce river flows downstream, while waste disposal into watercourses reduces their receiving capacity, a natural asset which can be dealt with as an economic good in association with its environmental management. Accordingly, water conservation and demand management may be promoted through a water price that reflects its value, although how best to apply this concept in practice is under evaluation.

Notwithstanding difficulties in determining prices administratively, the trading of water-use allocations, if allowed as an alternative, will require increasing degrees of control depending on whether trading is permitted within a user sector, or between sectors or water management areas. Serious problems and arduous administrative requirements are associated with trading water between disparate locations which are subject to climatic extremes, dissimilar river regimes, high infrastructure costs and social imbalances. Traded prices will not necessarily reflect the resource's full value under these conditions, and many South Africans would not accept that landowners who had acquired rights to use water under SA's historical land tenure system could possibly gain profits. It could also be seen as improper to subject the available water to allocation by tender or auction if wealth is concentrated in minority hands.

Fiscal capital expenditure on water development will be restricted to providing for basic human needs, assuring the environmental reserve and meeting commitments on internationally shared rivers. On government water schemes tariffs will be adjusted over a reasonable period to cover their full operation, maintenance, capital and resource management costs. Where water is not supplied from government schemes the price will reflect the resource management costs and an appropriate resource conservation charge. Water prices will vary according to location, being derived on a locally appropriate sub-catchment, catchment or system basis. Disadvantaged individuals and communities will be assisted by specific measures for beneficiaries of land restitution, land reform or other corrective action programmes, including periods when the full cost will not be charged in order to support new enterprises. Income will be shared between operating agencies, water management authorities and the national government in accordance with their contributions and responsibilities. The principles for disposing of income need to be widely accepted so as not to introduce new distortions or hidden taxes.

5   The history and unsettled future of irrigation in South Africa

The golden age of irrigation development in SA gained momentum after about 1906. Inspired by the archetype of Empire, with the population still small and not yet appreciating the hydrological extremes and dearth of suitable soil, Britain brought engineers from Egypt and India to SA to promote widespread irrigation schemes which today absorb more than half of SA's usable river flow. But SA entered an economic recession during the first world war and rural difficulties were intensified by drought and economic depression during the 1920s and 30s. Accordingly, the government built several labour-intensive irrigation settlements to employ and settle destitute whites from drought-stricken areas. Blacks were barely considered. In many areas, drought commissions recommended the writing-off of irrigation loans.

A negative consequence of the government's provision of infrastructure and the cancellation of debts was the entrenchment of a sense of entitlement to State assistance for irrigation water at charges which rarely covered the operating costs. Also, the Director of Irrigation, when visiting Italy in 1927, was persuaded that subsidies for irrigation were warranted by their "direct and indirect benefits". This ringing phrase, while sound in principle, was used in an unquantified manner for decades in White Papers to gain parliamentary approval for the fiscal financing of individual irrigation schemes. After the mid-1930s, when farmers were discouraged by drought, salinity and pests, capital subsidies were introduced to speed up the construction of irrigation schemes by individuals and statutory irrigation boards. So many farm dams were built that a permit system had to be introduced in the 1970s to control their expansion. Owing to their presence, many of the country's large dams now receive water only after very protracted rainfall.

Irrigation in SA is now approaching its limits to growth. Since 1979, irrigated areas have grown by 62 per cent from 800 000 ha to 1 300 000 ha. The scope for new irrigation development is now limited by a lack of water and the high cost of schemes to about 178 000 ha, which is only about 14 per cent above the present extent. Expansion will be possible in certain areas only, which require prioritisation.

Faced with increasing competition for water and the need to keep people at the centre of concern, in 1996 the Ministers of Water Affairs and Forestry, and of Agriculture, began a process of public participation to develop a new, sustainable, irrigation policy[8]. It is intended to enhance the country's economic welfare as well as its social development by promoting efficiency, fairness and stability. The policy has to cater for a wide range of circumstances, ranging from specialised exports to community vegetable gardens. Beyond prioritising the remaining possibilities for expansion, it must achieve the best use of water, for which the requirements are demanding not only in terms of infrastructure and equipment, but also for genetic engineering, management, training, counselling, research, and the use of indigenous knowledge on small plots. The irrigation sector will have to find balanced solutions to water use both within agriculture as well as in competition with other users. For instance, large mechanised irrigation projects are not as effective in creating jobs as some areas of commerce and industry for the same capital investment.

Many aspects of irrigation farming have now to be debated[8]. For instance, to what extent should SA import water-intensive products rather than produce its own food, wood and power? How practical is it to upgrade to higher-value crops, considering climate, pests, diseases, varieties, fertilisation, transport and marketing? Are there sufficient international markets for them? Would it be wise to reduce the third of the irrigated area which is under profitable, labour-intensive export fruit and wine? Does the constitutional right to water for basic human needs also encompass household food production and how could it best be implemented? How important is the 14 per cent of the area under crop-rotated wheat in stabilising 30 per cent of SA's production and the 6 per cent under cotton and oilseeds which cannot meet the demand? Is the 9 per cent under maize which yields only 10 per cent of SA's production justified, or the 5 per cent under sugar cane which is sold below domestic prices on the world market? Is it reasonable to have 25 per cent under pastures for milk and meat production? How will the various changes affect processing, packaging, transport, marketing, equipment production, fuel, fertilisers, and biocides? How can schemes designed for night work be kept going under growing trade union pressures and new labour legislation? How can skilled farm employees be helped to acquire land and working capital? How can farmers eliminate the harmful effects of salinity, agricultural chemicals and feedlots to assist in maintaining the ecological reserve from return flows?

With tariffs moving towards full cost recovery, high-value users ought to be able to obtain the water they need at an economic price. However, particular attention will have to be paid to the efficiency of water use for irrigation in terms of location, soil suitability, crop selection, cultivar development, transmission losses, scheduling, application techniques, mechanisation, fertilisation, salination, drainage, research and extension support. Years of experimentation with varieties, rotations, fertilisation and pest and disease control are likely to be needed to counter the risks of converting to higher value crops, particularly perennials requiring large investments. For formulating integrated catchment management plans, irrigation farmers will have to develop the technical and administrative capacity to be well represented in water user associations and catchment authorities.

A diversity of farm sizes is considered desirable for SA, with the achievement of success by small farmer-managed schemes being seen as having great potential to stimulate rural development, as large schemes have rarely achieved the anticipated level of development. In areas where maintenance facilities are poor, improved traditional methods have proven to be more profitable than sophisticated installations. Accordingly activities at ex-homeland, communal, subsistence, food-plot, employee and market-garden level will focus on the rehabilitation or improvement of schemes and the encouragement of self-sufficiency and entrepreneurship. To achieve these aims the role of irrigation is important to land resettlement projects. In comprising about half of the farmers and most of the labourers in small-scale rural irrigation, women will need to play an active role in planning gender-specific schemes. Some projects have been unsuitable, requiring heavy pipes to be moved in the mud at times when household chores are at a peak. Many rural women also need training.

Fortunately the positive spirit in which the irrigation policy process has commenced has gained the co-operation of government departments, NGO's, CBO's, organised agriculture, irrigation boards, industry, research institutions, active and aspiring small-scale farmers and process facilitators. Grass-roots consultations have been held with stakeholder groupings and irrigation action committees in SA's nine provinces. The SA National Committee on Irrigation and Drainage (SANCID) is arranging for commodity studies on SA's main irrigated crops, covering facets such as location, area cultivated, water use, production, type of farming, investment, employment, input and crop processing industries, economic multiplier effects and foreign exchange earnings. Specialist workshops are to consider the economic and social value of irrigation, the improvement of efficiency, effective catchment administration, the development of small farmers and the role of women.

5   Conclusion

The various activities are to culminate in a national irrigation policy conference to seek clarity on its unsettled future. Although much international thought, policy and practice can be recast for SA, the nation's success in handling other complex problems peaceably provides a promising backdrop for SA to find its own solutions to addressing its water management in a principled and structured manner. The law review process has shown that policy formulation, institutional reform and capacity building are as important as capital projects and that securing SA's water future will require special emphasis on the handling of conflicts. As the transformation will take many years, time alone will tell how well SA was able to surmount its difficulties.

The views expressed are those of the author, and not necessarily of the South African Department of Water Affairs and Forestry or of SANCID.

7   References

1.   Statutes of the Republic of South Africa. *Water Act No 54 of 1956, as amended.*
2.   RSA Government Gazette No 17678 (1996) *Act 108 of 1996, Constitution of the Republic of South Africa.*
3.   RSA Department of Water Affairs and Forestry. (1995) *You and your water rights - a call for public response.* ISBN 0-621-16752-5
4.   RSA Department of Water Affairs and Forestry. (1996) *Water law principles.* ISBN 0-621-17399-1
5.   RSA Department of Water Affairs and Forestry. (1997) *Fundamental principles and objectives for a new water law in South Africa.*
6.   RSA Department of Water Affairs and Forestry. (1997) *White Paper on a national water policy for South Africa.* ISBN 0-621-17707-5
7.   RSA White Paper N'-84 (1984) *Policy on water tariffs and related matters.*
8.   RSA Department of Water Affairs and Forestry and Department of Agriculture. *Towards an irrigation policy for South Africa.* ISBN 0 7970 3505 2

ECONOMIC CONSIDERATIONS FOR WATER USED IN IRRIGATION IN ISRAEL

Pricing of water for irrigation

Y. SHEVAH
TAHAL Consulting Eng. Ltd. Tel-Aviv, Israel
G. KOHEN
Water Commission, Tel-Aviv, Israel

Abstract
Water as a scarce economic input has become a major decisive factor in water pricing for the competitive sectors and especially for irrigation for which water is normally allocated on administrative and historic water rights. In Israel, agriculture is the major water consumer using about 65% of available water resources. A water quality factor and an Increasing Block Rate system related to the quality of water at source and the level of water used are applied for payment for the first 50, 80 and 100% of the allocation. The introduction of a progressive pricing system together with other efficiency measures have reduced government subsidy for water used in irrigation which was reduced from 40% in 1994 to 26% in 1996 and to a projected 19% in 1997. The new pricing system, although still subsidized has influenced water consumption pattern by making irrigation water a major expenditure (8-9%) of the total purchased agricultural production inputs. Further reform is being discussed in which the existing block rate and the non - tradable allocations are to be changed into a market negotiating systems. The Government is to further reduce direct involvement in water supply by privatization of the national water company and the creation of regional water systems which will operate as private corporations under government control. Shares allocation attracting dividends and voting rights will replace existing water rights. Some of the changes already experienced and those planned for the Israeli water Sector are discussed in this presentation.
Keywords: Israel, water, irrigation, prices, subsidies, water economics, agriculture

Water: Economics, Management and Demand. Edited by Melvyn Kay, Tom Franks and Laurence Smith. Published in 1997 by E & FN Spon. ISBN 0 419 21840 8.

1   Introduction

Water is a scarce commodity and as such there is a strong need for a sustainable use and economic pricing. Water prices do not normally reflect marginal economic values because they are largely determined by the Government rather than by the operation of free market. Governments' preferred policies include incentives and regulations that motivate adoption of best management practices, taxes or subsidies on inputs that generate or reduce pollutants and restriction on the use of selected inputs including water.   In Israel, water conservation is pursued by a wide scale adoption of low volume irrigation systems (e.g., drip, micro-sprinklers) which increase the average efficiency to 90% as compared to 64% for furrow irrigation [1] [2]. From 1975 to 1982 the use of low volume irrigation equipment increased over 700 percent [3]. Generally, water savings were mentioned as the primary reason for adoption despite the substantial fixed costs associated with purchasing the modern systems and Government assisted programs encouraging their adoption. Nevertheless, empirical studies concluded that profitability was also the major motive [4].   In parallel, a large scale wastewater renovation programme was also introduced leading to the current reuse of about 60% of the generated domestic wastewater [5].

A continuous growth of the population and a competitive demand for water by the urban sector have greatly increased the pressure on available water resources making it necessary to divert water from irrigation to industry and domestic use. Adding water by desalination at a cost of about one US C 80/cum is not economical as compare to the   marginal economic return of 55 US cents in agriculture and therefore measures to increase the price of water are put in place ahead of water using interest groups.   Initially, an increase in the flat rate price of irrigation water was adopted. This system has been modified to the adjustment of the price in discrete steps or blocks as the volume of the applied water increases. A single tiering level is imposed on all crops, although this could distort farm level cropping pattern decisions in favor of crops with relatively low water requirements, but sensible when the ultimate goal is water conservation.

The current measures are fully exploited and Israeli water economy is on the verge of a major reform in which the existing block rate and the non-tradable allocation are to be changed into a market negotiating systems. Israel has already embarked on the political process of positioning agriculture in the proper place vis a vis national water allocation and use and the various issues have been fully debated by a public hearing committee [6].

2   Water resources and water supply and demand

2.1   Water supply

Water is allocated by the Water Commissioner, empowered by the water Law 1959. The Water Commissioner is responsible for safeguarding water quantity and quality, issuing abstraction license and allocations to consumers. Water is supplied from local sources, stormwater, treated effluents and water conveyed from north to south via the National Carrier. Potential resources include also brackish water and domestic effluents (Table 1).

Table 1. Water Supply and Demand 1995 and Projected - 2010 (MCM/Year)

Water Supply - 1995		Water Demand by sector	1995	2010
Natural Fresh	1600	Urban	700	900
Natural Brackish	180	Agriculture:		
Domestic Effluents	220	Fresh Resources	900	700
		Marginal Resources	400	600
Total	2000	Total	2000	2200

Source: Water Commission and authors' estimates

2.2 Water consumption

Currently, the urban sector consumes about 700 MCM and the annual increase is about 20 MCM per year, about 4 %. Agriculture accounts for about 65% of water use, although less than 50% of the cropland is irrigated. Since the seventies, the irrigated land and the total water demand are almost stable, although the volume of production more than doubled over the same period (Table 2).

Table 2. Irrigation water demand and productivity 1975 -1995

Year	Cultivated land (000 ha)	Irrigated land (000 ha)	Irrigation Water (MCM)	Productivity Index (% of 1986)
1975	4200	1835	1325	55
1980	4245	2030	1210	72
1985	2380	2110	1435	95
1990	2360	2060	1215	139
1995	2350	2000	1210	196

Source: Central Bureau of Statistics, 1996

Despite the impressive productivity results and the undoubted water use efficiency, the Israeli agriculture which consume more than 60% of available resources contributes only 3% of the GDP. The remaining 97% is generated in industry and services using less than seven percent of the national water. The projected consumption figures (Table 1) indicate an increasing demand for the relatively fixed supply of water caused by population growth. Irrigated agriculture must therefore compete with other uses for increasingly scarce water resources.

2.3 Water suppliers

The Israeli water is characterized by a large National Water Company - Mekorot and several water associations and private producers. Mekorot supplies about 60% of the total consumption, in bulk, to local authorities and water associations who are responsible for the distribution of water to the individual consumers. The part of Mekorot in the total turnover is 80% because it operate the major systems and convey water through the National Carrier over a long distance from North to South. Mekorot supplies about 1350 MCM, while the remaining 650 MCM are supplied by small producers. All the producers including Mekorot are bound by production license issued by the Water Commissioner and allocated to consumers according to fixed quotas. In a sense, Mekorot is a monopoly which could dictate the course of development to the benefit of the company but not necessarily to the interest of the national economy, assuming that Mekorot will accelerate development irrespective of the economic timing of the proposed development.

Previously, Mekorot operated at a Cost Plus whereby all the expenses are covered by the consumers and the government which determine the development and the subsidies to cover the difference between water charges and costs. Since 1994, a reform was introduced in which the fixed cost (capital and labor) and variable costs (energy and materials) were defined and a 2.5% efficiency factor was imposed on Mekorot while the amortization cost, interest and cost of capital were also credited in order that the Company will become business oriented. The effects on Mekorot's performance and government support are shown in Table 3.

Table 3. Mekorot water cost and government subsidy 1994 - 1997

Year	Average Water Cost (US C/cum)	Government Support (%)
1993	36.4	40
1994	34.4	35
1995	32.8	26
1996	30.5	23
1997		19 (projected)

Source: Mekorot Financial Statements, various years.

A substantial   increase in water prices coupled with   increased performance efficiency (saving in energy cost and other variable and fixed costs) have resulted in a significant reduction in Government subsidy from 40 to 20% over the last four years. Current tariffs have totally eliminated the subsidy for water supplied to the urban and industrial sectors where their pay is slightly higher than the average cost.

3   Water cost, pricing and economic return

3.1   Water cost

The Israeli water system is characterized by heavy investments in pumping, pipes and treatment plants. There is also the   need to maintain a balanced supply comprising a mix of water sources in order to avoid over exploitation of relatively cheap resources. For the same reason, there is also a need to disassociate between economic returns and water allocation in order to avoid excessive use.   Information on the actual costs of the private corporations is not easily obtained, therefore, cost estimates are obtained from   Mekorot which supply   about   60% of the total consumption. The costs of Mekorot are assumed to be much higher if the cost of the National Carrier are included. Based on 1996 Mekorot's accounts, the average water cost is US C 31/cum, of which   capital costs (41%), fixed costs (26%) and variable costs (33%).

The marginal cost of water supplied to distant and elevated areas are much higher, especially if a shadow price of water currently estimated at US C 10/cum is added). Kislev [7], suggested the introduction of a charge for water at source (shadow price. The increased demand requiring the development of additional water resources poses a shadow price on the existing source which is equal to the difference between the costs of the two sources. The Shadow price is liable to increase to about US C 50/cum, if a large scale sea water desalination is to be implemented.

3.2   Water tariff

Water tariffs and water allocation are based on a quantitative allocation to groups of consumers, namely: towns, local councils, and water users associations. water prices for the various consumers   are fixed by a parliamentary committee based on recommendation made by the ministry of Finance and the water Commission. Recently, an Increasing Block Rate Prices system is applied for payment based on the amount of the allocation used by the consumers. Different block rates are fixed by the various authorities and differ in the various zones, except for water purchased from Mekorot which is uniform for all consumers, as shown in Table 4.

Table 4. Block Rate Prices for irrigation water supplied by Mekorot, 1995

Water Source	Part of Allocation (%)	Price (US C/cum)
Fresh Water	1 - 50	15
	50 - 80	18
	80 - 100	21
	Average Price	19
Tertiary Effluents	Low Season	14
	High Season	15
Secondary Effluents	Average Price	12

Source: Israel Water Commission Data

A discount of 10 to 40% is applied on the above rate for brackish water with chlorides content ranging between 600 and 1500 mg/l and above, while a penalty is levied on consumers exceeding their allocation. The above prices are still heavily subsidized and the average tariff for irrigation covers only 58% of the average cost. The current tariffs are however on an increasing scale, as shown in Table 5.

Table 5. Water cost and water tariff for irrigation supplied by Mekorot

Year	Cost (cum/ha)	Tariff	
		(US C/cum)	(% of Cost)
1986	14.3	10	70
1993	36.4	16.1	44
1994	34.4	17.1	50
1995	32.8	19.0	58
1996	30.5	21.1	69
1997 (Projected)	28.2	23.2	82

Source: Mekorot Water Co. Financial Statements, various years.

3.3   Economic return

Pricing, using Increasing Block Rates coupled with a continuous increase in water tariff as shown above   allows payment of an effective price without adversely influencing the farmer's income who   still determines his   total consumption according   to his   marginal benefit. However, current tariff policy is likely to drastically affect the total water consumption for irrigation. Over the years, the expenditure on water has steadily increased from about 5% in the eighties to the current 8 - 9% of the total purchased inputs in agriculture (Table 6)

Table 6.   Expenditure on purchased agricultural inputs

Year	Total	Of which Water	
	(US $ million)	(US $ million)	(%)
1990	1950	144	7.4
1991	1920	131	6.8
1992	1752	151	8.6
1993	1807	148	8.2

Source: Central Bureau of Statistics, 1996

The increase in expenditure on irrigation water is also reflected on the economic return of the major crops, shading light on some of the brand crops such as oranges and field crops which hardly sustain the current water tariff as shown in Table 7.

Table 7. Water application rate and economic return for selected crops - 1995

Crop	Average Application Rate (cum/ha)	Economic Return in Various Ecological Zones (US C/cum)			
		North (Hula)	Center (Hadera)	South (Negev)	Total average
Citrus	7500	30.3	25.5	0	18.2
Other Plantations	6000	41.4	104.3	55.0	75.9
Field Crops	4900	24.6	9.3	41.1	32.2
Vegetables		49.7	106.5	129.5	101.4
Total Average		36.7	64.1	71.8	64.1
Total (MCM)		80	155	115	1210

Source: Central Planning Authority, Ministry of Agriculture, 1996

4   Future development

It is anticipated that the current government policy aiming at further reduction of subsidy will continue. The reasoning behind this policy is that higher water rates are the most appropriate tool to use for obtaining the essential water required for urban use, avoiding exerted pressure for further development of marginal and extremely expensive resources such as sea water desalination. The use of market forces seems to be most suitable for   the reduction in fresh water used by farmers, while price incentive is used to encourage the use of reclaimed effluents. Accordingly, water pricing has to gradually change to adapt to a new and effective allocation system and the need to divert fresh water from irrigation to domestic use.

The vehicle for higher prices and rational use of water will be the water suppliers who will be defined as public service under the supervision of the Water Commissioner. The main principle of the proposed reform is to separate the functions of the National Corporation operating the National Carrier from those of the regional schemes that will be privatized and supervised by the public. The National Corporation will operate the main system based on a balanced budget, fixing efficient prices along the main system, selling water to the regional authorities and supervise the function of the regional authorities. The Regional water Authority will supply water to the consumers, purchase water from the National Corporation and treat the return flow.

Internal efficiency within the new water corporations will be achieved by the shareholders who will control the corporation management, while external efficiency will be achieved by the market forces and the value of the shares in the financial market. It is assumed that the shareholders will fix the water price which will be equal to the marginal cost, although the use of a fixed price, return on the invested capital, cost plus, concessions and opportunity cost of the product can also be considered. However, water suppliers will be obliged to charge for water value at source, a shadow price which will balance between supply and demand. This charge will render the historic allocations into a non - effective issue, but maintaining their regulation order in case of emergency or under a series of drought years. Accordingly, the price would include: the value of water at source - shadow price, the cost of production and conveyance, the cost of system optimisation and the cost of subsidy/tax. Subsidized prices if available will be fully indicated and calculated reflecting their portion of the full costs and budgeted for each specific system.

5   Discussion

Nowadays, it is generally accepted that seeing water as a cheaply available public good must be abandoned and its limited supply and competitive economic value should be fully considered. The price of water should not cover only the direct costs of production but also its scarcity value as well. attention should be paid to pollution and over exploitation (mining) and social aspects of equity concerning access to water and the ability to pay of low income groups. Therefore, demand driven management has also to be introduced as one of the most important components of integrated water resources management strategies. The others being physical measures and institutional arrangements.

This implies that water prices that were largely determined by the government rather than by the operation of the free market have to phased out. Recently the use of economic incentives that modify the price of water were introduced including an increase in the flat rate price of irrigation water and the adjustment of the price in blocks as the volume of the applied water increases. These trends follow other countries experience, in which new tools to mitigate problems of competition for water were introduced. A Water Banking system is already in use in California to facilitate the voluntary transfer of water to entities with critical need. Using the same arrangement, cities can sell their treated effluents or form an exchange arrangement with farmers whereby the cities will get drinking water in return for treated effluents [8].

The potential role of economic incentives in modifying farm level decisions to coincide with socially optimal choices were reviewed by Wichelns [9]. Tiered pricing was reported to achieve 10 - 15 % savings in water used for irrigation coupled with effective allocation [10].   Specific management and economic issues relevant to the Israeli water sector were also extensively discussed [7], [11] and [12], concluding that a trade in water rights and real water prices have to be introduced, differentiating between water pricing and (farmers) income allocation.

In the following, a   payment function which combines pricing criteria such as balanced budget,   individual and   group rationality,   historic equity rights and simplicity   with bargaining was suggested [13], assuming that the price and quantity can be effective tools in allocating water quantities to regional authorities which will be allowed to trade between them, while the price will reflect historic rights and actual use within a particular region. In reality, the individual will pay the average price for water used within his rights, while paying the marginal cost for additional quantity. On the other hand he will be credited the difference between the average and marginal cost for unused quantities.

On a larger scale, it was also suggested that the often feared water crisis can be averted through a system of cooperation which involves trade in water permits, licensing short term use of each other's water. Previous studies showed that trading in water rights is a feasible solution to water scarcity [14], while a sensible treatment of water problem could remove obstacles to peace negotiations [15].

The regional acute   water scarcity is further magnified by the continuous and rapid increase in water demand of the urban sector, requiring that more water have to be diverted from irrigation to industry and domestic use. The diversion of fresh water from irrigation for human consumption is fully justified because adding water by desalination at a cost of 80 C/cum and above is not economical. However, it should not be overlooked that water used for irrigation has an additional benefit of US 55 C/cum which justify accelerated development of water reuse schemes for irrigation in general and Israel in particular.

6   References

1.   Caswell, M.F. (1991) Irrigation Technology Adoption Decisions: Empirical evidence. in the economics and management of water and drainage in agriculture,(ed. A. Dinar and D. Zilberman), Norwell, Massachusetts, Kluwer Academic Publishers, pp. 295-312.

2.   Shevah, Y. and Cohen, G. (1995) Advances in Irrigation Technology under arid and semi-arid conditions, 16th European Conference on Irrigation and Drainage, ICID, Sofia, Bulgaria, pp.123-131.

3.   Abbott, J.S. (1984) Micro-Irrigation - Worldwide Usage, ICID Bulletin, 33 (I), pp. 4-9.

4. Fishelson, G. and Rymon, D. (1989) Adoption of Agricultural Innovations: The case of Drip Irrigation of Cotton in Israel, Technological Forecasting and Social Change, 35, p p. 375-382.
5. Shevah, Y. (1996) Israel National Policy for wastewater reuse. Proced. of the Int .Congress on Water Resources Management.   Israel Association of Architects and Engineers, Herzlia, Israel.
6. Arlozorov, S. (1997) Restructuring the Israeli Water Sector Report. Ministry of Finance, Israel (in Hebrew)
7. Kislev, Y. (1996). The Water Economy at a crossroads. Int .Congress on Water Resources Management. Israel Association of Architects and Engineers, Herzlia..
8. Bouwer, H.   1993,   Resolving Competition between urban and agricultural Water Use. ICID, 15th Congress, The Hague: 1429-1437).
9. Wichelns, Dennis 1991b. Increasing Block-rate Prices for Irrigation Water to motivate Drain water Reduction. in The Economics and management of water and Drainage in Agriculture, ed. A. Dinar and D. Zilberman, Norwell, Massachusetts, Kluwer Academic Publishers.
10. Tsur Y. and A. Dinar 1995. Efficiency and equity Considerations in Pricing and allocating Irrigation water. World bank, Policy Res. Working Group No. 1460.
11. Zusman, P. (1991). A conceptual Frame for the Restructuring of the Israeli Water Sector. Economic Quaterly Review, 440-464 (in Hebrew).
12. Yaron, D. (1991). Water Allocation Policy and water prices for irrigation. Economic Quaterly Review, 465-478 (in Hebrew).
13. Brill, E., Hochman, E. and zilberman, D. (1996). "Allocation of Water at a regional Level" Center for Agric. Econ. Res. The vhebrew Univ., Rehovot, Working Paper 96.01(in Hebrew).
14. Becker, N. (1996). Reallocating Water Resources in the Middle East through Market Mechanisms. Proceedings of Int. Conf. on water Resources Management Strategies in the Middle East Hertzlia.
15. Fisher, F. M. (1994). The Economic Framework for water Negotiation and management. The Harvard Middle East Water project. Cambridge: Kennedy School of Government, Harvard University.

APPROCHE DES USAGES GLOBAUX DES EAUX AGRICOLES EN BASSE DURANCE

A. GALAND, M. TERRAZZONI, C.FAYOLLET
Société du Canal de Provence et d'Aménagement de la Région Provençale
Aix-en -Provence, FRANCE.

Abstract

Since the fifties, modern infrastructure has been developed in the field of irrigation distribution in Provence, South-East of France. However, a part of this region, representing 70,000 ha, is still irrigated with traditional methods. This area requires 1,5 billions m^3 per year.

Rhône-Méditérranée-Corse Water Basin Authority asked the Société du Canal de Provence to review the different uses of agricultural water between direct (irrigation) and indirect ones (drinkable water, industries, vegetation consumption...) in qualitative and quantitative terms. Changes in the management of irrigation systems has been tested on a 5,000 ha perimeter with an hydrogeological model of the Durance aquifer, so that the water authorities can evaluate the impact of traditional irrigation system and their possible modernization on both direct and indirect uses. This approach also illustrates the social and economical role of irrigation water in Provence, and its related stakes.

1   Introduction

Le Sud-Est de la France, et notamment la Basse Durance, zone qui correspond à la vallée de la Durance sur les 80 derniers kilomètres de son cours avant sa confluence avec le Rhône et à la plaine de la Crau, est le lieu de nombreux périmètres irrigués selon des techniques traditionnelles, caractérisées par de très fortes consommation en eau.

A une époque où la ressource en eau constitue un enjeu économique majeur, les fortes mobilisations de ressources en eau que génèrent ces systèmes provoquent une remise en cause de ces pratiques agricoles. Ainsi, afin de mieux appréhender l'utilisation de l'eau véhiculée par les canaux agricoles de Basse Durance, l'Agence de l'Eau Rhône-Méditerranée-Corse a confié à la Société du Canal de Provence (SCP) en 1993 une étude pour une approche des usages globaux des eaux agricoles en Basse Durance, autour de deux objectifs:

Water: Economics, Management and Demand. Edited by Melvyn Kay, Tom Franks and Laurence Smith. Published in 1997 by E & FN Spon. ISBN 0 419 21840 8.

- le développement de la connaissance des effets secondaires sur le milieu et des usages des différentes pratiques en matière d'irrigation,
- l'identification d'objectifs pour des utilisations complémentaires à l'irrigation pour l'eau prélevée par les canaux.

2   Connaissances des interactions entre système d'irrigation et milieux connexes

2.1   Diagnostic de la situation locale en matière d'irrigation

2.1.1   Présentation générale des irrigations de la Basse Durance

La plaine alluviale se développe sur les deux rives de la Durance, et dans les deux départements des Bouches du Rhône et du Vaucluse. Elle comporte trois terroirs assez bien différenciés : la **Basse Vallée**, le **Comtat**, et la **Crau**.

L'exploitation des données du Recensement Général Agricole (RGA) de 1988 conduit à envisager pour l'ensemble de la région :

- une **superficie irrigable** de l'ordre de **71 000 ha** et une **superficie irriguée** au moins une fois dans l'année, de l'ordre de **54 000 hectares**.
- **La technique d'irrigation** reste à plus de 80 % l'irrigation de surface, largement dominante sur chacun des trois terroirs.
- **Les réseaux collectifs de distribution** constitués autour de 14 canaux principaux (Débits maximum cumulé: $99 m^3/s$) représentent l'essentiel de l'**origine des eaux** d'irrigation utilisées, mais on note que les irrigations individuelles par forages occupent aujourd'hui une place importante : environ 20 % sur l'ensemble de la zone.

2.1.2   Comparaison entre volumes dérivés et besoins en eau des irrigations collectives

Les séries de données disponibles font ressortir une moyenne observée des prélèvements entre 1972 et 1989 à environ **1,5 milliards de m³**.

Pour une année sèche, type 1991, ces volumes se répartissent comme suit:

- Saison d'arrosage (5 mois, mai à septembre)   **980 Mm³ (75 %)**
- Hors saison d'arrosage (7 mois)   **330 Mm³ (25 %)**
- TOTAL   **1310 Mm³ (100 %)**

En outre, compte tenu de la répartition des cultures irriguées dans chaque région, on a pu estimer entre 3800 et 5700 m^3/ha irrigués les besoins annuels en eau d'irrigation (évapotranspiration moins pluie efficace). L'ensemble de ces besoins annuels est en quasi-totalité réparti sur les mois d'avril à septembre. Sur la base de ces besoins unitaires, en année moyenne, et pour l'ensemble des superficies irriguées collectivement à partir des prises agricoles, les volumes correspondant aux besoins théoriques sont de 150 millions de m^3. **En année sèche, ils représentent globalement 195 millions de m³**.

L'analyse comparée des séries de volumes dérivés et des volumes théoriques consommés par les irrigations collectives sur plusieurs années met en évidence les points suivants:

- Pour le mois de pointe de juillet, les besoins théoriques représentent, pour les deux années considérées, un peu moins de 25 % des apports dérivés.
- Pour la pleine saison d'irrigation (mai-septembre), les besoins représentent moins de 20% des apports.

- En dehors de la pleine saison d'arrosage, les besoins théoriques sont quasi nuls, alors que les dérivations portent sur 300 à 400 millions de m^3.

2.2   Devenir des eaux apportés par l'irrigation gravitaire - impacts de ces pratiques

2.2.1   Aspects qualitatifs

Depuis l'origine du canal principal jusqu'à la plante irriguée, le système d'irrigation comporte trois niveaux :

- le transport par le canal principal (et éventuellement canaux secondaires),
- la distribution par les "filioles" (secondaires et tertiaires) à ciel ouvert qui alimentent les parcelles,
- l'irrigation à la parcelle, mise en oeuvre par les agriculteurs.

Chaque niveau du système génère des "pertes" de natures diverses, qui rejoindront pour l'essentiel le milieu naturel : canaux d'assainissement et ruisseaux, nappe phréatique.

- **Au niveau du transport, et pendant la saison d'irrigation**, le mode de régulation traditionnel par l'amont de ces canaux ne permet pas d'ajuster avec précision les débits dérivés en tête de l'aménagement à ces fluctuations. Le débit dérivé par le canal est en fait ajusté par excès, ce qui se traduit par des rejets aux exutoires ou déversoirs de sécurité.

  En **période hors pointe des irrigations**, les débits nécessaires sont très faibles en valeur relative, mais en l'absence d'équipements de régulation de niveau, il est nécessaire de transiter un débit important pour garantir un niveau suffisant pour l'alimentation des prises, ce qui se traduit par des dérivations en excès très importantes.

- **Au niveau de la distribution** par le réseau des canaux secondaires et tertiaires, les irrigants utilisent alternativement l'eau dérivée, selon une séquence parfois pré-établie dans le cas des "tours d'arrosage", ou prioritairement pour les agriculteurs les plus en amont dans le cas contraire. Il est bien rare, et en particulier en dehors des périodes de pointe des arrosages, que ces passages de "relais" s'effectuent sans temps mort : la prise de tête des filioles est rarement refermée systématiquement pendant ces temps morts, ce qui génère des rejets aux exutoires des filioles. On observera notamment ces rejets la nuit et le dimanche d'une part, en début et en fin de campagne d'autre part.

- **Au niveau de la distribution à la parcelle,** la pratique de l'irrigation de surface (raie ou calan) implique de "faire courir" l'eau de l'amont vers l'aval de la parcelle. Dans cette phase de mise en eau, l'amont de la parcelle reçoit ainsi un volume supérieur à celui reçu par l'aval de la parcelle. Dans la phase suivante, l'infiltration de la dose n'est pas facilement contrôlable avec une grande précision, ce qui entraîne des "colatures" en extrémité aval de la parcelle qui rejoindront les exutoires.

  Ainsi l'irrigation de surface génère structurellement des pertes par **percolation** dans la nappe et des **colatures** vers les exutoires. La technicité des irrigants et la diversité des conditions naturelles (pente, sol, longueur des parcelles etc...) expliquent que dans la pratique, l'efficience des irrigations de surface est extrêmement variable.

2.2.2   Aspects quantitatifs

On a tenté de quantifier, à titre d'ordre de grandeur, les "pertes" décrites ci-dessus en recensant les expérimentations et campagnes de mesures effectuées dans le passé pour la région étudiée.

- Au niveau de l'ensemble **transport et distribution**, c'est un rendement hydraulique de l'ordre de 55 % qui est à envisager pendant la saison d'arrosage. Ce chiffre global pourrait, à titre d'ordre de grandeur, se ventiler entre 85 % pour le transport et 65 % pour la distribution, mais cette répartition est très dépendante du mode de gestion adoptée pour chaque aménagement.   Les "pertes" de 45 % rejoignent le système d'évacuation des eaux de surface. Elles peuvent partiellement être réutilisées:
- pour l'alimentation en eau de la végétation de bordure des canaux et filioles,
- pour l'irrigation d'avaliers à partir d'eau de surface,
- en quantité négligeable pour l'évaporation, et faible en percolation profonde.
- au niveau de **l'irrigation à la parcelle**, et dans le cadre d'une "bonne conduite" des irrigations, on peut estimer que les besoins en eau des cultures représentent environ 45% des volumes entrant dans la parcelle. Les "pertes" se répartiraient entre percolation vers la nappe, à hauteur de 70 % de ces pertes, et colatures, pour les 30 % restants.

2.3   Les usages indirects des eaux d'irrigation gravitaire

2.3.1   Usages indirects correspondant à une activité économique

Les eaux distribuées pour les irrigations collectives donnent lieu à des "pertes" partiellement réutilisées, par les irrigations individuelles, et par les captages d'alimentation en eau potable ou à usages industriels.

Les irrigations individuelles

Les prélèvements dans la nappe, pour les irrigations individuelles par forages, correspondent à environ 12 000 ha et 65 millions de mètres cubes par an pour l'ensemble Basse Durance. Les prélèvements sur les canaux de drainage (irrigations "à partir des eaux de surface") correspondent à 3 000 ha et 17 millions de mètres cubes par an.

Alimentation en eau potable et industries.

Les prélèvements pour l'alimentation en eau potable et à usages industriels correspondent à des volumes annuels respectifs de l'ordre de 59 et 27 Mm^3. La population localisée sur la zone d'influence des canaux d'irrigation est de plus de 500,000 habitants.

2.3.2   Usages indirects environnementaux

Identification et importance des prélèvements d'eau par la végétation entretenue dans un système d'irrigation gravitaire de surface

La végétation rivulaire des canaux principaux, secondaires et tertiaires, des canaux d'assainissement, ainsi que les haies brise-vent en bordure de colature des parcelles s'alimente sur le système hyraulique.A chaque niveau, on peut évaluer la consommation de la végétation, afin d'estimer les besoins annuels des écosystèmes liés à l'eau agricole. Les consommations sont récapitulées Table 1 (cf. 1.4). On remarquera en particulier, les prélèvement des zones humides de Crau (10000 ha) de l'ordre de 5 000 m^3/ha, soit **50Mm^3** annuels.

Intérêt paysager de l'eau et de la végétation entretenue dans un système d'irrigation gravitaire de surface
Des linéaires de canaux d'irrigation et d'assainissement, largement arborés, structurent l'espace, apportent ombre et fraîcheur à travers de parcours variés ponctués par des ouvrages d'art (aqueducs, ponts, chutes, partiteurs, martelières...) souvent remarquables.
Intérêt écologique de l'eau et de la végétation entretenue dans un système d'irrigation gravitaire de surface
    De nombreuses zones du territoire de Basse Durance font l'objet de classements faunistiques et floristiques nationaux et européens (ZNIEFF, ZICO, ZPS, Arrêté de Conservation du Biotope) illlustrant parfaitement l'intérêt écologique de ces zones dans lesquelles on trouve plusieurs types d'écosystèmes:
- Ecosystèmes liés aux cultures,
- Ecosystèmes des haies bordant les parcelles,
- Ecosystèmes des zones humides, principalement en Crau,
- Ecosystèmes aquatiques, de fort intérêt pour la faune notamment.

Appréciation de l'impact de restructurations hydrauliques
Au cours d'une étude menée par la SCP en 1992 sur l'impact de la modernisation des systèmes d'irrigation sur les paysages de Basse Durance, divers exemples illustrant des abandons ou des évolutions du système traditionnel ont été analysés.
Il en ressort que les réactions de la végétation et l'appréciation de l'impact de restructurations hydrauliques sur le paysage et les écosystèmes dépendent de la quantité d'eau qui va arriver au niveau de la plante après le changement. Il est évident que si la végétation n'est plus alimentée d'une façon ou d'une autre en eau, la quasi-totalité des espèces ne supporteront pas le changement pour deux raisons :
- d'une part parce que beaucoup des essences se développant actuellement sur les terroirs sont des essences hygrophiles, et ne peuvent croître sans apport d'eau régulier,
- d'autre part celles qui pourraient supporter la sécheresse en d'autres circonstances ne seront pas capables de s'adapter au changement, compte tenu de leur âge ou de leur état physiologique lié essentiellement à l'entretien.

2.4   Conclusion

Les analyses qui précédent ont permis d'identifier les différents usages directs et indirects des eaux agricoles de Basse Durance. Les estimations proposées conduisent au tableau suivant pour la pleine saison d'arrosage. Comparés aux apports de l'année 1991, **le total des "emplois" représente globalement 40 % des ressources**. (Cf.Table 1).

Table 1: Total des emplois des eaux agricoles

Volumes en jeu en Mm^3	Val de Durance	Comtat	Crau	Ensemble
Apports par canaux d'irrigation	**123**	**453**	**404**	**980**
Besoins en eau des irrigations collectives	20	91	84	195
Besoins en eau des irrigations individuelles				
par forages	1	49	15	65
par eaux de surface	-	7	10	17
Sous-total	**1**	**56**	**25**	**82**
Prélèvements				
AEP	1	19	10	30
Industrie	1	9	4	14
Sous-total	**2**	**28**	**14**	**44**
Végétation rivulaire et haies	2	10	6	18
Zones humides	-	-	50	50
Sous total	**2**	**10**	**56**	**68**
TOTAL DES EMPLOIS	25	185	179	389
(pourcentage usages/apports)	(20 %)	(40 %)	(44 %)	(40 %)

3   Appréciation de l'impact d'une politique de gestion des canaux d'irrigation

3.1   Simulation de divers scénarii de modernisation - l'exemple du Canal St-Julien

Le bilan hydraulique a été établi grâce à un outil de modélisation de la nappe alluviale de Basse Durance. Calé sur les données de la période 1980-1989, soit 10 ans d'historique, il permet d'envisager la simulation d'un certain nombre de changements et d'évolution en matière de modernisation des systèmes traditionnels d'irrigation.

Afin de fournir un champ d'application concret et déterminé, le secteur du canal de St Julien, en Vaucluse dans la région de Cavaillon, a été retenu. Ses 3200 ha irrigués, et son débit maximum conventionnel de 7 228 l/s en font une zone de réflexion intéressante, d'autant plus que la zone d'influence du Canal St Julien intègre également des cours d'eau et canaux d'assainissement, lesquels peuvent être sensibles à certains scénarios d'évolution, et donc à ce titre se révéler d'opportuns indicateurs.

Les simulations d'évolution envisagées ont porté sur les divers modes d'alimentation et techniques d'irrigation envisageables, les mêlant à des degrés divers, afin d'obtenir avec 4 scénarios distincts une palette de résultats assez significatifs:

Le scénario 1 que l'on qualifiera d'extrême, suppose l'abandon total du système collectif et envisage la généralisation des irrigations sous pression à partir de forages dans la nappe. Les volumes totaux libérés représentent environ 125 Mm^3, tandis que le bilan de nappe est réduit de 40 Mm^3. La baisse maximale en été est alors d'environ 3,75 m dans la zone la plus sensible. Les canaux de drainage subiraient de sérieux dommages sur les plans quantitatifs et qualitatifs.

Le scénario 2 suppose une modernisation du transport et de la distribution, notamment par recuvelage, mises sous canalisations et régulation adaptée offrant une efficience de 90 %, et une généralisation des irrigations sous pression à partir du système collectif. Les volumes totaux libérés représentent environ 105 Mm^3, le bilan de nappe étant réduit de 25 Mm^3. La baisse maximale en été est ici de 1 m environ dans la zone la plus sensible. Les canaux et cours d'eau seraient peu touchés.

Le scénario 3 suppose une modernisation du transport et de la distribution permettant la desserte de 80 % de la surface irriguée selon les techniques gravitaires de surface modernisées (i,e : gaines souples, tubes à vannettes). les 20 % restant sont irrigués sous pression depuis des forages dans la nappe. Les volumes totaux libérés sont alors de 87 Mm^3, le bilan de la nappe étant réduit de 11 Mm3. La baisse maximale est limitée à 50cm, les cours d'eau et canaux d'assainissement n'étant alors pas ou très peu affectés par cette évolution des échanges nappes - rivières.

Le Scénario 4 envisage la modernisation du transport et de la distribution pour la desserte de toute la surface dont 50 % pratiquent l'irrigation sous pression et 50 % pratiquent l'irrigation gravitaire modernisée. Les volumes totaux libérés sont de l'ordre de 91 Mm^3, avec un bilan de nappe réduit de 14 Mm^3. La baisse maximale de 0,75 m ne parait pas trop affecter les débits des cours d'eau et reste acceptable.

En conclusion, ces simulations indiquent clairement que la **généralisation**, sur un secteur tel que St Julien, des **forages** pour l'irrigation sous pression en l'absence de système collectif traditionnel orienté vers le gravitaire, causerait de graves dommages à la nappe, aux cours d'eau et à la végétation qui en dépend.

En revanche, le panachage entre irrigation gravitaire modernisée et irrigations sous pression d'une part, forages et alimentation collective d'autre part peut donner des résultats encourageants en terme d'économie d'eau tout en faisant encourir un risque très faible au milieu naturel. Les scénarios 3 et 4 en sont une bonne illustration avec des baisses maximales de 50 à 75 cm et de faibles variations dans les échanges nappe-rivière.

4   Conclusion

Au-delà de leurs usagers directs, les systèmes collectifs d'irrigations traditionnelles de Basse Durance jouent un rôle déterminant :

- dans l'équilibre de la nappe alluviale et les usages qui lui sont associés : alimentation en eau potable des agglomérations et de l'habitat isolé, irrigations individuelles par forages, prélèvements à usage industriel. **Ce sont des usages indirects correspondant à une activité économique.**

- pour le maintien d'une végétation dont l'intérêt paysager et écologique a été souligné, pour le soutien d'étiage de cours d'eau et l'alimentation de certains étangs: **Ce sont des usages indirects environnementaux.**

Les scénarios de modernisations ont permis de mettre en évidence les points suivants:

- Une modernisation des irrigations qui privilégierait des **pompages généralisés dans la nappe** (conduisant à l'abandon du système collectif d'alimentation à partir des eaux dérivées de la Durance) aurait pour conséquence **une baisse importante de la nappe et des impacts dommageables sur les usages indirects et l'environnement**.
- Par contre, des politiques de "modernisation équilibrée",
- privilégiant le maintien d'un système collectif modernisé de transport et distribution à partir des eaux dérivées de la Durance d'une part,
- et promouvant un équilibre entre les nouvelles techniques d'irrigation sous pression d'une part, et le maintien sous une forme modernisée des techniques d'irrigation de surface, d'autre part, permettraient une évolution peu dommageable respectant les équilibres actuels.

De telles politiques de "modernisation équilibrée" libéreraient des volumes d'eau considérables par rapport aux volumes actuellement dérivés. En hiver où les besoins sont insignifiants, c'est la quasi-totalité des volumes actuellement dérivés qui se trouveraient disponibles, dans la mesure où le système actuel de transport et distribution serait modernisé: revêtement et régulation des canaux principaux, modernisation du réseau des filioles remplacé par des réseaux de canalisations fermées. En été, les volumes dérivés seraient ajustés sur les besoins effectifs en tête des parcelles arrosées. Dans le cas de scénarios "équilibrés", c'est environ la moitié des volumes actuellement dérivés qui se trouveraient potentiellement "libérés". Il conviendra alors d'examiner les perspectives de gestion de ces volumes qui seraient rendus disponibles.

Au delà de ces perspectives, cette approche permet de recadrer les enjeux économiques et environnementaux liés à l'usage des eaux dans ces régions de tradition de production agricole, en dépassant le simple cadre de l'agriculture. Elle fournit au monde agricole une légitimité, d'une part pour mettre en place des modernisations permettant les économies d'eau dans le respect de la satisfaction de tous les usages directs comme indirects, et d'autre part pour demander aux pouvoirs publics de mettre en place les dispositifs financiers et institutionnels actant d'une réelle solidarité de bassin.

COSTS OF WATER IN KUWAIT

M ABDAL
Kuwait Institute for Scientific Research.

Abstract
Water is very essential for agricultural development in the state of Kuwait, particularly for the urban landscape and greenery aspects. Kuwait is an arid country with scarce rainfall and limited water resources. Fresh water consumption from desalinated seawater units was calculated in 1995 to 212.4 million cubic meter with total price cost of 539.496 million dollars.Brackish water production was also estimated in 1995 to 99.60 million cubic meter with a total price of 35.86 million dollars. The treated wastewater available for the greenery and urban landscape was determined with efficient piping systems at 107.80 million cubic meter for the price of 19.404 million dollars. The price cost for desalinated seawater is relatively high for agriculture development use, whereas the brackish and treated wastewater can be cost effective for agricultural production in Kuwait.

A desert country like Kuwait with large resource of oil reserve desire better living for its citizens with more beautification of the urban area, and this can be achieved by efficient management of water resources, and training of manpower.

1   Introduction

Kuwait is an arid country with limited water resources for agricultural development and landscape use. The temperature is very high in summer accompanied by frequent sandstorms, and the rainfall does not exceed 115 mm/yr. The people of Kuwait had explored the artistic urban landscape and fascinating regional parks and recreational areas of other countries of similar environment, which encouraged the country to spend more time, effort, and resources on the country's landscape enhancement. Water scarcity is one of several limiting resources to Kuwait's greenery. Underground aquifers, seawater desalination, and sewage wastewater treatment are the major resources for irrigation water. Underground water is hardly recharged from the low rainfall, thus enforcing the full use of the other methods, in addition to

Water: Economics, Management and Demand. Edited by Melvyn Kay, Tom Franks and Laurence Smith. Published in 1997 by E & FN Spon. ISBN 0 419 21840 8.

researching new methods and sorting ways to maximize the utilization of seawater desalination and wastewater treatment.

Sewage treated wastewater proved to be the most convenient resource for plant irrigation and being the least saline, and the cheapest in cost. Other methods of improving the desalinated seawater production is by installing a reverse osmosis system with the existing or new MSF plants.

Training of manpower in agricultural and landscape fields, especially on the maintenance and water management levels, using appropriate irrigation methods with suitable equipment, promoting the greenery plan for public awareness, and installing more network pipelines to the existing wastewater treatment plants, contribute to full execution of the greenery plan.

2   Water resources In Kuwait

There are a few water resources for agricultural development and urban landscape. The rainfall is very limited to around 115 mm/yr. The major resources available are brackish groundwater with salinity content of 3000-10,000 milligram per litre (mg/l),desalinated seawater, and treated sewage wastewater. The brackish water resources are limited due to the low rainfall recharge, and the quality is deteriorating because of the excessive water pumping throughout the years. The brackish water production was estimated in 1995 to be 99.6 million cubic meter with a total price of 35.86 million dollars. The average daily consumption per capita was 0.15 cubic meter of brackish water, and the daily production was 0.273 million cubic meter [1]. The treated sewage wastewater is one of the best water resources for the landscape ornamental use. The salinity level of the treated water is relatively low (1000-2430 mg/l). Furthermore, the treated wastewater produced had met the world health organization irrigation water standards. The treated water available for the greenery and urban landscape and efficient connecting piping system were determined at 107.80 million cubic meter for the price of 19.404 million dollars. Future expansion will eventually increase the total design capacity to 172.5 million cubic meter per year, of which most of the treated wastewater produced can be used for the ornamental landscape [2].

Fresh water production capacity of the existing multi-stage flash evaporation (MSF) desalination plants in Kuwait is approximately 421.4 million cubic meter; however , in 1995 the gross maximum consumption was estimated to be 212.4 million cubic meters with a total cost of 539.496 million dollars. In recent years, technical advances in the area of reverse osmosis (RO) have made this process economically comparable to the widely used MSF technique. By combining an RO plant with the existing or new MSF plant, utilization of the MSF desalination brines could provide substantial quantities of high quality water [1].

High subsurface water tables (within a three meter depth) have been recorded beneath approximately 98 square kilometer of the urbanized areas within Kuwait. Studies are presently underway to evaluate the feasibility of installing extensive underground drainage systems and storage facilities that would enable this water to be used fo rlandscape irrigation. It is estimated that 19.1 million cubic meter per year could become available in the near future. However, due to the extensive collection, storage, and distribution facilities that would be required, it is doubtful if this will prove feasible [3].

Treated industrial wastewater had been utilized in limited quantities for the irrigation of isolated areas of afforestation, motorway landscaping, and selected agricultural pursuits. A large amount of the industrial wastewater originates within the oil refinery industrial area and contains both industrial and municipal components. The majority of this effluent is disposed through existing sewer network, or mixed with seawater and returned to the Gulf. There are studies underway to evaluate the feasibility of expanded treatment and use of this wastewater as a viable source of greenery irrigation water supply.

3   Water requirements for the ornamental landscape

Preliminary landscape planting projections based upon total implementation of the greenery plan, indicate future water consumption could approximate 663.64 million cubic meter per year. As presently envisioned, the water requirement for the greenery program within the urban and coastal area is 268.77 million cubic meter per year. The water needs to support the expanded planting within the desert area green belt are an additional 394.86 million cubic meter per year. The previously established water needs are now being evaluated, based on current studies of the effects of the forced neglect resulting from Iraqi's invasion various plant species. It is anticipated that this study may indicate the feasibility of reduction in the projected water requirements for various mature established species [4].   There are other factors beyond individual plant needs that have an impact on projected water demands, including climate, soil, density   of   plant   population,   and   methods   of   irrigation.   Furthermore, "hardscape"amenities, such as decorative paving and rock, and much surfacing, can be utilized to provide visual enhancement while reducing the area of landscape planting and related consumption of irrigation water.

4   Conclusion and remarks

Ground water had been identified as the primary water resource for agricultural and landscape pursuits in Kuwait. It is probable that future consumption will exceed the water supply. Efficient resources management procedures will be required to ensure the compatibility of water consumption and aquifer recharge.

Greater utilization of treated wastewater is necessary. Strict quality control ensuring public health, in addition to educational programs promoting the the value and safety of effluent use for greenery will be required. The installation of a reverse osmosis treatment with the existing or new MSF system, could provide most of the greenery water needs in the country.

Continued research will be required to identify and evaluate new water resources and prospective adaptable plant species. This activity, supplemented with effective water resource management, appropriate landscape planting, and skilled ground maintenance operations, will ensure successful implementation of the proposed greenery master plan and it's contribution to Kuwait's environmental enhancement.

5   References

1.   KISR (1992). Water Availability for Greening, Final Technical Report, Water Resources Division, Kuwait Institute for Scientific Research,.

2.   I.Y. Hamdan, Puskas , and N. Al-Awadi, (1989). Treated Wastewater for Greenery Plan, Final Report, Biotechnology Department, Kuwait Institute for Scientific Research, Kuwait.

2.   L. Hamdan, (1997). Study of Subsurface Water Rise in the Residential Areas for Kuwait, Final Report, Volume 1, Kuwait Institute for Scientific Research, Kuwait.

4.   KISR (1990) Medium and Long Term (1990-2005). Strategic and Master Plan for Greenery Development and Environmental Enhancement of Kuwait. Second Technical Report, Kuwait Institue for Scientific Research, Kuwait.

VALUING THE MULTIPLE USES OF IRRIGATION WATER

R. MEINZEN-DICK
Environment and Production Technology Division, International Food Policy Research Institute, Washington D.C., USA

Abstract

With growing demand from municipal and industrial uses, there is increasing pressure to transfer water away from irrigation to cities and factories. A large part of the argument for this is based on the relatively low output per unit of water in agriculture (especially grain production), compared to its value for domestic water or manufacturing. But focusing only on the grain output of irrigation systems ignores the many other uses of water in irrigation systems, such as fish, livestock, and garden production, as well as domestic uses. Thus policies to improve the "efficiency" of water use in irrigation or transfer water out of agriculture can cause substantial losses in overall system output and rural livelihoods. Policies to improve the full economic returns to water need to recognize the value--and the implicit water rights--of non-agricultural users of irrigation systems.

Keywords: aquaculture, domestic water supply, environmental benefits, gardens, irrigation, livestock watering, valuation of water, water rights.

1   Introduction

Increasing demands for water for irrigation, municipal, industrial, and environmental uses are creating scarcity and competition for vital water resources in many parts of the world. Developing new water sources to meet these demands is increasingly expensive. The result has been a move to transfer water from agriculture to cities and factories. This has largely been viewed as moving water from "low-value" use in irrigation to "high-value" uses in other sectors. This paper argues that, in many cases, perceptions of the low value of water in irrigation systems are overstated as a result of focusing only on the grain output of irrigated fields without recognizing the value of the many uses of water in irrigation systems. Such a narrow focus can lead to interventions that produce only marginal improvements in the productivity of water, or even reduce the full economic returns to water.

Although the success of irrigation systems is predominantly evaluated in terms of

Water: Economics, Management and Demand. Edited by Melvyn Kay, Tom Franks and Laurence Smith. Published in 1997 by E & FN Spon. ISBN 0 419 21840 8.

their ability to provide water for agricultural production, irrigation systems also provide water for other uses. The quantities of water used in these activities may be small, but these uses have high value in terms of household income, nutrition, and health. The importance of non-agricultural uses of irrigation water in livelihood strategies has implications for irrigation management and water rights, especially as increasing scarcity challenges existing water allocation mechanisms. The following sections of this paper identify some of the multiple uses of water in irrigation systems, who the users are likely to be, and the implications for efforts to improve the "efficiency" of water use, using illustrations from Sri Lanka and other areas.

2   Recognizing the multiple uses of water

Water is an essential element in biological, social and economic systems. Humans, animals, and plants require predictable intake of water to survive. Household livelihood strategies also depend upon water at the most basic level, both for domestic consumption and for productive uses. Water plays an environmental role in maintaining aquatic ecosystems and protecting endangered species. In ecological systems water cycles through the atmosphere, soil, plants and animals, providing a critical input in the food chain.

Although the major purpose of irrigation systems is to supply water at the appropriate time and quantities to agriculture, these systems also provide water for a wide variety of less documented agricultural and non-agricultural needs. An understanding of these uses is essential in the development of flexible water management systems which can meet the needs of a variety of water users [1, 2].

2.1   Agricultural production

The large majority of irrigated areas around the world produces staple cereals. Throughout history, this has been considered vital in feeding populations and averting famines. The success of the green revolution in raising agricultural productivity and achieving food security in Asia would not have been possible without extensive irrigation systems. But the very success of irrigation and the green revolution in raising output and lowering world prices of cereals has contributed to the appearance of low economic returns to irrigation today. The current trend toward globalization of food markets has led many to believe that local and national food security are less important, further reducing the value of irrigation of staple foods. Including the effect of irrigation on world food prices would raise the value of irrigation systems somewhat, but not enough to justify much further investment in irrigation [3]. In this context, irrigation often does not appear economically viable unless it is producing high-value crops such as vegetables and fruits, often for international markets.

In fact, irrigation systems often supply water for a great deal of high-value horticultural and other agricultural production, but because most of this does not take place within the official command area, it is not accounted for. People living in or near the command of an irrigation system use its water for small-scale gardens, often near their homesteads, in addition to production in the formally irrigated fields. The water for these gardens may be diverted from the irrigation system through informal ("illegal") offtakes, or carried to the gardens, or come from household waste water (with the latter originating in the irrigation system). The irrigation water also supports trees and other permanent vegetation around the fields, either by raising water tables or through direct application of water. For example, in Kirindi Oya system in Sri Lanka, the trees and other permanent vegetation around homesteads consume a substantial portion of total water supplies. These homestead sites produce coconuts, vegetables, fruits, and other products, plus environmental benefits (including shade) that raise the value of the property. Because they grow horticultural crops, gardens can be a significant source of income and household nutrition (especially micronutrients)[1]. Yet none of this is accounted for in the output of the irrigation system, because gardens and homesteads are usually small (even if they cover large areas, in aggregate), lie outside the formal command, and much of the produce is used for direct household consumption. The fact that much of the produce of "kitchen gardens" is under the control of women further reduces its visibility, even when the products are sold [5].

2.2   Livestock and aquatic production

Irrigation supples water not only for plants, but also for a wide variety of animals and aquatic life. As in the case of gardens, these can be significant sources of micronutrients and incomes. In addition, livestock, fish, and other aquatic resources are a prime source of high quality protein.

Livestock in irrigated areas depend on the irrigation systems for drinking and bathing water. Cattle and water buffalos are the most obvious consumers, but sheep, goats, pigs, ducks, and other poultry are also important users. The water from irrigation makes a significant contribution to milk production. In areas of India and Pakistan where groundwater aquifers are saline, fresh water from canals increases milk yields, and production levels are noticeably lower when the canal waters are not flowing [6].

The focus on crop production in irrigation is not the only reason livestock water uses are often overlooked. Livestock may be raised primarily outside the command. In some areas, their production is also less visible because they are under the control of women or groups outside the community of designated "beneficiary" farmers [2]. Herders are often from a different ethnic group (caste or tribe) than settled farmers. However, there are important complementarities, with herders grazing their animals on harvested fields, and supplying manure in return.

The living aquatic resources of irrigation systems are often underutilized. Fish provides 70 percent of animal protein resources for the world's population, and demand is growing. At the same time, marine fisheries are reaching the point of overexploitation [7]. Meeting the growing demand therefore requires turning attention to inland waters, and irrigation systems provide a number of habitats for fish, crustacians, and molluscs. These range from the large reservoirs through the flowing canals to small tanks or ponds and includes even the paddy fields, drainage ditches, and pools of standing water along canals.

The value of aquatic resources has not been associated with irrigation systems for a number of reasons. Irrigation and fisheries are usually under different government departments, rather than a single authority. Fishing in large reservoirs is often leased to commercial fishermen, and any income the government receives from leases is rarely credited to the irrigation system or its authority. The same may be true in village tanks, or the fish in tanks and channels may be used for direct household consumption, without any cash "value." The traditional extraction of small fish, crabs, snails, and other aquatic resources in fields and other small bodies of water has, in many parts of Asia, decreased with agricultural intensification, as pesticide use has killed many species.

Yet the total value of fish and other aquatic resources can be substantial. Combined fish-rice production in paddies can produce 30 kg of fish per hectare, up to 300 kg per hectare with trenches in the field, and 5 to 10 tons per hectare of tank area [8, 9]. The biggest barrier to the use of living aquatic resources is the use of pesticides on crops in irrigation schemes. Thus, it may be necessary to trade off some crop production for greater fish output. The economic gain overall may be greater, as the added income through fish production is usually higher than that of the sacrificed rice production.

Aquatic plants such as lotuses or reeds are important sources of income in some areas. In Kirindi Oya, lotus flowers, seeds, and roots are a source of income in the dry season, especially for women. Reeds are used for mats for domestic consumption and sale. Rather than recognizing the value of these products, system managers often view the plants as "weeds" that interfere with efficient irrigation.

2.3 Non-agricultural production
Within the boundaries of irrigation systems, water is used for a wide variety of non-agricultural production, ranging in scale from cottage industries to substantial factories. One common example is the use of irrigation water (and silt) for mixing mud for bricks. Brewing, food processing, dyeing, or retting coconut fiber for coir are other possibilities. Hydropower generation is commonly included in large reservoir systems, but is also possible in smaller systems. In Kirindi Oya, there is increasing demand for water for hotels to support a growing tourist industry along the banks of the irrigation tanks. Whether this water comes directly from canals or from groundwater, it is the irrigation system that supplies the water--either directly or through recharge--to these rural industries.

2.4 Domestic water supply
One of the highest-value uses of irrigation water lies not in agricultural, nor in non-agricultural production, but in domestic water supply to rural populations. Irrigation systems which are channeled through villages provide a convenient water supply for a variety of household consumption needs. Direct withdrawals of irrigation water for drinking and cooking may be declining as protected water supplies (from hand pumps or piped systems) are provided. However, the source of these water supplies is often ultimately the irrigation system, either from a reservoir or from higher water tables that hand pumps and drinking water wells can tap. In Pakistan, hand pumps along canals are a major source of sweet water for drinking, where aquifers are otherwise too saline.

Even if drinking water systems are available, they may not supply enough for all household needs. If water needs to be pumped by hand or carried to the house, it is easier to bathe and wash clothes in the tank or irrigation canal. The closer water is provided to the house, the more people will use it for washing, thereby significantly reducing hygiene-related disease.

The full economic returns for domestic water supplies provided by irrigation are even harder to compute than the value of irrigation water for other types of production. These benefits include reduction in mortality and morbidity (especially of children), as well as savings of labor and time. Time savings come about not only when irrigation brings water supplies close to the home, thereby reducing the time to fetch water, but also through less time to care for sick children. Because in many societies, fetching water and caring for the sick falls to women and girls, they are often the greatest beneficiaries of improved domestic water availability.

Irrigation systems also subsidize municipal water supplies for larger towns and cities in or near the command areas. Dams built for irrigation often provide the storage for supplying cities with water in dry seasons. The Lunkaransar Lift Irrigation System in Rajasthan, India incurs heavy capital and ongoing energy and maintenance costs that may appear to be higher than the productivity of the irrigated area would justify on economic grounds. But at the tail end of the canal, it supplies water to the city of Bikaner, a growing industrial center in the desert, with no alternative source to expand its water supply to meet growing needs. The supply of irrigation water could therefore be regarded as a by-product of a municipal supply system, instead of the reverse.

2.5   Environmental benefits

The negative effects of irrigation systems on the environment (e.g. from large dams flooding forest areas) have received far more attention in recent years than the positive environmental effects of irrigation. While negative effects are certainly substantial in many cases, irrigation systems also provide habitats to a large number of species. The wetlands created in irrigation systems are especially important for migratory and other waterfowl. The tanks of the Kirindi Oya scheme in Sri Lanka, for example, support many species of birds, and the number and variety increased when the system was expanded. Elephants also come to the irrigation tanks within the system to drink and bathe when sources in the nearby wildlife reserves dry up. Thus, irrigation systems can help conserve biodiversity. However, as in the case of fisheries, the use of pesticides to increase crop production creates a substantial threat to wildlife, creating a trade-off between crops and wildlife.

3   Recognizing the rights of multiple users

Because many of the uses of water have been overlooked, those users who have depended on water for gardens, fishing, livestock, or domestic uses have also been overlooked. Although people may have enjoyed customary rights of access, use, and even management of water resources for certain purposes, those rights have rarely been codified. Many of the "marginal" uses (e.g. fishing, harvesting aquatic plants, livestock watering) depend on common property such as tanks, channels, or reed beds. Others, such as gardens and permanent vegetation, rely on water flows that are not easy to observe, such as aquifer recharge, return flows, or "waste" water. This makes it more difficult to document the use of water resources, as well as the formal and informal rights.

As demand for water has increased due to population growth and economic development, there is greater emphasis on formalizing water rights, and especially market allocation of tradable water rights. This is usually associated with assigning all rights to a single holder of the right. In this context marginal users are most likely to lose rights to water. Women have especially experienced difficulties in securing water rights [5]. Accommodating multiple uses requires acknowledging that different users may hold different bundles of rights. There are considerable difficulties in identifying overlapping domains of property rights which are defined by space, time and product, especially for most legal systems and water markets. However, recognizing the rights of multiple users is essential for the development of equitable, as well as efficient, systems of water use.

4   Implications for irrigation system design and management

Managing irrigation systems to maximize the value of all uses, and to accommodate all users, requires a different approach than the conventional attention to irrigation deliveries. Ironically, many measures to improve the "efficiency" of irrigation water use can reduce the total output of the irrigation system by restricting other uses and users. In the Minipe Naga Deepa system in Sri Lanka, introducing canal lining to reduce seepage losses lowered the water table beyond the reach of the hand pumps people used for drinking water. An expensive piped water supply system would be required to replace the drinking water supply. In other areas of Sri Lanka, use of tubewells to supplement surface supplies lowered the water table below what the trees could tap, and killed valuable permanent vegetation. In systems in Pakistan and India, rotational deliveries have been introduced instead of continuous supplies to "save" water in irrigation [6]. But while the crops can survive for up to several weeks between waterings, the people and livestock who depend on that water for drinking, bathing, and other purposes cannot.

Recognizing the multiple uses of water in irrigation systems provides a potential means of improving their overall productivity. Rather than maximizing the output or efficiency of a single activity (e.g. grain production), it is better to optimize the joint production of field crops, gardens, livestock, fish, and other activities. This may mean reserving water for certain uses (e.g. bathing humans and animals), or giving water releases when the main field crops do not require irrigation. More difficult than deciding the quantities of water needed for various uses is dealing with the impact of each use on water quality. Reducing pesticide use on field crops (and gardens) is likely to be the most important change required to improve the use of irrigation systems for domestic water supply, fish, and wildlife, but greater attention to the quality and pathways of drainage flows from all uses is called for.

The biggest obstacle to improving the total value of water in irrigated areas lies in the artificial sectoral boundaries between agencies involved in water resources. While the state as a whole has responsibility for overall water use, the executing agencies have neither mandate nor incentive to create integrated projects or to balance the needs of various users [1]. Most agencies dealing with water resources have responsibility only to deal with irrigation or drinking water or industry or the environment. Thus, they operate within strict limits on the quantity of water use, or respond only to single constituencies (e.g. farmers, industrialists). Organisations such as the Water Resources Council established in Sri Lanka in 1995 can provide the framework for interagency coordination [10], but to be effective these need to reach beyond the national level, down to the actual implementation of projects.

Although they do not explicitly encourage alternative water uses, irrigation departments often take steps to ensure that the physical structures do not suffer from repeated non-agricultural uses. Lining or steps may be built in canal banks to control damage to irrigation structures by washers, bathers, and livestock [1]. In some systems special water outlets have been constructed which divert water from canals into pools for livestock drinking water and bathing. When grazing is scarce, water may be diverted out of the main canal system to promote the growth of grass for fodder.

Where irrigation development has also involved resettlement, the executing agencies have had to pay attention to the full spectrum of users' needs for water. For example, the Indira Gandhi Nahar Pariyojana (IGNP) in Rajasthan, India, has integrated ponds for domestic use into the system design because there was no other source of water for humans and livestock when the canal was not running. Financially autonomous public agencies may also turn to other uses of the irrigation systems as a means of recovering their costs. Cross subsidization from fishing, recreation, river pearls, tourism, and other uses has become a significant source of revenue for irrigation projects in China [11, 12].

As demand for water in industry and municipal systems grows, it has often been easier to look to irrigation systems to give up water than to develop other sources, especially as the readily exploited sources of water supply have been used up. But when transfers out of agriculture and rural areas are considered, it is essential to evaluate the *total* output foregone, and the implications of *who* gains and loses when water is diverted for irrigation systems. For example, there is a proposal to build an oil refinery near the Kirindi Oya system in Sri Lanka, that would require 25,000 cubic meters of water per day.[2] This would need to be evaluated not only in terms of the amount of paddy production foregone to divert water for the industry, but also the impact that this diversion would have on gardens, domestic water supply, livestock, and aquatic plants and animals. With all these costs included, diverting water out of irrigation systems may not be the cheapest way to meet growing needs in other sectors.

5   Discussion and conclusions

The consideration of full economic and social returns to all water uses is necessary to increase the accuracy of evaluation of alternative water development strategies. Identifying all the uses and users of water in irrigation systems, and the connections among them, is a critical first step. But we also need more information on the full economic value of irrigation. This paper has provided illustrations of the range of interconnected uses, but few studies have quantified the values of these uses, or explored the extent of competition and complementarity between them. Although cost-benefit analyses should consider all costs and benefits associated with resource use, in practice most secondary uses remain invisible. There is a need for considerably more research on the multiple uses of water in irrigation systems, and to develop methodologies to estimate the values of those uses.

It is important not only to identify and value all the uses of irrigation water, but to identify who the users are. Where water rights are being formalized, maintaining access to water is critical for the livelihoods of not only the primary irrigators, but of all water users. To do this requires recognizing the rights of marginal users, including women as well as men, farmers, herders, fishers, and other groups.

When we begin to examine the many ways in which water is used in irrigation systems, it becomes clear that intersectoral competition for the resource does not only occur between cities and rural areas, but can also occur at the local level. The quantity, quality, and timing of water supplies are all areas of potential conflict. While sources of competition are not hard to find, the challenge is to identify complementarities between uses of water resources within irrigation systems. This requires changes in the institutions governing water resources, including government agencies, user organisations, and property rights. Maximizing the output of a single product (e.g. grain) may not provide the optimal solution, from a productivity or equity standpoint. Rather, learning to accommodate different uses and users offers a way to raise total output per unit water, and to improve the viability of irrigation systems in an era of water scarcity.

6   Acknowledgement

This paper is part of an ongoing study on Valuing the Multiple Uses of Irrigation Water, under the CGIAR System-Wide Initiative for Water Management. I am grateful to Randolph Barker, Ramesh Bhatia, Lawrence Haddad, John Hoddinott, Lee Ann Jackson, Waqar Jehangir, Peter Jensen, Flemming Konradsen, Kusum Athukorala, Yutaka Matsuno, Rekha Mehra, Mark Prein, Daniel Renault, Claudia Ringler, Mark Rosegrant, and Margreet Zwarteveen for their comments and insights. All responsibility rests with the author.

7   References

1.   Yoder, R. (1981) Non-agricultural uses of irrigation systems: Past experience and   implications for planning and design. Prepared for the Agricultural Development Council, New York (mimeo).
2.   Salem-Murdock, M. and Niasse, M. (1996) Water conflict in the Senegal River   Valley: Implications of a "no-flood" scenario. Dryland Programme Issue Paper   No. 61. International Institute for Environment and Development, London.
3.   Svendsen, M. and Rosegrant, M. W. (1994) Irrigation development in Southeast   Asia beyond 2000: Will the future be like the past? *Water*

International Vol. 19;   pp. 25-35.

4   Meinzen-Dick, R., Sullins, M.   and Makombe, G. (1994) Agroeconomic performance of smallholder irrigation in Zimbabwe, in *Irrigation System Performance in Zimbabwe*, (eds. M. Svendsen, M. Rukuni, and R. Meinzen-Dick)   UZ/IFPRI Irrigation Performance Project, Harare

5.   Zwarteveen, M.   (1997) Water: from basic need to commodity. *World Development* (forthcoming).

6.   Bhatia, R. (1996). Personal Communication.

7   FAO Fisheries Department.   (1995).   *The State of World Fisheries and Aquaculture*. Food and Agriculture Organisation of the United Nations, Rome.

8.   Muir, J.F. (1995). Aquaculture Development Trends: Perspectives for Food Security.   Paper presented at International Conference on Sustainable Contribution of Fisheries To Food Security, Kyoto, Japan, 4-9 December.

9.   Cruz, C.R., Lightfoot, C., Costa-Pierce, B.A., Carangal, V.R., and Bimbao, M.P., eds. (1992) *Rice-Fish Research and Development in Asia*. International Center   for Living Aquatic Resources Management, Manila.

10.   Arriëns, W.L., Bird, J., Berkoff, J., and Mosley, P. (1996) Towards Effective Water Policy in the Asian and Pacific Region... Volume 1: Overview of Issues and Recommendations. Asian Development Bank, Manila.

11.   Gitomer, C.   (1994) Price regulation in the reform of irrigation management in rural China. Paper presented at the Political Economy Workshop, Berkeley, California, 10-13 February.

12.   Svendsen, M. and Changming, L. (1990) Innovations in irrigation management and development in Hunan province. *Irrigation and Drainage Systems*, Vol. 4. pp. 195-214.

1 In a comparative study of irrigation systems in Zimbabwe, small-scale gardens on dambo wetlands were found to have significantly higher productivity per unit land and per unit water than formal smallholder irrigation systems [4].

2 Supplying that volume of water every day would mean setting aside considerably more, to allow for seepage and evaporation losses, especially during the dry season--a fact that is too often overlooked in calculations of municipal and industrial demand.

AVAILABILITY OF COLORADO RIVER WATER FOR TRANSFER IN SOUTHERN CALIFORNIA

K. M. BALI and R. A. GONZALEZ
University of California Desert Research and Extension Center, Holtville, CA, USA

Abstract

Colorado River water is the only available source of irrigation and drinking water in the Imperial Valley. As much as 345,375 ha-m (2.8 million ac-ft) of Colorado River water are used every year to irrigate more than 202,342 ha (500,000 acres) of land in the Imperial Valley. The recently proposed water transfer draft between Imperial Irrigation District and the San Diego County Water Authority calls for transfer of up to 49,337 ha-m (400,000 ac-ft) annually of Imperial Valley's allotment of Colorado River water. This study was conducted to evaluate the potential availability of water for transfer and the economic value of conserved water from two on-farm water conservation measures; 1) implementation of tailwater return systems and 2) implementation of pressurized irrigation systems. Implementation of tailwater return systems is limited to field crops grown on heavy clay soils and implementation of low-volume irrigation systems is only feasible on specific vegetable crops. The implementation of tailwater return systems will reduce surface runoff from the current Valley-wide average of 16.8% (of delivered water) to approximately 5% and may yield a maximum potential savings of approximately 24,454 ha-m (198,253 ac-ft) of water. The implementation of low-volume pressurized irrigation systems may yield a maximum potentials savings of approximately 12,237 ha-m (99,209 ac-ft) of water. The use of low-volume irrigation systems requires high capital investment and significant changes in cultural practices. It also yields higher value of conserved water as compared to tailwater recovery systems. Costs associated with decline in yield and increase in soil salinity should be considered in determining the economic value of water.

Keyword: Irrigation management, water availability, water conservation, water transfer.

1 Introduction

Water is a limited resource in California. Water demands among agricultural and

Water: Economics, Management and Demand. Edited by Melvyn Kay, Tom Franks and Laurence Smith. Published in 1997 by E & FN Spon. ISBN 0 419 21840 8.

urban users in California require strategies for water transfer and management that are based on practical water conservation practices. Water transfer and marketing is playing a major role in water resources management in Southern California. Water transfers from agricultural to urban regions in California have been used as a partial solution for water shortages in the state through implementation of the state water bank during the drought period of 1987-92.

In 1989, the Imperial Irrigation District (IID) and the Metropolitan Water District (MWD) of Southern California signed a long-term agreement to transfer 13,075 ha-m (106,000 ac-ft) of water to MWD annually for 35 years. In exchange for water, MWD is providing funds for water conservation projects in the Imperial Valley. The price of transferred water is about $811 per ha-m ($100 per ac-ft, 1989 cost basis). The annual payment from MWD to Imperial is about $350 million over the 35-year period of the contract. Most of the conserved water came from district-level conservation projects such as building new reservoirs and from on-farm water conservation projects.

In 1996, the IID and the San Diego County Water Authority (SDCWA) started negotiating another water transfer agreement. The recently proposed water transfer draft between IID and SDCWA calls for transfer of up to 49,339 ha-m (400,000 ac-ft) annually of Imperial Valley's allotment of Colorado River water [1]. Water available for transfer must be conserved water from on-farm water conservation programs and district-level structural conservation projects. Most of water available for transfer may come from on-farm water conservation projects. This study was conducted to evaluate the potential availability of water for transfer and the economic value of conserved water from two practical on-farm water conservation measures; 1) implementation of tailwater return systems and 2) implementation of pressurized irrigation systems. This study considers only the above conservation measures and does not consider other on-farm or district-level conservation measures that might be implemented nor does it include the possibility of a reduction in farmable land due to urbanization or land fallowing. These additional measures are expected to yield additional water savings.

2   Location and history

The Imperial Valley is located about 193 km (120 miles) east of San Diego, CA in the Northwestern Sonoran Desert. The Colorado River water was diverted to the Valley at the turn of the twentieth century. Colorado River water is diverted to the Valley through the All American Canal. Approximately 75% of California's agricultural water entitlement of Colorado River water is used in the Imperial Valley, the other 25% is used in Bard, Palo Verde, and Coachella Valleys in the southeastern desert area of California. The Imperial Valley has the largest irrigation delivery systems in Southern California and its agriculture contributes about one billion dollars in gross income to the California economy [2].

3   Irrigation practices

The Colorado River is the lifeblood of the low desert of Southern California. It is the only available source of irrigation and drinking water in the Imperial Valley. As much as 345,375 ha-m (2.8 million ac-ft) of Colorado River water are used every year to irrigate more than 202,342 ha (500,000 acres) of land in the Imperial Valley. The cost of water in the Valley is $97.29 per ha-m ($12 per ac-ft) in addition to a water availability charge of $12.36 per ha per year ($5 per acre per year). Surface irrigation, the primary method for irrigation in the Imperial Valley, is used on more than 95% of the irrigated land. Drip irrigation is used on less than 5% of the irrigated area and mostly on vegetable crops. Field crops are grown on 80% of irrigated land [2]. Surface and subsurface drainage water from irrigated fields enters the Salton Sea which serves as a drainage sink for the Imperial and Coachella Valleys (formed in 1905). The Salton Sea continues to exist because of the drainage water from agriculture in Imperial and Coachella Valleys as well as flow of agricultural drainage and untreated and partially treated sewage from the Mexicali Valley. Several water quality issues exist in the Imperial Valley because of surface and subsurface drainage and its impact on the Sea, for which water conservation and transfer play a major role.

4   Water conservation

4.1   Implementation of tailwater return systems

In 1985, IID initiated a demonstration project to evaluate the potential use of tailwater recovery systems (TRS) in the Valley [3]. The estimated cost of a runoff recovery system designed by IID was about $1,186 per ha ($480 per acre, 1990 cost basis). The annualized pump back system cost was approximately $173 per ha ($70 per acre, 1990 cost basis) which includes capital and operation/maintenance costs. The impacts of such systems on soil salinity were not conclusive [3]. Changes in soil salinity ranged from 29% higher to 38% lower when compared to conditions prior to the construction of these systems. Increases in soil salinity may result in decline in crop yield and result in an increase in the economic value of conserved water. The economic value of conserved water is defined here as the costs associated with conserving a unit of water (ha-m or ac-ft) and includes capital, operational, and maintenance costs in addition to costs related to decline in yield as a result of certain water conservation measure. Based on data available to date, the costs associated with decline in crop yield are negligible with such a system. The effect of expected increase in soil salinity and potential reduction in crop yield on water value is also considered here.

Tailwater recovery systems yielded a 12% potential savings of delivered water [3]. The average agricultural tailwater in the Valley is approximately 16.8% of applied water [4]. The average depth of applied water for each of the three major field crops [5] in the Imperial Valley and the average acreage [2] are shown in Table 1. The weighted average depth of applied water for these major field is 1.67 ha-m/ha per year (5.48 ac-ft/ac per year). Therefore, tailwater recovery systems yield an average of 0.200 ha-m/ha per year (0.6576 ac-ft/ac per year) with an average value of conserved water at \$864.87 per ha-m (\$106.45 per ac-ft, 1990 cost basis). Summary of the value of water saved and the amount of water that can be saved are shown in Table 1. All values are adjusted to account for inflation and reported in 1996 cost basis (\$1 in 1990 is equivalent to approximately \$1.23 in 1996).These values are based on the assumption that costs associated with increase in soil salinity and decline in crop yield are negligible. The effect of salinity on crop yield and the economic value of conserved water is discussed in section 4.2.

Table 1. Average acreage, water use, and value of conserved water for major field crops in the Imperial Valley (1990-1995).

Crop	Area (acre)	Avg. water use (ft)	Vol. of cons. water (ac-ft)	Water value* (1996) (\$ per ac-ft)
Alfalfa	181,134	6.5	141,285	110.38
Sudangrass	63,106	4.8	36,349	149.48
Wheat	57,275	3.0	20,619	239.17
Total	301,515	5.48**	198,253	131.22**

*Based on negligible changes in soil salinity
** Weighted average

Another method was recently proposed to reduce or eliminate surface runoff water in heavy clay soil [6]. The surface runoff reduction (SRR) method is based on a relatively simple technique that predicts the cutoff time necessary to minimize or eliminate runoff and to improve water use efficiency. While the method is applicable for all soils, it works best in heavy clay soils. The method is a combination of a volume balance model [7] and a two-point measurement method [8]. The objective in heavy clay soil is to have enough water to fill soil cracks with little or no runoff. Preliminary results of research conducted at the University of California Desert Research and Extension Center in the Imperial Valley indicate that the method is suitable for clay soils, however, an increase in soil salinity and a reduction in yield were observed at the tail end of the field. While there are no capital costs associated with this method, there is the additional time needed for irrigation planning and management and other costs associated with decline in yield as a result of salinity buildup in soil. The advantage of this method is that it can be implemented in relatively short period of time in response to short-term water needs by urban regions in Southern California.

The implementation of tailwater return systems is limited to field crops grown on heavy clay soils. Approximately 60-70% soils in the Imperial Valley are classified as heavy clay soil [9]. The IID estimates that approximately 40% of heavier soils will benefit from tailwater return systems [2]. We estimate that at most 121,406 ha (300,000 acres) will benefit from the implementation of tailwater return systems and surface runoff reduction method. Our estimates is based on the average acreage of alfalfa, sudangrass, and wheat between 1990 and 1995 [2].

The implementation of tailwater return systems will reduce surface runoff from the current Valley-wide average of 16.8% (of delivered water) to approximately 4.8% and yield a maximum potential savings of approximately 24,454 ha-m (198,253 ac-ft) of water. Because of the large acreage of alfalfa in the Valley and its relatively high consumptive water use, water saving from alfalfa is the highest and the value of conserved water is the least when increases in soil salinity are negligible. In addition, implementation of the above methods will reduce surface elevation of the Salton Sea and minimize the amount of agricultural chemicals in runoff water.

4.2 Effect of salinity on crop yield and water value

We consider here the effect of increases in soil salinity on crop yield and economic value of conserved water for the three major field crops in the Imperial Valley. The effect of soil salinity on crop relative yield can be estimated with following equation [10]:

$$Y = 100-b(EC_e-a) \tag{1}$$

where Y is the relative yield for soil salinities exceeding the threshold of any given crop, a is threshold value (dS/m) or the maximum root salinity at which no reduction in yield is observed, b is % reduction in yield per dS/m above the salinity threshold value, and EC_e is the average root zone salinity. Soil salinity levels below the threshold values will not affect crop yield or quality. The values of a and b are; 2.0 dS/m and 7.3 % per dS/m for alfalfa, 2.8 dS/m and 4.3 % per dS/m for sudangrass, and 5.9 dS/m and 3.8 % per dS/m for wheat [10].

The value of conserved water increases as the average root zone salinity exceeds a threshold level for a given crop. The value of conserved water can be estimated with the following equation:

$$V = (C/W)+[(100-Y)*Z/100] \tag{2}$$

where V is the value of conserved water ($ per ac-ft), C is cost of water conservation measure ($ per ac), W is the amount of conserved water (ac-ft/ac), Y is as defined above, Z is the crop value ($ per ac) when the average root zone salinity below the threshold level. Crop values for the three major field crops in the Imperial Valley (obtained from the 1995 Imperial County Agricultural Crop and Livestock report and adjusted for inflation to reflect 1996 cost basis,$1 in 1995 is equivalent to approximately $1.038 in 1996) are shown in table 2. Fig 1. shows the economic value of conserved water at various salinity levels for major field crops in the Imperial Valley.

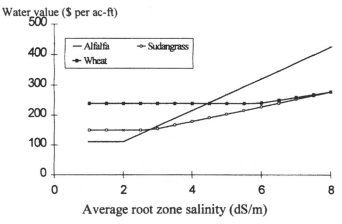

Table 2. Average yield and crop value for major field crops in the Imperial Valley.

Crop	Yield (tons per ac)	Value per unit* (1996) ($ per ton)	Value per ac*(1996) ($ per ac)
Alfalfa	7.94	90.79	720.87
Sudangrass	6.50	88.23	573.50
Wheat	2.75	176.46	485.27

* 1996 value basis

4.3   Implementation of low-volume irrigation systems

The implementation of low-volume irrigation systems is only feasible on specific vegetable crops (approximately 20% of the irrigated area). The value of the potential decline in yield and increase in soil salinity should be considered. We consider here the additional amount of water needed for leaching to maintain soil salinity at levels optimum for crop production. The amount of water used by vegetable crops varies widely. The amount of water saved from implementation of low-volume irrigation systems on vegetable crops will vary since vegetable crops are grown at various times during the year. The average acreage of major vegetable crops grown in the Imperial Valley and their current consumptive water use [11] are shown in Table 3. These crops represent more than 80% of vegetable crops in the Valley during the period between 1990 and 1995.

The cost of conversion to drip irrigation system varies considerably from one location to another and depends on many variable related to crop, soil, climate, type of system, and water source. Here, we use the average cost of a typical drip irrigation system in the Valley of $1532 per ha ($620 per acre, 1996 cost basis) [11] which includes costs of the system and labor. We assume the system will last for three years or three growing season. The average seasonal cost associate with removal of drip tape, chemical for maintenance, and additional costs is $193 per season [11].

Drip irrigation is relatively new in the Imperial Valley. Crop water use figures for drip irrigated crops in the Imperial Valley are not available and vary widely depending on system type (surface or subsurface), soil type, growing period (planting and harvesting dates), and location (variability in reference evapotranspiration). Several studies have shown that while crop water use under subsurface may be lower than under furrow or sprinkler irrigation, crop water use is virtually the same under surface drip and furrow systems [12]. Therefore, we considered the above variables and estimated the minimum, maximum, and average crop water use under drip irrigation for each of the major vegetable crops in the Valley independently from crop coefficients and reference evapotranspiration [13, 14, &15] during the growing season. The additional water needed for leaching was calculated and added to crop water use [10]. The predicted water use and amount of water that can be saved from converting to drip irrigation systems are shown in table 3. These figures include a leaching fraction of 15% or a minimum of 0.10 m (4 in) of water for salt leaching after each season. The frequency of leaching depends on crop type, salinity of irrigation water, and salinity level in the root zone. Our assumptions are based on the leaching requirements of the most sensitive vegetable crop in the region. Leaching could be done after each season using a sprinkler irrigation system or after several seasons using flood irrigation. To maintain salt balance, we estimated that a minimum of 1 ac-ft/ac is needed after three seasons of continuous planting. These values are included in the predicted water use figures in table 3.

Table 3. Average acreage, current water use and predicted water use figures for major vegetable crops in the Imperial Valley (1990-1995).

Crop	Area (Acres)	Water use (ac-ft/ac) Current*	Predicted** Avg. Range	Avg. Cons. water (ac-ft/ac)	(ac-ft)
Head lettuce	28,376	3.00	1.70 (1.1-2.3)	1.30	36,889
Cantaloupes	20,377	3.75	3.15 (2.7-3.6)	0.60	12,226
Carrots	13,993	3.75	2.10 (1.4-2.8)	1.65	23,088
Onions	10,924	4.50	3.10 (2.4-3.8)	1.40	15,294
Broccoli	7,808	3.50	2.00 (1.4-2.6)	1.50	11,712
Total	81,478				99,209

*Surface irrigation, **Drip irrigation

We estimate that approximately 40,469 ha (100,000 acres) will benefit from the implementation of low-volume irrigation systems. In this analysis, we consider the number of acres that could be converted to drip irrigation (vegetable crops). The implementation of low-volume pressurized irrigation systems will save approximately 12,237 ha-m (99,209 ac-ft) at an average value of $2,661 per ha-m ($328.24 per ac-ft, 1996 cost basis). The amount of water that can be saved and the value of conserved water are shown in table. 4.

Table 4. Volume and value of conserved water for major vegetable crops in the Imperial Valley (1990-1995).

Crop	Vol. of cons. water (ac-ft)	Water value (1996) ($ per ac-ft)
Head lettuce	36,889	307.44
Cantaloupes	12,226	666.11
Carrots	23,088	242.22
Onions	15,294	285.48
Broccoli	11,712	266.47
Total	99,209	328.24*

* Weighted average

In general, the value of saved water using drip irrigation systems is at least 250% higher than the value of conserved water from TRS. However, the value of conserved water from TRS increases as soil salinity increases due to decline in yield. The use of low-volume irrigation systems requires high capital investment and significant changes in cultural practices. It also yields higher value of conserved water as compared to tailwater recovery systems.

5   Conclusions

The implementation of tailwater return systems will reduce surface runoff from the current Valley-wide average of 16.8% (of delivered water) to approximately 5% and yield a potential savings of approximately 24,454 ha-m (198,253 ac-ft) of water. The implementation of low-volume pressurized irrigation systems will save approximately 12,237 ha-m (99,209 ac-ft). The use of low-volume irrigation systems requires high capital investment and significant changes in cultural practices. It also yields higher value of conserved water as compared to tailwater recovery systems.

6   References

1.   Imperial Irrigation District (1996) *Water requirements and availability study-Executive summary* (Draft). Imperial Irrigation District, pp1-8.
2.   Birdsall, S. L. (1996) 1990-995 Imperial county agricultural crop and livestock reports. Imperial County, California.
3.   Boyle Engineering Corporation (1990) *Tailwater recovery demonstration program study.* Special technical report. Imperial Irrigation District.
4.   Boyle Engineering Corporation (1993) *On-farm irrigation efficiency.* Special technical report. Imperial Irrigation District, CA
5.   Mayberry, K. S. and Holmes, G. J. (1996) *Guidelines to production costs and practices for Imperial county-field crops 1996-1997.* University of California Cooperative Extension, CA.
6.   Wallender, W. W. (1986) Furrow model with spatially varying infiltration. *Transactions Of Asae,* Vol. 29, No. 4. pp. 1012-1016.
7.   Elliot, R. L. and Walker, W. R. (1982) Field evaluation of furrow infiltration and advance functions. *Transactions Of Asae.* Vol. 25, No. 2. pp. 396-400.
8.   Bali, K. M. and Grismer, M. E. (1997) Evaluation of Surface Irrigation Systems in Heavy Clay Soils, Submitted to *California Agriculture,* 12 pp.
9.   United States Department of Agriculture Soil Conservation Service (1981) *Soil survey of Imperial county, California, Imperial valley area.* USDA.
10.   American Society of Civil Engineers (1990) Manuals and Reports on Engineering Practice No. 71. *Agricultural Salinity Assessment and Management* (ed. K. K. Tanji) , ASCE, New York.
11.   Mayberry, K. S. and Holmes, G. J. (1995) *Guidelines to production costs and practices for Imperial county-vegetable crops* 1994-1995. University of California Cooperative Extension, CA.
12.   Hanson, B., Schwankl L., Grattan R., and Prichard T. (1994) *Drip Irrigation for Row Crops.* University of California Water Management Series Publication No. 93-05.
13.   Goldhamer, D. A. and Snyder, R. L. (1989) *Irrigation scheduling: A guide for efficient on-farm water management.* University of California. DANR publication 21454.
14.   Snyder, L. R., Grattan, S. R., and Schwankl, L. J. (1989) *Drought Tips for Vegetable and Field Crop Production.* University of California. DANR publication 21466.
15.   Food and Agriculture Organisation of the United Nations (1984) *Crop Water Requirements. FAO Irrigation and Drainage Paper 24.* FAO, Rome.

EFFECTS OF WATER MANAGEMENT ON WATER VALUE RELATIVE TO ALFALFA YIELD
A case study for the low desert of California

M.E. GRISMER
LAWR, Hydrology, UC Davis, Davis, CA, USA
K.M. BALI and F.E. ROBINSON
UC Desert Research & Extension Center, Holtville, CA, USA
I.C. TOD
Independent Irrigation/Drainage Consultant, Cambridge, UK

Abstract
The value of water conservation to irrigated agriculture can be linked to the capital investment required for improved irrigation water management, or costs associated with loss in yield due to smaller annual water applications. In this study, we consider the value of water attributable to loss in alfalfa yield in the Imperial Valley, California resulting from reducing the frequency of irrigation during peak evapotranspiration demand periods, and use of a cut-off time border irrigation method that eliminates tail-water runoff. Reducing the number of annual water applications by two, four and seven times as compared to an optimally watered control resulted in yield losses of 1530, 3840 and 5230 kg/ha, or an annual water saving of 53, 362, and 627 mm/ha, respectively. At a market value of $0.11/kg ($100/ton), the average value of each increment of water saved for the progressively fewer annual irrigations was $3308, $1194 and $940 /ha-m, respectively. Following reduced irrigation treatments, salt leaching was required to restore satisfactory soil conditions. The cut-off time irrigation method resulted in similar annual water savings (108 mm/ha), yield losses (1431 kg/ha) and slightly smaller water values ($538 /ha-m) but incurred only limited salinization at the low end of the fields after one year.
Keywords: Water conservation, soil salinity clay soils, irrigation management, crop yield

Water: Economics, Management and Demand. Edited by Melvyn Kay, Tom Franks and Laurence Smith. Published in 1997 by E & FN Spon. ISBN 0 419 21840 8.

1   Introduction

Water conservation is often considered an economic good that is associated with more efficient use of a limited resource that is subject to competing demands. In semi-arid or arid regions, the economic value of water to a growing urban population is linked to the costs of developing additional water resources. Similarly, the economic value of water to the agricultural sector associated with water conservation, or improved irrigation management can be linked to the capital investment required for improved application efficiency, *or* costs associated with loss in crop production income due to lower yields resulting from limited water availability.

In this study, we determine the value of water associated with the effects of reducing irrigation frequency during peak evapotranspiration (ET) periods and use of a cut-off time border irrigation method on alfalfa yield and soil salinization of clay soils in the Imperial Valley, California. We consider alfalfa because it is a major crop in the Imperial Valley that has a high consumptive water use. Reducing irrigation frequency during high ET periods was originally proposed by Lehman (1979) when he found that the ratio of alfalfa yield to water use declined substantially during periods of high water use. Robinson et al. (1994) noted that the net reduction of one 120 mm irrigation per year on the Valley's over 80 000 ha of alfalfa would yield an annual water savings of over 10 000ha-m that could potentially be reallocated for urban use. We note that approximately 40% of the Valley alfalfa is planted on clay to clay loam soils that are typically border irrigated. Grismer and Tod (1994) developed a simple cut-off time method for border irrigation of cracking clay soils that eliminates tail-water runoff. Tail-water runoff typically comprises 10 to 25% of the individual water application, or about 17% of the total applied water on a Valley-wide basis. We consider both reducing irrigation frequency and use of the simple cut-off time method of irrigation because these are water management practices that can be readily incorporated into existing cultural practices in the Valley without additional capital investment by the growers.

2   Materials & methods

We conducted this study over the period 1991 through 1996 at the Desert Research and Extension Center near Holtville, CA on an alluvial clay loam soil (hyperthermic Typic Torrifluvent) referred to as Area at the Center. The reduced irrigation portion of the study was a randomized block design that included several varieties of alfalfa (only CUF 101 will be considered here) that was conducted from fall 1990 (planting) through 1993. The block design consisted of 9.1 m wide by 83 m long plots. The four irrigation treatments (together with frequency of harvesting) considered in the design are summarized in Table 1 below. Each irrigation treatment was conducted in triplicate and each irrigation of the treatment was sufficient to replenish soil moisture to field capacity.

Table 1. Summer irrigation frequency and annual harvest frequency during study period.

Irrigation	Number of Irrigations			Number of Harvests		
Treatment	July	August	Sept.	1991	1992	1993
Optimum	3	2	2	8	9	9
Minimum Stress	3	1	1	8	9	9
Short Stress	3	0	0	6	7	7
Long Stress	0	0	0	5	6	6

Because of the extensive soil cracking as the soil dried, 3.7 m wide fallow strips were maintained on either side of the plots and kept moist so as to prevent water movement from one irrigation treatment to the next. Soil moisture was monitored at 0.15 m depth intervals in each of the plots via access tubes using a neutron probe calibrated for the field (Grismer et al.,1995). Soil salinity and chloride concentration were measured for each plot from saturation extracts taken at 0.3 m depth increments to a depth of 1.2 m in March, June, August and October of each year. The short and long stress treatments were not harvested during the time they were not irrigated so as to allow the standing stalks to shade (protect) the crowns from direct solar radiation. Following the last harvest of fall, 1993, the field was disked and leached repeatedly (then planted to Sudan grass in 1995) until soil salinity levels were similar to pre-experiment conditions in anticipation of alfalfa planting for the second phase of the study.

For the cut-off irrigation method portion of the study, the field was divided into two sets of four border checks 20 m wide by 340 m long. One set of borders was planted to alfalfa (CUF 101) in November, 1995 and the other set to Sudan grass (not considered here). Again, soil moisture was monitored using a neutron probe in 3 m deep access tubes, 8 per border check, at regular intervals prior and immediately following irrigation. Similarly, 8 observation wells per border check were used to monitor shallow groundwater depth, salinity and chloride concentration. Soil salinity was determined from saturation extracts collected in November of 1995 and 1996. The cut-off irrigation method relies on calculation of the time required for the surface wetting front to just reach the end of the field based on the onflow rate and the time required for the advance front to reach approximately 30% of the border length. Borders were irrigated as necessary to meet the soil moisture deficit (approximately monthly) and averaged approximately 120 mm per irrigation for 1.43 m total water application in 1996. During harvesting, each border check was divided into 8 equal segments along the length of the border so that yield measurements along the border could be correlated with potential soil salinization along the border due to progressively decreasing intake opportunity times. As with the first part of the study, pest and weed control and harvesting procedures followed the commercial practices in the Valley.

3   Results & discussion

Table 2 and Figure 1 summarize the yield and water use efficiency data, and Figures 2 & 3 summarize the soil salinity data resulting from the two different water management strategies, respectively. Alfalfa yield obtained in the cut-off irrigation plots exceeded that

Table 2.   Annual yield and water use efficiency for the two water management strategies.

| | Irrigation Treatment or Method | | | | | | | | |
| Year | Optimum | | Minimum | | Short | | Long | | Cut-off |
	(kg/ha)	(kg/ha-m)	(kg/ha)	(kg/ha-m)	(kg/ha)	(kg/ha-m)	(kg/ha)	(kg/ha-m)	(kg/ha)	(kg/ha-m)
1991	13 188	10.78	12 390	10.51	9560	9.65	7374	8.85		
1992	11 456	9.50	9959	8.34	8300	8.90	7601	8.95	-	-
1993	10 113	8.92	8263	7.51	6376	6.70	5560	6.99	-	-
1996	-	-	-	-	-	-	-		21 460	15.0

obtained from even the optimum irrigation plots in the previous years because of a severe sweet potato whitefly infestation that caused great damage to agriculture in the entire Valley beginning in 1991 until it was brought under control in 1994.   The infestation resulted in decline in yields during this period for even the optimum treatment and is reflected in the smaller yield to water use values (approximately 10 kg/ha/mm for the optimum treatment versus 15 kg/ha/mm for the cut-off method). Figure 1 illustrates the variability in yield from one cutting to the next and the decline in yield along the border due to increasing soil salinization (Figure 3) during the year for the cut-off method.   Average yield from the cut-off method compared favorably with the Valley-wide average for the year.   Soil salinity also increased in the reduced irrigation plots, but was only a statistically significant increase for the short and long treatments(Fig.2).

Figure 1. Alfalfa yield along border check for each cutting in 1996

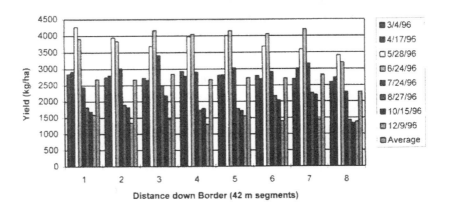

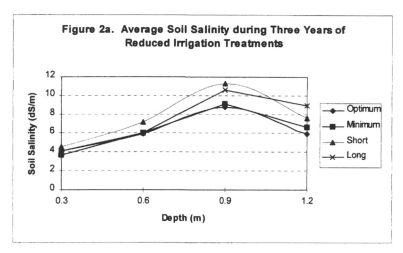

Figure 2a. Average Soil Salinity during Three Years of Reduced Irrigation Treatments

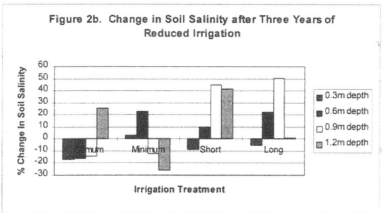

Figure 2b. Change in Soil Salinity after Three Years of Reduced Irrigation

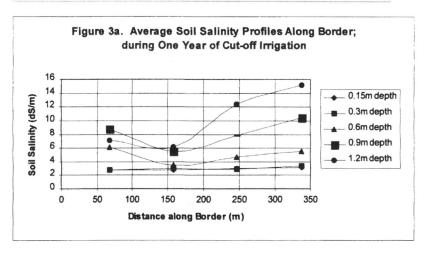

Figure 3a. Average Soil Salinity Profiles Along Border; during One Year of Cut-off Irrigation

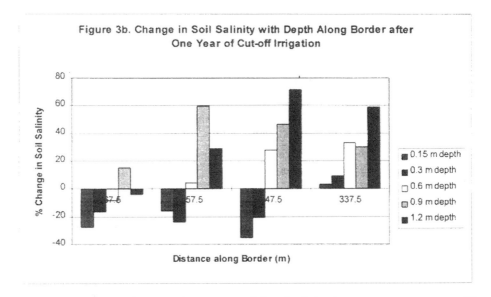

Figure 3b. Change in Soil Salinity with Depth Along Border after One Year of Cut-off Irrigation

A comparison of the total water saved by the irrigation treatments, or cut-off method, to the total yield lost over a one, two, or three year period produces a yield equivalent in water that can then be related to the value of that water in terms of lost yield. Table 3 summarizes the net yield loss, water savings and water value for the two water management strategies assuming an alfalfa value of $0.11 /kg ($100 /ton). Net water savings and yield losses for the different irrigation treatments were determined by comparing results of each reduced irrigation treatment with corresponding results from the optimum treatment. Similarly, water savings for the cut-off method were calculated from the onflow rate and projected continuation of the irrigation beyond the cut-off time as normally practiced. Yield loss was taken to be the difference in yield at the top (first 84 m) and bottom (last 84 m) of the field. No attempt was made to account for the costs of the leaching irrigations required to restore the reduced irrigation short and long treatments plots which would have the net effect of reducing the water value.

It is clear that there is significant yield loss as well as water savings resulting from use of the two water management methods, however, the water value associated with the reduced irrigation method is roughly twice that of the cut-off method. This larger water $/ha-m value is due largely to the plant damage caused by the whitefly infestation that suppressed yields beyond that induced by the water stress alone. The whitefly effect on yield can be accounted for in part by considering that the yield per water use values for the reduced irrigation method is roughly half that of the cut-off method, especially if the leaching irrigation requirement for the short treatment is included (Table 2). Correcting the alfalfa yields of the short and long treatments for the 1991-1993 period using the ratio of the yield/water use values of the cut-off method to the reduced irrigation method results in water values for the short and long treatments that are comparable to that for the cut-off method. Of the water management methods considered here, the one most likely to be adopted by the

Valley growers would be the short treatment. With additional education and support of the Imperial irrigation district, the cut-off method for clay fields could also be readily adopted. Use of these two methods suggests that the value of water associated with these water conservation measures in the Valley would be approximately $600 /ha-m for yield losses on the order of 1500 kg/ha.

Table 3. Annual yield loss, water saved and water value based on yield loss and an alfalfa value of $0.11 /kg ($100 /ton) for the different water management methods.

Treatment, or Method	Net Yield Loss (kg/ha)			Net Water Saved (ha-m)			Water Value ($/ha-m)		
	1st yr	2nd yr	3rd yr	1st yr	2nd yr	3rd yr	1st yr	2nd yr	3rd yr
Minimum	876	2540	4585	0.031	0.038	0.062	1265	2952	3308
Short	4023	7394	11529	0.159	0.338	0.430	1127	976	1194
Long	6428	10698	15732	0.268	0.511	0.744	1070	933	940
Cut-off	1431	-	-	0.797	-	-	538	-	-

4   Summary & Conclusions

The purpose of this study was to assess the value of water conservation related to alfalfa production in the Imperial Valley of California. We considered two water management strategies; reduced irrigation frequency during high ET periods of July through September, and a cut-off time border irrigation method that eliminates tail-water runoff. Measurements of alfalfa yield, water use and soil salinity enabled calculation of production water use efficiency, loss in yield and net water value on an annual basis. Of the three irrigation treatments considered, minimum, short and long stress, the method most likely to be acceptable to growers is the short stress treatment that eliminates irrigations during August and September, but maintains sufficient plant stand such that the crop recovers for the fall harvesting period. The long stress treatment nearly depleted the crop stand completely resulting in eventual soil salinity problems, weed control problems, and loss of plant vigor and poor recovery for the fall harvest period. The minimum stress treatment continues irrigation during August and September, but the net yield per volume of water used is not significantly greater than that for the short stress treatment, suggesting that water use during these months is relatively inefficient in terms of crop production. The cut-off irrigation method results in yield losses near the end of the borders due to increasing soil salinity along the border, but overall yields remained high, comparable with Valley-wide averages. With additional education and support of the irrigation district, this method could be used on clay soils throughout the Valley, or combined with the short stress method of reducing or eliminating August and September irrigations for even greater water savings. In either case, the value of water from use of the short stress or cut-off irrigation methods is approximately $600 /ha-m. This value can then be compared with costs associated with additional water development for urban use and the economic externalities associated with loss of crop production in the Valley. Such a method for evaluating the value of agricultural water use may have application to other crops and other arid regions faced with competing demands for water resources.

5   References

1. Grismer, M.E., Bali, K.M. and Robinson, F.E.. (1995) Field-scale neutron probe calibrations and errors in water content estimates for a heavy silty-clay soil. *ASCE of Irr. & Drn. Engr.* 121(5):354-362.
2. Grismer, M.E. and Tod, I.C.. (1994) Field procedure helps calculate irrigation time for cracking clay soil. *Cal. Ag.* 48(4):33-36.
3. Lehman, W.F. (1979) Alfalfa production in the low desert valley areas. UCBerkeley Pub., Berkeley, CA 94720.4   Robinson, F.E., Teuber, L.R. and Gibbs, L.. (1994) Alfalfa water stress management during summer months in Imperial Valley for water conservation. Final Report to the Metropolitan Water District and the UC Water Resources Center. 34p.

ASSESSMENT OF IRRIGATION INVESTMENTS IN BULGARIA

R. PETROVA and S. CHEKHLAROVA
Institute for Irrigation, Drainage and Hydraulic Engineering, Sofia, Bulgaria

Abstract
Vitality of irrigation depends on the effective use of the water sources in the country. Making irrigation financially attractive for agricultural producers is of national importance. In this respect the value of water for irrigation is essential.
An economical and financial appraisal of two representative irrigation schemes in Bulgaria has been made on the basis of system analysis. The model irrigation schemes are in Plovdiv region (low-pressure scheme with gravity irrigation) and in Vratza region (high-pressure scheme with sprinkler irrigation), where the pilot extension services were founded.
The present condition and the organisation of water supply of these schemes are also estimated. The impact of the irrigation scheme parameters such as the cropping plan, the specific irrigation equipment, the market prices etc., on the value of irrigation water are evaluated.
The benefits of rehabilitation of the two model irrigation schemes are estimated using some well known methods for appraisal of investments. The results can be used in practice.
Keywords: Agriculture, financial and economical analysis, investments, irrigation, price of irrigation water, appraisal

1   Introduction

The political and economical reforms in Bulgaria have led to great changes in agriculture and particularly in the irrigation sub-sector. If in a national context, the economic viability of irrigation depends on the effective use of water resources, then for individual water users financially attractive irrigation is important.

During the period 1989-1996 the quantity of supplied water has decreased 10 times. The percentage of irrigated area in 1995 became 10% from the whole equipped for irrigation area and during 1996 it has increased to 20%. The grain crops (maize) are irrigated only once or twice per a season; vegetables and tobacco are irrigated two or three times. The cropping plan has changed. In regions which are

Water: Economics, Management and Demand. Edited by Melvyn Kay, Tom Franks and Laurence Smith. Published in 1997 by E & FN Spon. ISBN 0 419 21840 8.

traditional vegetables producers such as Plovdiv and Pazardjick, the area cropped with cereals has increased compared to the vegetables which demand more irrigation. The correlation between the different kind of irrigation equipment has also changed. The part of gravity irrigation (traditional furrow irrigation) increased on the account of sprinkler irrigation (center pivots, mobile sprinkler machines etc.) and drip irrigation.

The reasons for this are complex and include, on one hand the serious problems related to the land reform and on the other credit policy, markets and final price of farm products.

At present, when most of the agricultural products prices are below the free market ones (which does not apply for input costs), the organisations involved in irrigation investments and management are making only partial analysis of the input costs and their recovery.

2   Objective and methods

The objective of this case study is on the basis of financial and economic analysis to assess irrigation investments in two representative irrigation schemes in Bulgaria: the Plovdiv branch and the Vratza branch. This preliminary estimation will allow to determine the area that can be maintained and developed further and the area that shall be abandoned keeping in mind the time factor and the economic conditions in the country.

It should be noted that the study was accomplished under conditions of transition to market oriented economy with characteristics such as: 30 - 35% decrease in GDP, budget deficit, very high inflation and devaluation of Bulgarian currency. Under these conditions the problems of investment efficiency in irrigation and decision-making are very complex. Different methods for efficiency estimation of the investments are applied in the international management practice.

For the purposes of this case study the authors have used a method for financial and economic analysis developed by Silsoe College, Cranfield University, UK and demonstrated in Bulgaria during a project financed by PHARE [1]. An appraisal of irrigation investments was done on irrigation schemes' using this system analysis. The system diagram is presented in Fig.1.

According to the principles of financial and economical analysis, the study goes through the following stages:
- definition of inputs and amount of production, including the needs of irrigation water; estimation of their prices;
- determination of financial gross margin for irrigated and non-irrigated conditions based on international border prices;
- transition from financial to economical prices is made by exclusion of taxes and subsidies.
- the influence of cropping plan and the irrigation intensity on the water requirements at irrigation schemes' level is defined;

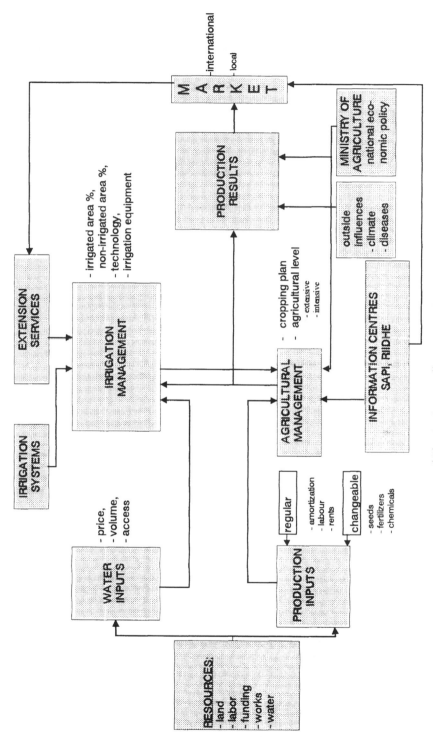

Fig. 1 System diagram

- the strategy for future development "with" or "without project" contains an estimation of investment costs for the scheme construction as well as operation and maintenance (O&M) costs.
- the discounted cash flow method is used with criteria: net present value (NPV) and the internal rate of return (IRR).
- sensitivity analysis is done in order to determine the critical variables which influence the price of water.
- the average unit cost of the delivered water is estimated on the basis of a model which includes different assumptions regarding the extent of input recovery. They may vary from full recovery of all capital and operational costs to recovery of the energy costs only.

The sources of information are: the budget of Irrigation Systems Company, farmers' business plans, producer's prices of inputs, international FOB prices, etc.. The discount rate is 10 %.

3   Model Irrigation Schemes

The model irrigation schemes are situated in Plovdiv and Vratza region. Both schemes are well known for their excellent practice referred to their high technical and technological characteristics. These regions have high potential for development.

3.1   Vratza Irrigation Scheme

The Vratza scheme is representative of high pressure irrigation schemes. It is situated in the Northern part of Bulgaria where the water is delivered on 2-3 to 4-5 pumping stages. The total irrigated area is 50,000 ha, 38,200 ha of which are water secured and 11 800 ha are temporally not in use and need investments for rehabilitation.

Three main dams and 26 pumping stations are built. The scheme efficiency is 0.8 for gravity and 0.9 for sprinkler irrigation. Irrigation depth varies from 120 mm to 70 mm.

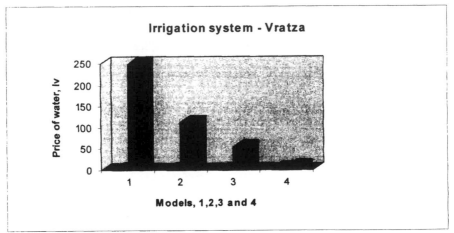

Fig.2. Price of water in Vratza Irrigation scheme

The cropping plan is as follows: maize 72%, sunflower 1%, tobacco 17%, soy beans 3%, alfalfa 1%, vegetables 4%, sugar beat 1%, orchards 1%.

At present the rehabilitation of the irrigation scheme consists of reorganisation of pump stations, i.e. some of the pumps are replaced by ones with lower-pressure.

3.2 Plovdiv Irrigation Scheme

The Plovdiv irrigation scheme is representative of the low-pressure irrigation schemes and is situated in South Bulgaria. The total irrigated area is 176,000 ha: 154,000 ha (88%) are for gravity irrigation, 5,000 ha (8%) - stationary sprinkler irrigation, 2,000 ha (1.5%) - drip irrigation. The canal network on about 40 000 ha is in extremely good condition, 10 000 ha needed reconstruction investments, 30 000 ha permanently are not in use and 6 500 ha are not good for irrigation (according to the Water Users Associations).

There are 18 main reservoirs with a total capacity of 285,000,000 m^3 and 280 smaller reservoirs managed by the farmers co-operatives and water associations in the region.

Since the Plovdiv irrigation scheme is too large to be considered as a whole and in order to obtain comparative results, the model is applied for an area of only 50,000 ha. The cropping plan is as follows: maize 24%, alfalfa 10%, rice 19%, orchards 7%, wines 4%, wheat 10%, soybeans 2%, sugar beat 1%, vegetables 10%, tobacco 17%.

At this stage the investments for rehabilitation of the scheme and the equipment shall be estimated. The essential issue in the project rehabilitation is that the efficiency of the gravity scheme shall be increased with regard to both technical and management parameters. That is why equipment for water measurement is also included in the project.

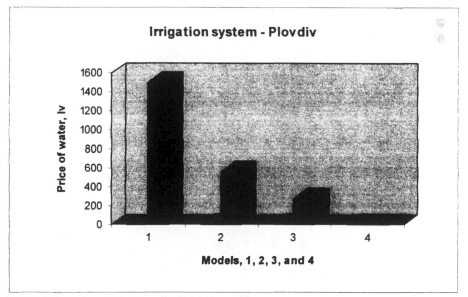

Fig.3. Price of water in Plovdiv Irrigation scheme

4   Results and discussion

The data about the schemes (capital costs for recovery) are from 1996; the crop prices and inputs are from December 1996 when the exchange rate was 1 $ = 500 Leva.

The results obtained are graphically represented in Fig.2 and 3. The method for estimation of investment efficiency allows the following analysis:

1. Despite the extremely adverse economic conditions in the country the investments in the field of irrigation might be successful.
2. Based on the criteria net present value (NPV) and internal rate of return (IRR) under the adopted discount rate it was concluded that in both representative schemes the efforts for rehabilitation should be continued. NPV in the gravity scheme is much bigger than the one in the high-pressure scheme, although the values of IRR are very close(14% for gravity scheme and 11% in a sprinkler irrigation scheme);
3. The price of water in Vratza Scheme varies from 9.0 Lv (5 cents) - for model-4 to 250 Lv (50 cents)- for model-1. Respectively for Plovdiv Scheme: from 4.0 Lv (8 cents) to 1500 Lv (3 $) at the same level of the cost recovery.
4. The factors of possible risk: percentage of irrigated land, prices of agricultural production, energy prices and expenses for repairs influence the economic results to different extent. The changes of the percentage of irrigated land and the prices of agricultural products have the greatest influence. The smallest influence is caused by changes in energy price and the maintenance costs.

5   References

1.   PHARE Programme for the Ministry of Agriculture Bulgaria (1995) *Technical Assistance to Land Reform*, Ref. Project BG 9001-01-04, Cranfield University, Silsoe, UK
2.   Bulletin of Irrigation Systems Company (1996), Bulgaria

RENT-SEEKING BEHAVIOUR AND WATER DISTRIBUTION
A Southern Punjab irrigation scheme, Pakistan

J-D. RINAUDO
Irrigation Division, Institute of Agricultural and Environmental Engineering Research (Cemagref), Montellier, France
S. THOYER
Agricultural Economics Department, Ecole Nationale Supérieure d'Agronomie de Montpellier, France
P. STROSSER
International Irrigation Management Institute, Lahore, Pakistan

Abstract
The paper describes how farmers in a protective irrigation scheme of Punjab (Pakistan) interfere in water distribution to privilege their own supply and therefore appropriate water rents. On a specific case study, water rents are quantified and their economic values are estimated.
Keywords: water distribution, outlet, interference, rent-seeking, Pakistan.

1 Introduction

Built by the British more than a century ago, the Indus Basin Irrigation System was designed as a protective irrigation system: limited water resources are spread over an area as large as possible, thereby imposing a relative water scarcity on farmers through the implementation of specific rationing procedures. Since cropping intensities have more than doubled over the years, the farmers have developed a number of adaptive strategies to cope with the increasing water scarcity: they have augmented the total water supply by tapping groundwater resources via the installation of private tubewells; they have developed informal water markets to exchange surface and ground water, thereby increasing the flexibility of the supply; also farmers have developed specific mechanisms to appropriate more than their due share of canal water, either by interfering directly in the distribution process or by influencing the decisions of the irrigation agencies. This interference provides rents (extra water supply illegally appropriated) to some farmers and induce reduction of the supply received by downstream water users.

Water: Economics, Management and Demand. Edited by Melvyn Kay, Tom Franks and Laurence Smith. Published in 1997 by E & FN Spon. ISBN 0 419 21840 8.

This paper is a first attempt at gaining a better understanding regarding the impact of rent-seeking behaviour on water distribution. It mobilises data and the simulation tools developed by the International Irrigation Management Institute and Cemagref as part of a 5-year research effort in the Chishtian Sub-division of the Fordwah Branch irrigation system. The study area is located in the cotton-wheat zone of Punjab and serves a cultivable command area of 71,000 hectares.

After describing the gap between official and actual distribution of canal water (Section 2), the different types of interference are related to farmers' constraints, so as highlight rent-seeking strategies (Section 3). Section 4 presents some preliminary results of a case study conducted in one secondary canal (or distributary) to quantify rents created by one type of interference and to estimate their economic value.

2   From design to actual water distribution

The Indus Basin Irrigation System was designed as a run-of-the-river system. A succession of main canals and distributaries convey surface water to more than 89,000 tertiary units (*watercourses*). Design objectives included a minimum of human intervention for operation and maintenance. Canals were to be run at their full capacity all of the time (supply based system). And few gated structures have been installed to control flows (mainly at the head of distributaries or along main canals). In case of severe water shortages, operational requirements are increased with the implementation of specific rotational schedules between distributaries. Within each distributary, water is automatically distributed to all watercourses through non-gated structures *(outlets),* having dimensions designed so as to allow a discharge proportional to the watercourse command area *(authorised discharge).* Outlets are design so as to guarantee a relative equitability of water distribution (e.g. an equal distribution of shortage or excess of supply between watercourses). While the main canal and distributaries are operated and maintained by staff from the Provincial Irrigation and Power Department (PIPD), water management below the outlet is the responsibility of the farmers who share canal water following a fixed weekly roster of water turns *(warabandi).* The duration of each turn is proportional to farmer's land holding size.

Research studies have emphasised the significant gap between official and actual canal water distribution [1]. Discharges at various points in the irrigation system are at variance from their official figures. Also, equitability is not conserved, with some areas bearing the bulk of the water shortages. This results in highly inequitable water distribution among farmers.

In addition to technical issues frequently cited by actors involved in irrigation management to explain the gap between the official and actual distribution, interference by farmers themselves in the operation of the irrigation system is of significant importance [2] though seldom documented in field studies. Such interference is increasingly acknoledged by PIPD officials as one of the major reasons explaining the poor performance of canal water supply [3].

To increase their water supply, farmers try to modify one of the three variables that determines the quantity of water they receive: (i) the duration of their water turn;

(ii) the discharge at the watercourse head; and/or (iii) the discharge at the head of the distributary.

(i)   The modification of the water turn duration may range from a simple water turn confiscation to a long term modification of the *warabandi*, which favours some farmers and deprives others. A limited number of farmers also trade partial or full water turns. Bandaragoda and Rehman [4] confirmed that, in practice, water is not shared equitably between farmers of a given watercourse. Power relationships between different groups of farmers may explain the pattern of water distribution within the watercourse command area. Also, inequitable water distribution may be socially accepted and compensated by input and credit supply or employment relation [5].

(ii)   Farmers have also developed a number of strategies to increase water flows to their watercourse.   The most simple strategy consists in obstructing the flow downstream of the outlet and provoke a raise in the distributary water level with direct (positive) impact on the watercourse discharge. Flexible siphoning pipes or direct cuts in the canal banks are also used by farmers either to increase the outlet discharge during their water turn or to irrigate directly their fields when those are located close to the distributary. Although this practice generally lasts for a few hours only (mostly at night, when the risk of being caught is almost nil) a few permanent illegal pipes have also been observed. Another frequent practice is to damage the outlet structure and enlarge its dimensions. It might take a few months for the PIPD staff to notice and repair the outlet. Other modifications, such as lowering the crest of the outlet, cannot be carried out without the consent of the lower rank PIPD staff. To obtain a permanent increase of the watercourse head discharge, some farmers prefer to deal with PIPD staff higher in the hierarchy and obtain an official modification of the size of their outlet.

(iii)   The PIPD staff in charge of the operation of the gated structures which are used to control the distribution between distributaries may be pressurised by influential farmers or groups of farmers so as to privilege one distributary at the expense of the others. The increase of discharge to selected distributaries may be temporary and last for a few hours or days only (during periods of severe shortage) or be more permanent and be considered as a *de facto* change in the allocation of the distributaries.

Each type of interference provides different benefits to farmers and is also associated with a specific set of constraints.   In order to understand farmers' interference strategies, the following section investigates the characteristics of the different types of interference, with a particular focus on the actors involved, the costs of intervention, and the type of benefits which are sought.

3   Rent-seeking strategies

3.1   Farmers and PIPD interaction

Although the irrigation laws and regulations clearly specify the rules and procedures for water allocation and distribution, these activities are permanently contested and negotiated among the actors of the irrigation system. In this negotiation, PIPD staff is in a rent-providing situation because of its key role in the monitoring, operation and maintenance of the system, but also in the resolution of conflicts and enforcement of law and order. Although some forms of interference do not involve PIPD staff, farmers often negotiate with PIPD officials in order to: (i) obtain more water through a direct intervention of the PIPD staff; (ii) influence the judgement pronounced by the canal magistrates for an offence committed with regard to the Canal and Drainage laws; and/or (iii) avoid paying the financial penalty *(tawan)* imposed by the canal magistrates.

In its simplest form, the negotiation between rent-seekers and PIPD officials consists of exchanging a payment (most of the time in cash) in return for an increase in canal water supply to their distributary or watercourse. Putting pressure on the PIPD officials through a connection with a politician is a second alternative. Politicians, such as Members of the Provincial or National Assemblies, have a strong influence on the transfer of Government officials. They utilise this influence to obtain from PIPD officials increased water supply to localities which constitute their electoral base. This distribution of privileges contributes to securing their electoral support in their constituencies[1].

In practice, the negotiation between water users and the PIPD is a combination of pressure from politicians (which often remains implicit) and financial incentives provided by farmers.

3.2 Interference characteristics

Each type of interference can be characterised by: (i) the benefits it provides (e.g. the value of the rent); (ii) the costs of rent-seeking activities; and (iii) the type of externalities it induces for (usually) downstream water users.

(i) The benefits gained by rent-seekers can be measured by the increase of agricultural income permitted by the extra water supply appropriated illegally. The four major variables that determine the value of the rents are the timing (when) the duration (how long), and the level (how much) of the increase in canal water supply, along with the rapidity of the negotiations. The benefits can be individually appropriated (pipes, cuts) or shared between different water users (modification of the outlet, increase of the distributary head discharge).

(ii) Theoretically, the cost of interference can be expressed as the amount of the legal fine imposed to the offender multiplied by the probability of being caught and prosecuted. Costs of interference often derive from different activities which aim at reducing this probability; organisation of a collective action, negotiation with the PIPD staff (including possible bribes), and lobbying activities to obtain the support of politicians. Part of these costs correspond to a redistribution of the rents to rent-providers. The modification of the physical infrastructure may also be required and induce labour and material costs.

(iii)   By illegally appropriating extra water supply, rent-seekers deprive downstream water users of their due share. The negative externalities may be direct (one farmers gains, one loses) or diffuse (the externalities are shared between several water users). They may be transparent (the deprived farmers are able to identify the origin of the negative externalities) or hidden (the origin is not identified). These characteristics determine the difficulty for the rent-seekers to secure the privileges they obtained.

Farmers engage in rent-seeking activities only if the net expected benefit of the interference (defined as the value minus the costs of the rent) is positive. However, obtaining these benefits is subject to certain constraints, such as the capacity to pay, the potential for collective action, or the access to political support, but also the capacity to compete with the groups that may suffer from, negative externalities. The interference strategy adopted by each group of farmers is determined by their capacity to overcome these constraints.

3.3   Interference strategies

An individual strategy is generally chosen by farmers who have a high personal bargaining power with the authorities (high capacity to pay and/or obtain political support). The type of intervention chosen may provide a short or long-term increase in canal water supply. The benefits may be individual or shared; such farmers are prepared to support the totality of the rent-seeking costs as long as their net benefits remain positive, whether in the context of this very transaction or in a larger socio-economic context. Individual action can also be chosen by farmers who are prepared to support a high individual risk of penalty. In this case, the offender acts secretly and seeks not to be caught by PIPD staff. The choice of such a risky strategy often corresponds to emergency situations when the crops can only be saved by an immediate irrigation. Since the offender does not share the risk, he most often chooses an intervention which allows him to appropriate the benefits entirely.

Collective interference is often a response to a low individual capacity to negotiate with the authorities, either because of financial constraints or because of a lack of political connection. Forming a coalition is then the only solution for being able to compete with other rent-seekers. This represents an effective strategy when the organisation costs (including costs of free-riding control) are low. Collective action is also required when the benefits or the costs of interference cannot be individualised. This is mostly the case for actions which aim at modifying the outlet discharge, or at restoring water distribution along the distributary when combined externalities from upstream water users have significantly reduced available flows to downstream water users.

4   Quantification of rents: The Fordwah Distributary case study

A quantification of rents requires a detailed hydraulic analysis of each type of interference to quantify the volume of water appropriated by rent-seekers. This section focuses on the rents due to unauthorised modifications of outlets (change in dimensions or crest level). These rents are quantified and their economic value is estimated. The study is undertaken in the Fordwah Distributary located at the tail of the Chishtian Sub-division. This distributary has a design discharge of 4.47 m^3 for a Culturable Command Area of 14,800 hectares. Rents due to other types of interference are not quantified in this preliminary study.

4.1   Deterioration of the canal or interference?

A hydraulic model (SIC) is used to describe the actual distribution of water to all of the watercourses of the Fordwah Distributary [7]. The model takes into account the outlet modifications (measured in the fields) and also the actual geometry of the canal (slope, sections). It was calibrated with a 1996 set of discharge measurements. It computes the discharge received by each outlet when the distributary receives its design discharge. The comparison for each outlet of the actual discharge (simulated value) with the authorised discharge (official value) emphasises the distortions in distribution along the distributary (see Fig. 1).

The analysis of the physical characteristics of the canal shows that its deterioration partly explains the distortions shown in Fig. 1. For instance, outlets in the first reach of the distributary draw less than their authorised discharge as water levels in the distributary are low due a wider than design bed of the distributary in this reach. Since most of the outlets in the first reach draw less water than their due share, the discharge in the distributary progressively exceeds its design value. This excess of discharge in the distributary is rapidly absorbed in the second reach by a number of outlets which receive more than their authorised discharge. Some of these outlets have not been modified but directly benefit from the excess of supply (increase of water levels). Others (shown in black in Fig. 1) have been modified (observed and measured modifications) and the extra water supplies they draw may be considered as rents.

Sediment deposition in the canal may also disturb water distribution. For instance, the presence of 20 cm of sediment in the third reach of Fordwah Distributary is responsible for a raise of the water level. This increases the discharge received by the Jiwan Minor off-taking at the tail of the second reach. The excess supply received by this minor worsens the water shortage downstream; when it enters the third reach, the discharge in the distributary drops 10% below its design value. However, as a consequence of the raise in water level, the outlets of the third reach get their due share despite the relative deficit of water in the distributary. Finally, the water shortage due to the deterioration of the canals and to the water thefts upstream is almost entirely transmitted to the last reach where it is largely shared between the last few outlets.

This analysis of water distribution along the distributary shows that the gap between the actual and authorised discharges is not only due to unauthorised modifications of the outlets, but also to be deterioration of the canal. Sediment deposition and scouring modify the canal width, depth, and slope, leading to a change in the water level in the distributary and ultimately to a modification of the discharge received by each outlet.

4.2   Quantification of rents

In order to quantify the marginal impact of farmers' interference on water distribution, the same hydraulic model is then used to simulate the new water distribution that would be obtained if the outlet dimensions and crest levels were set back to their design values. The rents and externalities due to unauthorised modifications of outlets are computed comparing the output of the simulations with and without outlet modification. They are presented in figure 2, expressed in extra volumes of canal water received per hectare and per month.

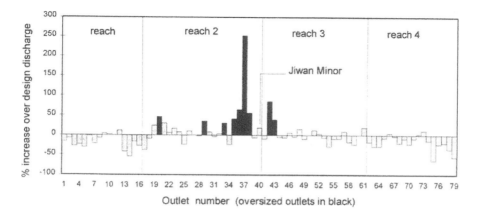

Fig. 1. Actual water distribution in the Fordwah Distributary.

Only 9 outlets, mainly located in the second reach have been significantly modified and receive important surplus of canal water supply, i.e. from 110 to 1930 m³/ha per month. As a comparison, the average design canal water duty is equal to 750m³/ha per month for the Fordwah distributary. The negative externalities affect almost all of the outlets located downstream from the oversized outlets.

Fig. 2 shows that the gains related to rent-seeking are concentrated, whereas the negative externalities are spread over a large number of farmers located in the tail outlets, and for whom organisation costs to form a counter coalition would probably be higher than the expected benefits of collective action.

4.3   Economic value of rents

The economic value of rents is defined as the increase of agricultural income permitted by the extra canal water supply illegally appropriated. To illustrate the impact of rents on income, a simple approach is proposed in this paper, based on the estimate of additional area that can be cultivated with extra water supply. As the cotton-wheat rotation is dominant in the area, it is assumed that water represents a major constraint during two periods only, corresponding to the wheat and cotton peak water requirement periods. And farmers' production decisions are based on the availability of water during these two periods. The additional area under wheat (cotton) permitted by the excess of water supply are estimated by the ratio "*excess canal water supply during the wheat (cotton) peak demand period*" over "*wheat (cotton) maximum water requirements*".

The additional income generated by these additional areas under such crops are then calculated using average gross margin estimated through a farm survey in the Chishtian sub-division. The results are presented in Table 1. The comparison with the average price for the rent of land (US$ 180 per hectare and per year) reveals the importance of the rents.

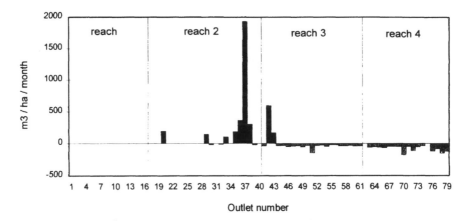

Fig.2. Rents and negative externalities due to unauthorised modifications of outlets along Fordwah Distributary

Table 1. Economic value of rents and negative externalities along Fordwah Distributary

	extra supply (m3/ha/month)	% variation cultivated area	of variation of income (US$/ha/year)
Rents : 9 outlets			
average	445	59	105
minimum	110	15	25
maximum	1930	257	455
Externalities : 40 outlets			
average	-57	-8	-13
minimum	-14	-2	-3
maximum	-181	-24	-43

The increase of water supply received by some watercourses is such that it can only be absorbed by increase in areas under crops with high water requirements (rice sugarcane). This stresses the need to develop more complex approaches to refine estimates of the economic value of the rents. For example, to develop micro-economic (linear programming) models for individual farms would provide more complete answers on the impact of an increase or reduction of canal water supply on farm income.

5   Conclusions

The present paper identifies a large range of rent-seeking mechanisms that are currently in use in irrigation systems in Pakistan. The paper shows that the gap between official allocation and actual distribution is partly due to interference by water users. The benefits obtained through illegal appropriation of water significantly increases rent-seekers income. Interfering in the distribution of water represents a real alternative to other strategies for increasing the cropping intensity and crop yields, such as ground water use (tubewell investment) or participation in water markets.

The important irrigation management reforms which are being proposed in Pakistan are expected to lead to a redistribution of rents between water users and to changes in coalition groups. The groups negatively affected by these changes are likely to resist to the implementation of these reforms. A good understanding of rent mechanisms and of the impact of rents in terms of water supply and farm income will facilitate the identification of potential resistance from different actors. It will contribute to design reform scenarios more likely to be accepted by (all) actors involved in irrigation system management.

The present study represent a first effort in this direction. Further emphasis will be given to the variability of the observed rent-seeking mechanisms under different water supply and social conditions. Also, investigations of the impact of the different rent-seeking mechanisms on canal water supply, farm income and externalities will be pursued.

6   References

1.   Kuper, M. and Strosser, P. (1996). *Canal water management in Pakistan: lessons from current practices for proposed policies and management changes.* Draft document. IIMI, Lahore.
2.   Lowdermilk, M.K. (1990). *Irrigation water control and anarchy.* In: *Social, economic and institutional issues in third world irrigation management.* Sampath, R.K. and Young, R.A. (eds). Studies in water policy and management, No. 5. Westview Press, Boulder, CO.
3.   Mian Haffiz Ullah. (1994). *Participatory Irrigation Management : a case study of the Azim distributary.* Punjab Irrigation and Power Department, Bahawalpur.
4.   Bandaragoda, D. J. and Rehman, S. U. (1995). *Warabandi in Pakistan's canal irrigation systems: widening gap between theory and practice.* Country Paper No.7. International irrigation Management Institute, Colombo, Sri Lanka.
5.   Mollinga, P. (1996). *The politics of water distribution: negotiating resource use in a South Indian canal irrigation system.* In: Water policy: allocation and management in practice. P. Howsam, and R.C. Carter (eds). E&FN Spon, London, U.K.
6.   Wade, R. (1982). *The system of administrative and political corruption: canal irrigation in South India.* The Journal of Development Studies. Vol. 18, No. 3.
7.   Hart, W.W.H. (1996). *Research into the relationship between maintenance and water distribution at the distributary level in the Punjab.* M.Sc. Report. University of Technology, Delft, The Netherlands.

---

1. The similar role of politicians in water distribution has been intensively documented in South Indian canal irrigation schemes by Robert Wade [6] and more recently by P Mollinga [5].

FOOD SECURITY IMPLICATIONS OF RAISING IRRIGATION CHARGES IN DEVELOPING COUNTRIES
RAMESH BHATIA

Abstract
This paper outlines the operational implications of the concept of water as an economic good. Four components of value of water in irrigation include: direct crop benefits to the farmer; non-irrigation benefits(e.g. drinking and bathing water for people and livestock, recharging of ground water by canals, water for micro-enterprises) from irrigation infrastructure; benefits from return flows; positive economic and environmental externalities and premium for societal benefits. Costs have been defined in terms of Supply Costs which include financial costs of O&M plus capital charges. Economic Costs consist of Supply Costs plus opportunity cost of water (i.e. benefits foregone of water use in other sectors) plus negative economic externalities. Full Costs are Economic Costs adjusted for environmental externalities. The impact of raising irrigation charges to cover various components of costs have been analysed using farm-budget data from field surveys in Haryana,India.

1   Introduction

There are several general principles involved in assessing the economic value of water and the costs associated with its provision . This paper suggests an approach for estimating value of water in agriculture and provides empirical estimates for Haryana, India.

2   Value of water in irrigated agriculture

Value to Users of Water: For industrial and agricultural uses, the value to users is at least as large as the marginal value of product . For domestic use, the willingness-to-pay for water represents a lower bound on its value. For, water used in agriculture, value to users is equal to the Net Value of Output in Irrigated Agriculture as discussed below.

Water: Economics, Management and Demand. Edited by Melvyn Kay, Tom Franks and Laurence Smith. Published in 1997 by E & FN Spon. ISBN 0 419 21840 8.

2.1 Volume of water diverted for irrigation

The Net Value of Output in Irrigated Agriculture: If water markets were functioning, the value of water in irrigated agriculture could be obtained from the prices paid by the farmers in the market. In the absence of water markets (particularly in surface irrigation), the value of water in irrigated agriculture can be derived as the Net Value of Output (NVO) attributed to the use of water diverted for irrigating crops. It is defined, as the Value of Water in Agriculture, which is the Net Value of Output with Irrigation minus that without Irrigation

The Net Value of Output (NVO) is estimated as the gross value of output minus the cost of cultivation. The value of water in the numerator is estimated as the difference between the NVO with irrigation and the NVO without irrigation. Thus, the numerator represents the additional value of net value of output that is attributable to the use of irrigation water Volume of water in the denominator refers to the quantities diverted for irrigation and not to the volume of water used by the crops or the evapotranspiration needs of the crops. This is because the costs of supply of water are determined by the volumes of water stored and/or diverted by irrigation structures and not by the volume of water used by crops. To the extent there are positive return flows from water diverted from irrigation, these should be explicitly accounted for in terms of externalities as discussed later in this section. Similarly, rainfall is not included in the volume of water in the denominator but is accounted for when net value of output without irrigation is quantified .

Table 1 provides some data at the farm level which can be used to estimate the value of water in agriculture (VWA). The data are for a one year rotation of wheat and paddy in Haryana in North-West India. The gross value of output with irrigation is estimated at US $1290 per hectare (ha) while the gross value of output without irrigation is estimated as only US $220 per ha. Thus, irrigation enables the farmer to increase the gross value of output by a little over than US $1070 when two water-intensive foodgrain crops are grown per year. However, in the intensive farming adopted in the arid region, the cost of inputs, including the cost of irrigation, fertilizers and labor, account for US $870 per ha. This leaves only US $390 per ha as the Net Value of Crop Output.

Table 1: Estimates of Additional Net Value of Output in Agriculture in Haryana, India

	With Irrigation	Without Irrigation	Additional Value/Costs
Gross Value of Output (US$ /ha/year)	1290	220	1070
Cost of Cultivation (US$ /ha/year)	870	190	680
Net Value of Output (US$ /ha/year)	420	30	390
Estimated Water Input(m^3 of water diverted per year per ha)	20,600	0	20,600
Net Value of Output per Unit of Water Input(US cents per cubic meter)	1.9		

In view of the low rainfall in this arid region, and high evapotranspiration needs of crops, the irrigation water diverted from surface irrigation and pumped from the ground is very high. The estimates of irrigation requirements for Haryana are: 1640 mm for paddy and 420 mm for wheat. This is equivalent to 16400 cubic meters of water diverted for paddy and 4200 cubic meters of water diverted for wheat—i.e. a total of 20,600 cubic meters of water per year. Given the crop output of 2.9 tons of rice and 3.6 tons of wheat (Table 1), these figures show that about 20.6 tons of water diverted for irrigation in a year resulted in a foodgrain output of 6.5 tons—i.e. a ratio of a little over three to one. However, this figure will be lower when return flows are taken into account as discussed below.

Adjustment for Societal Objectives: For water use in the household and agricultural sectors, there may be an adjustment made for societal objectives such as poverty alleviation, employment and food security (particularly in rural areas, where foodgrain prices tend to be high in the absence of the additional food output gained from irrigated agriculture, and where it may be difficult to supply imported foodgrains). Such adjustments are over and above the value of water to the user and should be added to reflect various societal objectives. To reflect this, a premium of 50 percent has been added to the value of agricultural output in Haryana. To reflect the objective of employment creation, a shadow wage rate equal to one-half of the market wage rate has been used in view of the prevalence of unemployment in areas from where labor migrates to Haryana. These adjustments increase the economic value of irrigation water by 2.8 cents per cubic meter.

In view of the absence of empirical studies in this crucial sector, the estimates used here are to be considered as the best judgments available and emphasize the need for empirical work in this critical area. The estimates of these values are not to be arbitrarily set, but should be determined on the basis of the best available methods and empirical studies which give the real gains to society of using water in irrigated agriculture as compared with its use in other sectors.

Non-Irrigation Benefits of Irrigation Supplies: Irrigation schemes provide water for domestic (drinking and personal hygiene) and livestock purposes which can result in improved health and higher incomes for the rural poor. For example, in areas of northwest India (Haryana and western Uttar Pradesh) where groundwater is saline, irrigation canals not only provide water for domestic and livestock uses, water in these canals recharge the groundwater table, thus enabling the pumping of sweet water from handpumps and shallow tubewells. In the absence of this sweet water, use of saline groundwater by animals is reported to result in about a 50 percent reduction in the output of milk [1]. In many arid regions of Haryana, the Indian Punjab and the Pakistani Punjab, income from livestock accounts for a significant proportion of the income of poor households, particularly in the drought season. In Haryana, there are no empirical studies where the additional value of these benefits have been quantified. In the absence of such data for Haryana, an estimate of US 1.0 cent per cubic meter is used for additional benefits to the value of water diverted for irrigation. In addition to livestock, irrigation canals provide water for wildlife, flora and fauna and provide in-stream benefits. In some canals in South India, canal drops are known to be used for installation of small and mini hydro plants. These indirect benefits have to be included while estimating the value-in-use of water that is diverted for agricultural purposes. Ignoring these benefits will result in a serious under-estimation of societal benefits available from the volume of water that is diverted for irrigation. Irrigation is also known to have some adverse environmental and social impacts which result in hardships for poorer households. Such adverse consequences include, inter alia, waterlogging and salinization of soils, declining groundwater tables (which result in dry handpumps and shallow tubewells), and pollution of water from agrochemicals and water-borne diseases. These environmental impacts can be adjusted either in terms of the negative benefits in estimating the value of water in agriculture, or added to the environmental externalities component of the Full Cost of water.

Net Benefits from Return Flows: Return flows from water diverted for urban, industrial and agricultural uses constitute a vital element of many hydrological systems, thus the effects of these flows must be taken into account while estimating the value and cost of water [2] [3] [4] and [5]. For example, a part of the water diverted for irrigation recharges the groundwater table in the region and/or increases the returns to the river/canal downstream. However, the benefits from the return flows will critically depend on the proportion of water that is lost to evaporation (due to open drains and canals) or to other sinks. As pointed out by Briscoe [2], it is useful to consider three different (not necessarily exclusive) situations, each of which is quite common, and two of which require modifications in ways in which water needs to be managed as an economic good. These three situations are:

Situation I: Return flows are removed from the system, in that they are discharged to the ocean or into salt sinks (as in some districts of Haryana in India). Under these conditions, there is no need to adjust the values for return flows.

Situation II: Return flows impose negative externalities to the users, in addition to the usual opportunity costs. For example, in areas where groundwater tables are high (such as in the Punjab in Pakistan), return flows exacerbate problems of salinity and waterlogging, thus imposing negative externalities on other users [2] [6].

Situation III: Return flows are beneficial, either to other downstream users or because they recharge groundwater aquifers on which others depend. In the Indo-Gangetic Plain, for example, an important function of leaky surface irrigation systems is the recharge of the groundwater aquifer.

Net Benefits from Return Flows in Haryana: It is known that a part of the return flows in Haryana go to the sink of saline groundwater while the rest charge the groundwater. Although water tables in Haryana have been declining, mining of groundwater is occurring only in one or two districts. Hence, on average, it is assumed for illustrative purposes only that net benefits from return flows are 25 percent of the net value of output in agriculture. This gives an estimate of US 0.5 cents per cubic meter of water diverted for irrigation purposes.

Total Economic Value in Irrigated Agriculture: The estimated total economic value of water diverted to irrigated agriculture is estimated at US 6.2 cents per cubic meter.

3   Full economic costs of water used in irrigation

Full Economic Costs of water diverted for agriculture are: US 0.2 cents for O&M costs plus 3.8 cents per cubic meter for capital costs and plus 1.5 cents per cubic meter for pumping costs plus 3.0 cents as opportunity cost of water (in the urban household use). These components add up to 8.5 cents as the Full Economic Costs of supplying water for irrigation in Haryana.

It may be emphasized that the opportunity cost of water used in irrigation would depend on the opportunities and costs of transferring large volumes of water used in irrigation to the potential users in other sectors. Given the fact that irrigation is the predominant user of water in most river basins, the opportunity cost would be zero(or close to it depending upon ecosystem demand for water) after the demands for other sectors /users have been met.

4   Irrigation charges, cost recovery and subsistence farming

If Irrigation Charges are levied so as to cover O& M costs of irrigation schemes, these charges will form only a small (about 10 per cent) proportion of the Net Value of Output (US 1.9 cents per cubic meter of water diverted) from irrigation water. However, if the farmer has to pay O&M costs plus annualized capital costs as irrigation charges, these charges alone (equivalent to US 4 cents per cubic meter) will be almost twice the Net Value of Output the farmer obtains by using irrigation. Apart from the political infeasibility of raising irrigation costs to such high levels, there is a need to evaluate the disincentive effects of raising irrigation charges on the reduction in the use of water for irrigation, lower crop yields, higher foodgrain prices and adverse effects on the rural and urban poor. These adverse impacts would be reduced over time because higher irrigation charges will, hopefully, result in improved and reliable irrigation services, use of other inputs and upward adjustments in output prices. However, small and marginal subsistence farmers whose consumption of foodgrains depends on their own output, will be adversely affected since they will have to sell a larger proportion of their crop output to meet additional costs of irrigation charges.

5   Conclusions

Unlike in other user sectors of water, direct economic benefits to the farmer (from crop output) reflect a small proportion of the total benefits to society of using water in irrigated agriculture. Irrigation infrastructure provides non-irrigation benefits to other user sectors and results in substantive benefits to subsistence farmers, agricultural labourers and the urban poor. Ignoring these benefits will result in a serious under-estimation of societal benefits available from the volume of water that is diverted for irrigation.

In view of additional societal benefits (generally not quantified due to estimation problems), irrigation charges need not reflect Full Economic Costs of irrigation supplies. If irrigation charges are raised to cover O&M costs, this will not impose any burden on the farmers. However, there will be substantial disincentive effects for farmers and adverse impacts on subsistence farmers , agricultural labourers and urban poor if irrigation charges are fixed to cover O&M plus capital charges plus opportunity cost of water.

6   References

1.   Bhatia, R. and S.K. Raheja, (1996). *Multiple Uses of Water: A Research Proposal.* Submitted to IIMI, Draft.
2.   Briscoe, J. (1996) Water as an Economic Good: The Idea and What it Means in Practice, *Proceedings of the ICID World Congress,* Cairo.
3.   Seckler, D. (1996) New Era in Water Management. IIMI, Colombo.
4.   Sinha, B. and Ramesh Bhatia. 1992. Economic Appraisal of Irrigation Projects in India. Agricole Publishing Academy, New Delhi. Vaidyanathan, A. 1993. Second India Series Revisited: Water. Madras Institute of Development Studies; Madras, India.
5.   Bhatia, R. (1989) Financing Irrigation Services in India: A Case Sudy of Bihar and Haryana States, in Financing Irrigation Services: A Literature Review and Selected Case Studies from Asia by L.E. Small, M.S. Adriano, E.D. Martin, R. Bhatia, Y.K. Shim and P. Pradhan. International Irrigation Management Institute, Colombo, Sri Lanka.
6.   Kijne, J. and R. Bhatia. 1994. Conflicts in Water Use: Sustainability of Irrigated Agriculture in Developing Countries. Proceedings of the Stockholm Water Symposium, Stockholm, Sweden. pp. 361-375.

SECTION B

THE VALUE OF DRAINAGE AND FLOOD CONTROL

THE ECONOMIC BENEFITS OF FLOOD WARNING AND FORECASTING

P.J. FLOYD, J.R.V. ASH and M.S. HICKMAN
Risk & Policy Analysts Ltd, Loddon, Norfolk, UK.
P.F. BORROWS
Environment Agency, Thames Region, Reading, Berkshire, UK.

Abstract

The Ministry of Agriculture, Fisheries and Food has policy responsibility for flood and coastal defence in England and Wales. The Environment Agency, as the relevant operating authority, exercises a general supervision over all matters relating to flood defence. The provision of flood warning systems is a high priority of both organisations. In the promotion and operation of flood warning systems, there is a need to demonstrate that any flood warning scheme is not only technically sound but also economically viable. However the benefits arising from flood warning are not well understood nor well documented. The tangible benefits are often small, relating mainly to the moving of personal possessions to above the predicted flood level. The time needed to undertake such actions is obviously a factor and can depend on age, mobility, etc. The provision of timely warnings may also give intangible benefits such as a reduction in the level of concern or stress experienced by the individual. This paper describes an approach for quantifying both the tangible and intangible benefits of flood warning and forecasting and for assessing the effectiveness of the warning process.
Keywords: Decision analysis, economic benefit, event tree, flood forecasting, flood warning flood trauma value stress.

1   Introduction

Sufficient water is a prerequisite for life but in excess in the form of floods, water can be life threatening and cause catastrophic damage. The United Nations has estimated that flooding is responsible for one third of all damage caused to the world's economies by natural disasters. It follows therefore, that effective flood warning offers the prospect for reducing damage.

Water: Economics, Management and Demand. Edited by Melvyn Kay, Tom Franks and Laurence Smith. Published in 1997 by E & FN Spon. ISBN 0 419 21840 8.

A significant proportion of the population of England and Wales live in areas at risk from flooding. It has been estimated that over 5% of the population are at risk from tidal flooding. Additionally, in England and Wales, there are some 600 km^2 of urban area and 10,000 km^2 of non-built up area at risk from a 1 in 100 year return period fluvial flood event.

The Ministry of Agriculture, Fisheries and Food (MAFF) has responsibility in England and Wales for policy in respect of flood defence and coast protection and has lead department responsibility for co-ordinating the response to river, sea and tidal flooding emergencies in England. The aim of the Ministry's policy [1] is to reduce the risks to people and the developed and natural environment by:

- encouraging the provision of adequate and cost-effective flood warning systems;
- encouraging the provision of adequate, technically, environmentally and economically sound and sustainable flood and coastal defence measures; and
- discouraging inappropriate development in areas at risk from flooding or coastal erosion.

The safeguarding of lives has the highest priority and the provision of flood warning systems is the highest of the priorities for grant-aid as published by the Ministry.

A number of Acts of Parliament are administered by MAFF in relation to its flood defence responsibilities and these Acts also empower the relevant authorities to undertake flood defence measures. The Environment Agency (the Agency), under these Acts, has a duty to exercise a general supervision over all matters relating to flood defence and also, in respect of flood warning, has powers enabling it to:

- provide and operate flood warning systems;
- provide, install and maintain apparatus required for the purposes of such systems; and
- carry out other engineering or building operations so required.

In promoting or operating a flood warning system, there is a need to demonstrate that any scheme is not only technically sound but also economically viable. The methodology used for this is cost-benefit analysis. There are published guidelines for the appraisal of flood alleviation schemes [2] covering the estimation of tangible, and, to a lesser extent intangible benefits accruing from the avoidance of damage by flood water.

The benefits arising from flood warning are not so well understood nor well documented. Tangible benefits are often small, relating to the moving of removeable items to a level above the anticipated flood water level. The time to allow for preparations to be made is not well understood and can depend on timely receipt of warning, knowledge of actions to take, and other response factors such as age, mobility and type of dwelling.

The assessment of intangible costs or disbenefits, of a flood and their reduction which can be achieved through effective warning is under-researched, especially in the area of human health effects.

It was against this background that the Agency commissioned a research and development contract to evaluate the tangible and intangible benefits associated with the provision of a flood forecasting and warning service and to establish a methodology which could be assessed by the evaluation of specific case studies (to be carried out as a separate phase). This first phase has now been completed by *En*tec in association with Risk & Policy Analysts Ltd, and is the subject of this paper.

2   Benefit assessment

2.1   Tangible benefits

The techniques for assessing the tangible damages (i.e. dis-benefits) of flooding are well researched. Three manuals have been published by Middlesex University's Flood Hazard Research Centre (MUFHRC) [3] [4] [5] which deal with the evaluation of damage caused by a flood event. The overall methodology is in accordance with UK Government guidelines in that damage costs are evaluated in terms of national loss.

The assessment of tangible effects can be subdivided into direct and indirect effects. Direct effects are those relating to restoration to the condition preceding the flood. Indirect effects are those occurring as a consequence of the flooding, such as disruption to traffic, loss of industrial production, etc.

The other direct cost to be taken into account is the beneficial effect a warning can have in reducing the loss of life and number of accidents caused by panic responses. This aspect has to be approached with certain caution as there is little information on injury statistics and there have been few deaths in the UK since the 1953 flood, probably as a result of improved communications and warnings given.

Assessing the reduction to tangible losses as a result of a flood warning can, to a large extent, use similar methods to those used for flood defence. The benefit however, is only accrued if certain actions are taken to reduce the effects (which is usually the relocation of moveable objects to a position of safety and minor works to reduce the severity of the flooding to one's property). Time to be able to respond effectively is, therefore, an important consideration.

2.2   Intangible benefits

The intangible effects of flooding are those for which costs cannot be directly assessed as there are no readily available market prices for them. The assessment of costs, or dis-benefits to recreation or the natural environment affected by a particular flood event can be assessed by the use of indirect methods such as the travel cost method or through methods which rely on hypothetical markets such as the contingent valuation method (CVM). However the reduction in this dis-benefit arising from a flood warning is small if, in fact, any reduction at all can be achieved.

The intangible effects to human health are not so well understood nor researched and their inclusion within any flooding or flood warning scenario has often been through subjective descriptive statements which allow the decision maker to assess the scale of effects under different scenarios. The effects of floods on those who actually experience the event at first hand include those impacting on the health of the individual [6]. Yet there has been little research into placing a monetary estimate on such impacts.

Part of this project involved developing a potential methodology for assigning a cost to the human health effects based on the evaluation of flood trauma following examination of a range of approaches and from stress research carried out in the US.

This was then costed through comparison with values used by the Department of Transport in relation to the human effects of accidents in the UK. This is described in the following section.

3   Stress model

Given the possible acute and chronic effects of stress induced by a disaster, the costs attributed to stress can be a major factor in not only the area of flood warning and forecasting, but also in flood defence scheme justification. It was proposed, therefore, to develop a stress model that provided actual costings per person following (or, indeed, prior to) a flood event. The starting point for this model was the Social Readjustment Scale [7]. This scale defined stressful events in relation to other stressful events and examples are given in Table 1.

Table 1.   Examples of the social readjustment scale

Life event	Mean value
Death of spouse	100
Divorce	73
Personal injury or illness	53
Marriage	50
Pregnancy	40
Death of a close friend	37
Change in responsibility at work	29
Change in living conditions	25
Change in residence	20
Change in schools	20
Change in recreation	19
Change in sleeping habits	16
Change in eating habits	15
Christmas	12

Although flooding may be responsible for some of the life events given in the table, it was hypothesised that flooding as a life event itself would rank about 25 on the scale. This does not mean that actually being flooded would be the same as a 'change in living conditions', but the relative impact upon an individual's life would be similar.

The next stage in the analysis was to calculate a flood trauma value (FTV), this value lies on a scale from 0 to 100, where 0 reflects 'no trauma' and 100 'maximum trauma'. It is this value which forms the basis for assigning an economic value to stress. The FTV is composed of four elements:

- a demographic factor;
- an urgency factor;
- a panic factor; and
- a 'hassle' (or nuisance/inconvenience) factor.

The sum of the four factors has a maximum value of 100. Within each factor group there is a maximum value, reflecting its relative importance in the analysis.

The four factors can move within a range of 0 to their maximum value. This is summarised in Table 2.

Table 2.   Summary of the components of the flood trauma value (FTV)

Component	Range	Maximum Total
Demographic		
- population reactivity	0 - 20	
- nature of housing	0 - 12	
- car ownership	0 - 8	40
Urgency		
- initial preparedness	0 - 5	
- "knowing what to do"	0 - 5	
- time left to flooding incident	0 - 10	20
Panic		
- reaction	0 - 3	
- severity of flood	0 - 7	
- level of preparedness	0 - 3	
- previous flood experience	0 - 7	20
Hassle		
- forced evacuation	0 - 6	
- length of time out of homes	0 - 6	
- lack of basic living necessities	0 - 4	
- extent of clean-up and associated problems	0 - 2	
- work missed	0 - 2	20
Total (FTV)	0 - 100	100

The demographic factor is considered the most important variable (and given twice the weight as the other factor groups) and is determined by the characteristics of the area flooded, i.e. it is constant throughout the analysis for a given area. The sub-components are placed on a sliding scale reflecting their weighting and importance for the particular area under consideration. When summed, the resultant figure out of a maximum score of 40 reflects the area's potential ability to minimise damages. Therefore, a score of 0 would indicate a high ability to cope and a score of 40 would indicate a very low ability to cope.

Urgency factors reflect the level of urgency prior to a flood event (such as when a warning is given) or during and after a flood event. The urgency factor is considered the rational reaction (as opposed to panic, see below) of a flood victim to the flooding event. The factor is further sub-divided into three areas, initial preparedness, "knowing what to do" and time left to flooding incident. Once summed, these give an urgency factor with a maximum score of 20, where 0 shows no level of urgency and 20 an extreme level.

The panic factor highlights a totally irrational response to either a flooding event or the warning of a flooding event. The sub-components of this factor when summed give a panic factor with a maximum value of 20. A score of 0 would show no panic, while a score of 20 shows extreme panic.

The hassle factor reflects the level of inconvenience or nuisance that a flood warning/event can cause. The source for such a factor has its roots in Allee [7]. It is dependent upon the severity of a particular flood event, and this value is therefore constant throughout an event. The sum of the sub-components have a maximum score of 20. A score of 0 shows that no hassle is attributable to the event (or a warning) and a score of 20 shows maximum hassle.

The next stage is to attribute a monetary value to the FTV. From Department of Transport figures [8], it is possible to determine average injury costs per person. Averaging the human costs (reflecting pain, grief and suffering) of a slight and serious injury gives a monetary value of £46,615. This is a cost per accident and requires conversion into per person costs, considering that, on average there are 1.23 casualties per road accident. The average injury cost per person, when rounded, is £38,000.

To combine this value with the social readjustment scale, it is assumed that this figure is equal to a mean value of 53 ('personal injury or illness' in Table 2). Given that it is also assumed that a mean level of 25 is attributed to flooding, the maximum value, in monetary terms, of the FTV is calculated as 25/53 x £38,000, i.e. the maximum value of FTV (100) is equal to £18,000 per person. The value for a particular flood event is then a percentage reduction of this figure.

4   Benefits of flood warning and forecasting

The modelling of benefits accruing from flood warning and forecasting was based on the addition of the tangible and intangible benefits and their comparison over different warning times. The model limited the analysis to lead times of 2, 4 and 8 hours relative to the no warning base of 0 hours. This was to allow the use of published data for the estimation of the tangible benefits and the reduction, given set warning times. The framework used for the benefits model is shown in its completed form in Table 3.

Table 3. Benefit model case study

	No warning (£)	2 hours (£)	4 hours (£)	8 hours (£)
Damage to property				
- damage to fabric	101,500	101,500	101,500	101,500
- damage to fittings	131,000	98,000	85,000	77,200
Agriculture/livestock	3,240	1,620	0	0
Roads/infrastructure	8,000	8,000	8,000	8,000
Environment	25,000	25,000	25,000	25,000
Human effects				
- stress	740,000	630,000	576,000	523,000
- physical	833,000	14,000	7,000	0
Total Damage	1.8m	880,000	800,000	735,000

This model was then tested as a desk top exercise using a number of recorded flood events. The analysis assumed that all those at risk received a warning and took appropriate action. The results, from an example of a coastal village experiencing a 1 in 100 year tidal event, are shown in Table 3.

5   Decision analysis

The final part of the analysis was to calculate the benefits for different levels of flood warning in annualised terms taking into account the range of expected flood events, the reliability of forecasting, the adequacy of dissemination of the flood warning, and the effectiveness of the response by individuals. This was proposed by the use of an event tree which would be the subject of detailed examination in the Phase 2 study in order to explore the significance of particular components in the chain, from the occurence of initial conditions through to flood forecast, the issue of flood warnings and response to the warnings. An example of a generic event tree is shown in Fig 1.

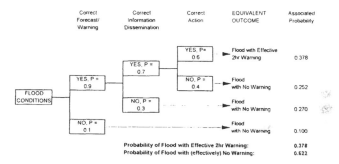

Fig. 1. Event tree for a 2 hour warning

From the assumptions underlying Figure 1, some of the benefits of the two hour warning are lost due to failures in forecasting, inadequate dissemination of information and, perhaps most importantly, failure to respond. Based on the event tree, it can be seen that there is only an estimated 38% chance of success in achieving a two hour warning (i.e in 62% of the cases the tangible losses would be effectively the same as for a flood with no warning, although this may not be the case for stress effects).

The decision analysis, together with the annualised costs of different scenarios for flood warning, can then be used to explore the cost-effectiveness of improving any of the components within the process (from forecast to avertive action taken) and hence be used to maximise effectiveness and resources available.

6   Conclusions

The effective delivery of flood warnings can offer substantial tangible and intangible benefits. Although the assessment of tangible benefits is well developed and documented, the intangible benefits in relation to human effects have not been researched to the same degree. The use of a costed Flood Trauma Value is proposed as one method for assessing the human health response to flooding and the relative changes that can be achieved as a result of flood warnings.

It is proposed that this methodology is further evaluated by its application to a number of case studies which will analyse in some detail the effect of lead times, the manner in which people react when warned and the human effects caused by flooding and flood warning The forecasting and warning procedure should also be explored by the use of decision analysis with the aim of improving the effectiveness and efficiency of the flood warning service.

7   Acknowledgements

This research and development project, *The Economic Benefits of Flood Warning and Forecasting* was carried out on behalf of the Environment Agency by *En*tec in association with Risk & Policy Analysts Ltd. The authors would like to thank all those who have provided information for the project and for the encouragement and support of the Project Steering Group.

8   References

1.   MAFF. (1993) Strategy for Flood and Coastal Defence in England and Wales,MAFF, London.
2.   MAFF. (1993) Flood and Coastal Defence - Project Appraisal Guidance Notes, MAFF, London.
3.   Penning-Rowsell, E.C. and Chatterton, J.B. (1977) The Benefits of Flood Alleviation, A Manual of Assessment Techniques, Gower, Aldershot.
4.   Parker, D.J., Green, C.H. and Thompson, P.M. (1987) *Urban Flood Protection Benefits,* Gower, Aldershot.
5.   Penning-Rowsell, E.C. et al. (1992) The Economics of Coastal Management, A Manual of Benefit Assessment Techniques, Belhaven, London.
6.   Bennet, G. (1970) Bristol Floods 1968: Controlled survey of effects on health of local community disaster. *British Medical Journal,* August 22 1970, pp454-485.
7.   Allee, D.J. et al. (1980) *The Impact of Flooding and Nonstructural Solution,* US Army Corps of Engineers, Army Engineer Institute for Water Resources, Fort Belvoir.
8.   Department of Transport. (1996) Highways Economics Note No 1: 1995 Valuation of Road Accidents and Casualties, DoT, London.

OPTIMIZING WATER MANAGEMENT IN POLDER FLEVOLAND

SAIFUL ALAM
Ministry of Water Resources, Dhaka, Bangladesh
E. SCHULTZ
International Institute for Infrastructural, Hydraulic and Environmental Engineering (IHE), Delft, The Netherlands, and Directorate-General for Public Works and Water Management, Utrecht, The Netherlands

Abstract

Flevoland (97,000 ha) is one of the polders of the Zuiderzee project. During reclamation, irreversible crack formation in the upper clay layer has taken place, resulting in a substantial increase in permeability and storage capacity. This additional storage reduces the rise of water levels during extreme wet periods.

The extent to which storage in cracks reduces the effects of closing one of the pumping stations was analysed. Based on 15 years data, rainfall - runoff processes in the rural and urban areas, operation and maintenance costs, reductions in crop yields and damage to buildings and infrastructure for different water management options were simulated with the optimisation model OPOL. Closing of one of the pumping stations would result in a reduction of the equivalent annual costs for water management of 8%. The estimated damages to buildings and infrastructure show an insignificant rise, while no reductions in crop yields were identified.

Keywords: polder, drainage, water management, optimisation in design.

1   Introduction

The polder Flevoland (97,000 ha) - consisting of Eastern and Southern Flevoland - is one of the reclaimed lands in the former Zuiderzee, The Netherlands. The land use is agriculture, urban, recreation and nature reserves. The water management system consists of field drains, main drains and four pumping stations (Fig. 1).

The upper layer (1.0 - 1.5 m) consists of ripened, cracked clay, which developed during reclamation. The cracks provide an additional storage during extreme wet periods, which was not taken into account in the design of the drainage system. Recently the electrical pumping stations were automated. In normal situations these pumping stations are used. Only during wet periods are the diesel pumping stations

Water: Economics, Management and Demand. Edited by Melvyn Kay, Tom Franks and Laurence Smith. Published in 1997 by E & FN Spon. ISBN 0 419 21840 8.

needed. Closing of the pumping station Wortman would reduce the costs of operation and maintenance. Before taking decisions, the effects of such a closure on the behaviour of the water management system and on possible yield reductions and damage to buildings and infrastructure would have to be investigated.

In this paper the analyses for the Low section with the existing pumping capacity and without the pumping station Wortman will be presented. The optimisation model OPOL, consisting of a hydrologic, an economic and an optimisation module, was applied to compute costs and damages for different water management scenarios in the rural and urban areas [1]. In this model, the relation of storage in the cracks with the open water level is taken into account. The hydrodynamic model DUFLOW was used to simulate water levels in the main drainage system during extreme wet periods [2]. The geographic information system ILWIS was applied to prepare parts of the input for the optimisation model and the hydrodynamic model [3].

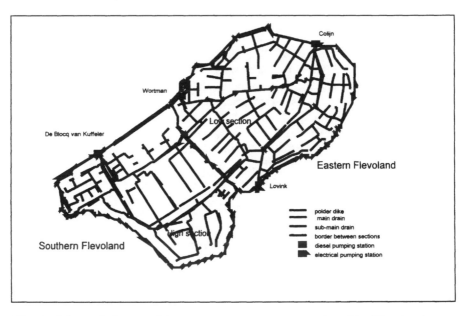

Fig. 1. Schematic layout of the water management system in polder Flevoland.

2   Physical situation

The Netherlands has an even distribution of rainfall over the year, mean annual rainfall is 750 mm. The evaporation from open water varies from 0 mm/day in winter to 5 mm/day in summer, mean annual evaporation is 700 mm. On average the rainfall deficit in summer is 120 mm and the surplus in winter is 300 mm. The 24-hour rainfall with return periods of 2 and 10 years is 31 mm and 42 mm.

The land surface is between 2.5 and 4.5 m-MSL (Mean Sea Level). The soil consists of 1 - 8 m holocene deposits, mainly consisting of clay, resting on pleistocene sand layers. When the polder emerged, the holocene sediments were loosely packed -pore volume around 75% -, saturated with water, and had a very low permeability ($1.7 * 10^{-4}$ m/day). Due to soil ripening during reclamation, among others, mainly vertical cracks have developed upto about 1.0 m-surface. Plowing has mixed the top layer, but the cracks under this layer remained open (Fig. 2). Due to the cracks in the clay soils the permeability increased to 50 - 100 m/day and the storage coefficient upto 0.15 - 0.18. When the water level in the main drainage system exceeds the crack depths, a nearly horizontal groundwater table can be observed, although the groundwater level may differ between the cracks and the soil columns.

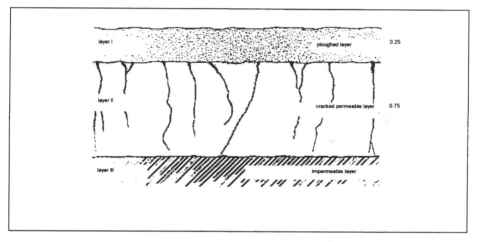

Fig. 2. Schematization of the soil profile.

3   Water management in the polder

Primary function of water management in the rural area is to create good soil moisture conditions. Characteristic features of the drainage system are: depth and spacing of field drains, normal water level, percentage of open water and pumping capacity.

Due to the high permeability, widely spaced sub-surface drain pipes (24 - 48 m) at 1.0 m-surface keep the groundwater table at an appropriate depth. After subsidence the outlets of the drain pipes are at 1.2 m-surface. The bottom of the collector drains lies just above normal water level, but at least 0.2 m below the outlets, in order to ensure non submergence of the drain pipes under normal conditions.

In the urban areas, landfill has been applied with 1.0 m of sand and sub-surface drains were installed at least at 1.30 m-surface. Due to the sand layer, the sub-surface drains react more slowly on rainfall than the sub-surface drains in the rural area. Separate sewer systems were installed. The storm water sewers discharge quicker than the sub-surface drains in the rural area. The sewer system and the sub-surface drains discharge through the urban canals and weirs or pumps into the main drainage system. Domestic and industrial waste water is discharged to treatment plants.

The main drainage system, consisting of collector drains, sub-main drains, main drains and structures, is divided into two sections: High section - normal water level 5.20 m-MSL - and Low section - normal water level 6.20 m-MSL. In the higher parts, weirs maintain the normal water level in the sub-main drains at higher levels, but at least at 1.4 m-surface. In the new towns, the water level is maintained at other levels with weirs, or pumping stations. The open water storage at normal water level is 1.2% in the rural area and about 3% in the urban areas.

The pumping capacities are 11.0 mm/day and 12.4 mm/day in the Low and High section. The overall capacity is 11.5 mm/day. Without pumping station Wortman, the capacity for the Low section reduces to 6 mm/day. Normally excess water in both sections is less than 2 mm/day, which for the Low section can be drained at Colijn (2.5 mm/day), and for the High section at Lovink (2.1 mm/day).

4   Model analyses

The model analyses involved the following steps:
- hydrological analysis of the polder;
- computation of storage in the polder in relation to water levels;
- determination of extreme hydrological situations;
- system optimisation for extreme hydrological situations and economic criteria;
- computation of non-steady flow conditions in the main drainage system for extreme situations;
- interpretation of various options of water management in terms of groundwater level, land use, and damage.

4.1   Preparation of GIS maps and results
Until a rise of the water level of 0.40 m there is only storage in the main drainage system, above that level storage in cracks plays an increasing role.   Spatial distribution of the storage, divided into open water storage and storage in cracks, was estimated with the geographic information system ILWIS. Maps were produced for: landuse, soil distribution, elevation, normal water levels in and lay out of the main drainage system, groundwater levels for various water management options, and reduction in crop yield. The depth map was combined with the soil map to produce a new image representing depth and storage capacity. This image was processed by GIS to produce a table listing the pixels with the same groundwater depth and soil properties. This table was processed by a FORTRAN programme which computes the storage related to the rise in the water level. The results are presented in Fig. 3.

4.2   Optimisation model for water management in a polder
To determine the effects of the closing of pumping station Wortman, the simulations for the rural and urban areas were treated separately. The simulations were done with time steps of two hours for the rural area and 5 minutes for the urban area. The actual optimisation was done with the Rosenbrock scheme [4]. The equivalent annual costs were determined by the annual interest and the life-time of project components. Other input data were derived from test fields and literature. Most field data were taken

from three experimental plots with representative soil types and field drainage intensities that existed in Eastern Flevoland from 1978 - 1985 [5].

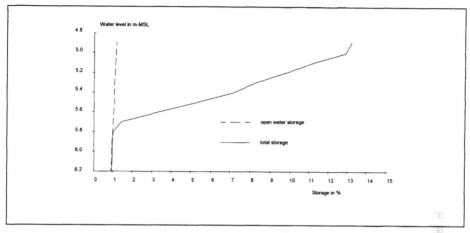

Fig. 3. Open water and soil storage related to the water level in the Low section.

In the hydrological model for the rural area, the surface and sub-surface environment were schematised into four layers, in which the water balances and relevant processes were determined for each time step.   The following data were used: soil characteristics, daily reference evapotranspiration, and hourly rainfall, groundwater levels and collector drain discharges. Although in the study on which this paper is based, simulations for three soil types were done, only the results based on data for one soil will be presented [6]. The crack width varies between 0.25 and 10 mm. The ploughed layer (layer I) of 0.25 m has a clay content of 30%, in the underneath 0.75 m of cracked soil (layer II) and beyond 1.0 m the clay content is 16% (layer III) [7]. The storage coefficient in the ploughed layer is 4 % and in the cracked layer 18 %. The model computes the unsaturated permeability based on an empirical relation between soil moisture tension and soil moisture content [8]. The crop factors - based on potato, sugarbeet and winter wheat - were taken from literature [1]. The basic values to schematize the drainage system were: sub-surface drain spacing 24 m and depth 1.0 m, collector drain spacing 400 m, parcel length 1,000 m, normal water level 1.4 m-surface, open water 0.9% and pumping capacity 11.0 mm/day.

The economic computation for the rural area consists of the determination of costs for water management, and costs related to yield reductions and damage to buildings and infrastructure. The installation, and operation and maintenance costs of the field and main drainage system were computed as functions of unit values reflecting the dimensions and frequency of activities. The areas covered by each crop were given as percentage of the total area. The value of crops, buildings and infrastructure was calculated based on the maximum and minimum investment costs per crop for each decade per ha, the area of buildings per farm, the value of buildings per farm, the percentage of facilities and the value of infrastructure. 20% of the value of the buildings has been taken for facilities, like shops and factories. The density of paved roads was determined at 10 m/ha of 5 m width. Reductions in crop yield may occur due to too wet conditions in the top layer during sowing or harvesting, or due to too high groundwater tables in winter. They were determined with empirical formulae [1]. The dates of planting were determined based on the earliest day, the required number of days and the condition of the top layer. The critical periods were determined and yield reductions resulting from untimely sowing were computed. Yield reductions during harvesting were calculated based on the lowest pH value at which harvesting can take place, the duration it takes, and the date after which yield reductions can occur. There may also be yield reductions due to insufficient air entry as a result of a bad soil structure in layer I, or II. Damage to buildings and infrastructure was assumed to be a percentage of the value in relation to the groundwater level.

In the hydrological model for the urban area, for the paved and unpaved surfaces, the transformation of rainfall into sewer inflow and flow to the subsurface drains, and the outflow from sewers and drains to the urban canals were determined for each time step. Input data were mainly based on research results in four urban areas [9]. Data on rainfall, sewer inflow and discharge are available per 5 minutes for the period 1970 - 1985. Construction and maintenance costs for the different elements of the urban water management system were based on fixed values per meter length, diameter and number, and unit prices. The data by which the water management system in the urban areas was schematized were: sewer diameter 0.30 m, canal spacing 410 m, canal width 11.00 m, crest level of weirs 1.80 m-surface, total crest width 11.00 m. Data on the value of buildings were used for houses with one or two floors, shops and offices, and industrial buildings. The value of infrastructure was determined based on: squares and paths, quarter roads and main roads. Public facilities and utilities were set at a fixed value per hectare. Direct damage is caused by the contact of the water with the goods that can be affected and was quantified based on the costs for repair and replacement. Damage values for the different types of buildings, roads and canals were expressed in percentages of the value of these elements, and were calculated based on: calculated volume of water in the streets, groundwater levels and exceedence of critical water levels. Damages of infrastructure were computed using an empirical relation with inundation or depth of the groundwater table. Indirect damages concerning lost services, working hours, industrial production, or inconveniences in traffic were expressed in percentages of the direct damages [10].

Selected wet periods were used to compute the discharge and the water levels. The urban discharge was added to the rural discharge. The operation of the pumping stations was handled based on prescribed levels to start and to stop.

The hydrological models were calibrated for one year and validated based on inputs for other years. Calibration of the hydrological model for the rural area resulted in the following optimal values: maximal interception 2.38 mm/day, the parameter giving the lower limit of soil moisture for potential evapotranspiration 133 mm, maximum moistening rates for layer I and II of 1.49 mm/m/hour and 0.75 mm/m/hour, and for the parameters 0.89 and 0.75 of the non-linear discharge model. Calibration of the hydrological model for the urban area resulted in the following optimal values: storage in depressions 0.80 mm, runoff coefficient 0.73, the two parameters of the non-linear models for the transformation of runoff into sewer inflow and from inflow to sewer discharge 2.34 and 0.300 and 1.92 and 0.950.

The land use and values in the Low section are shown in Table 1. The simulations resulted in a reduction of the annual operation and maintenance costs for the water management system in the Low section, with and without the pumping station Wortman, of 8%. In the present situation damage to buildings and infrastructure in the rural area was estimated at 0.07% and to the crops at 1.27% of the value. Without the pumping station Wortman there is an insignificant estimated increase in the damage of buildings and infrastructure.

Table 1   Values of crops, buildings and infrastructure in the rural area of the Low section

	Area in ha	Highest value in NLG/ha	in 10^6 NLG	Lowest value in NLG/ha	in 10^6 NLG
Buildings	132	13,800,000	1,822	13,800,000	1,821
Infrastructure	277	632,500	175	632,500	175
Potato	18,359	9,430	173	977	18
Sugarbeet	18,359	6,805	125	977	18
Winter wheat	18,359	3,968	73	977	18
Total	55,706		2,368		2,050

For the economic computation for the urban areas in the Low section the periods that may be important to assess damage in the urban areas were selected before the optimal values of the main drainage system were determined. The determination of these periods was based on the following criteria regarding periods of rainfall: a preceeding dry period of at least 24 hours, rainfall exceeding 15 mm and a minimum rainfall intensity of 0.5 mm/hour. A review of the values of buildings and infrastructure is given in Table 2. In the present situation the damage expressed in the values in the different areas were: living areas 0.10%, shops and offices 0.08% and industrial areas 0.05%. Without the pumping station Wortman there is an insignificant estimated increase in damage in the urban areas in the Low section.

Table 2   Review of total value in NLG of buildings and infrastructure in the urban area

	Area in ha	Value including furniture in NLG/ha	in 10^6 NLG
Houses, one floor	749	2,250,000	1,684
Houses, two floors	749	3,375,000	2,526
Shops and offices	96	16,200,000	1,555
Industrial buildings	573	8,500,000	4,870
Infrastructure	360	1,409,722	507
Public facilities	2,166	15,000	32
Public utilities	2,166	15,000	32
Total	2,526		11,206

4.3   Hydrodynamic simulations

With the DUFLOW programme package the effects of the closing of pumping station Wortman on the behaviour of the main drainage system under extreme conditions were simulated. Unsteady flow in the sub-main and main drains was described in the model by one-dimensional partial differential equations, which are discretized in space and time using the four-point implicit Preissmann scheme. For each drain section the discharge and the water level were determined for each time step. The pumps were assumed to be at full operation or not at all, dependent on the upstream water level. The storage in the cracks was incorporated in the main drain sections. At a return period of 10 years in the present situation a 2 day rainfall of 27.5 mm/day and without the pumping station Wortman a 6 day rainfall of 15.0 mm/day were input. The maximum rises in the water levels were 0.80 m and 1.05 m respectively after 48 hours and 210 hours. Control computations were made at a return period of 100 years.

5   Discussion

Storage in the cracks in the upper soil layer plays an important role in the water management of polder Flevoland. The storage variability due to topography and soil type has to be taken into account in the simulations. Based on the results the option of water management in the Low section without the station Wortman could be feasible, however, all aspects that may play a role would have to be considered in detail, before decisions can be taken.

6   References

1.   Schultz, E. (1992) *Water Management of the drained lakes in The Netherlands* (In Dutch). PhD thesis. Delft University of Technology, Delft.
2.   ICIM (1992) *DUFLOW* The Hague.
3.   ITC (1992) *ILWIS. Users manual.* Enschede.
4.   Kuester, J.L. and Mize, J.H. (1973) *Optimalization techniques with Fortran.* Mc Graw - Hill, New York.
5.   Ven, G.A. (1980) Runoff from arable land in the Flevopolders, in Proceedings of the symposium 'The influence of man on the hydrological regime with special reference to representative and experimental basins'. IAHS-AISH publication nr. 130. Helsinki.
6.   Saiful Alam (1994) *Optimizing water management in polder Flevoland, The Netherlands.* MSc thesis, Int. Institute for Infrastructural, Hydraulic and Environmental Engineering (IHE), Delft.
7.   Rijniersce, K. (1983) *A simulation model for physical soil ripening in the IJsselmeerpolders.* PhD thesis. Agricultural University, Wageningen.
8.   Staring Centre (1975) *Annual Report 1974* (In Dutch). Wageningen.
9.   Ven, F.H.M. van de (1989) *From rainfall to sewer inflow in a flat area* (in Dutch). PhD thesis. Delft University of Technology, Delft.
10.   Grigg, N.S. and Helweg O.J. (1975) State of the art of estimating flood damage in urban areas. *Water Resources Bulletin*, volume 11, nr. 2.

SECTION C

SOCIAL AND ENVIRONMENTAL VALUE OF WATER

ALLOCATING WATER FOR PRODUCTION AND RURAL CONSERVATION: CHOICES AND INSTITUTIONS

Allocating water for conservation

I.D. HODGE
Department of Land Economy, University of Cambridge, Cambridge, UK
W.M. ADAMS
Department of Geography, University of Cambridge, Cambridge, UK

Abstract

This paper discusses the problem of allocating water between different uses, with particular reference to the impacts of low flows in rivers for the conservation of aquatic and riparian ecosystems. It considers the lack of effective markets for water in rivers and shallow aquifers, and the role of planning in resource allocation. It explores the importance of an institiutional approach to water resource allocation between competing demands.

Key words: Conservation, economics, institutions, markets, rivers, wetlands, wildlife

1   Introduction

Water must be one of the most complex of economic commodities. It has numerous direct out-of-stream uses but is also drawn from a complex of aquatic environments that themselves offer numerous services. Economic growth stimulates the demand for water in respects of perhaps all of these uses. But the water available is already very extensively utilised. In a number of catchments demand for water outstrips local supply, and a significant number of rivers suffer from low flows [1, 2]. In this paper we outline some of the economic issues associated with water management. Our objective is to suggest an approach to the analysis of these issues, and potential directions for water management. Our aim is to widen the debate and suggest avenues for more detailed analysis, rather than to make specific policy proposals.

Water: Economics, Management and Demand. Edited by Melvyn Kay, Tom Franks and Laurence Smith. Published in 1997 by E & FN Spon. ISBN 0 419 21840 8.

2   Low flows and nature conservation

Low river flows represent a significant threat to conservation interests. A series of years of below-average rainfall have highlighted the problem of licensed abstractions exceeding natural supply, resulting in low river flows, and in extreme cases, dry channels. Low flows and depleted aquifers present problems for a range of water uses. Nature conservation interests are affected in three main ways. First, low flows can have impacts on habitats and species within the river channel itself. Second, they can affect water bodies supplied by flowing water and groundwater. Third, they can affect river bank and floodplain (riparian) habitats. Conservation values may also be threatened outside these areas, for example if species of restricted range (such as the native white-clawed crayfish) are present.

In river channels themselves, reduced discharges reduce channel size and hence populations of animals and plants, reduced velocities allow increasing sedimentation of fine particles and reduced scour of larger sediments. This in turn causes changes in benthic ecology, affecting plants, grazing animals and predators within the ecosystem. The most dramatic impacts may be on flagship species such as game fish, but ecological impacts can be rapid and significant on a range of taxa.

In riparian ecosystems, vegetation communities are adapted to a seasonal cycle of inundation, or high water tables. The ecology of meadows and peatlands can be profoundly altered by changes in water table depth and flood durations, impacts may be exacerbated by associated changes in water quality (such as high nitrogen of phosphorous loads due to reduced dilution of agricultural run-off or sewage). While uncommon species and communities may quite easily be lost from newly-desiccated wetlands, land exposed by shrinking water bodies is likely to be colonised by robust common species, particularly introduced species with little conservation value such as the Himalayan balsam, and phosphorous-loving plants such as the common nettle. Ecological changes may not be reversed in any simple way if wetter conditions subsequently return.

The adverse impacts of abstraction on conservation sites are well demonstrated by Redgrave and Lopham NNR on the Norfolk/Suffolk border [1], with six plant species and 77% of invertebrate species recorded in 1958 extinct by 1991, and many further species threatened, including the great raft spider. This is not unique. In 1996 English Nature identified 18 SSSIs presently at serious risk from abstraction, and suggested that the number could rise to 41 [3]. The values of these wetland sites are threatened by the demands of other water uses. Clearly, allocation decisions need to be made between conflicting uses. How are such decisions to be made?

3   Resource allocation in the water environment

While in a physical sense, water is completely utilised, in an economic sense it must be under-utilised. In many instances it is used inefficiently in economic terms. There must be at least some scope to increase the total value that is derived from the use of water and from the water environment, by reallocating water between uses and by managing the water environment differently. Indeed, given that it is widely agreed that demands for water are likely to go on increasing in the future, we face severe difficulties if this is not the case.

A logical objective for water management therefore is to direct water (and the environmental characteristics it supports) towards their highest value uses. This is not meant either in a narrow financial or in a short term sense. The term 'value' includes all potential sources of value, both those represented by monetary flows and those not reflected in market prices. It is important to consider both use and non-use values. This approach to water allocation may be interpreted as seeking to achieve the highest possible total value from the use of a resource in the long term. This is not dissimilar to the goal of sustainability, although this may involve placing an even higher priority on maintaining specific items of natural capital through time, even if this requires foregoing high value uses.

The relatively high pressure on resources indicates the importance of opportunity costs. If one use of water removes the opportunity of another use, the cost of resource use is reflected in the value of the uses foregone, which may or may not be reflected in the price that the user has to pay for the water. Typically, prices are not set in this way, being more likely to reflect the costs of supply and institutional arrangements for the determination of prices. For example, charges for water abstraction and for discharges are based on the expenditure, such as the cost of administering the consent system and related monitoring, and not on the economic value of the water consumed [4].

Further, the price of water often fails to reflect the full social opportunity cost because of the problem of externalities. Removing water from a source and returning it elsewhere, commonly in a less pure state, will often impose costs on others. These may be associated with the quantity of water remaining in the system or with water quality. Either way, these costs are borne by people who have not contracted to accept them, and who receive no compensation for them. This situation persists because of transactions costs, the costs associated with all economic transactions such as the cost of assembling information, of establishing and agreeing contracts and of enforcement. Where there are significant, clearly defined and well understood impacts (e.g. of pollution on fisheries), those affected may have recourse to legal action in order to protect their interests. Where the effects on each individual are relatively small, or where the evidence required to prove a direct link between an action and a consequence is difficult to obtain, such action is often not possible, and impacts persist. It should be noted that, of course, there may also be external benefits of water use, although they are less common.

The effects of water utilisation are transferred between potential beneficiaries in various ways. Water moving through the catchment carries chemical and biological impacts to other locations. Changes in the quantity or quality of water can, in particular, have impacts downstream. These may cause direct losses, either financial (e.g. the impact of chemical pollution on a commercial fishery) or non-financial (e.g. impacts on recreation or aesthetic values). They may also impose opportunity costs. In some circumstances these effects are unidirectional, whereby A's actions affect B. Alternatively, in the case of a common pool resourcee where several users draw supplies from a single acquifer, these impacts may be reciprocal, such that A and B's actions affect each other. Furthermore, in some cases impacts may be synergistic, where individual impacts combine to cause some greater overall impact. Thus low levels of pollution and water withdrawal may separately not be a problem, but together their combined impact could be significant, with the concentration of pollutants increased by low river flows.

The whole system is further complicated by the question of uncertainty. Levels of river flow vary seasonally and from year to year, and while these variations can (within limits) be modelled and predicted statistically, it is not simple to devise procedures for including such probabilistic data in arrangement for resource allocation. There can also be significant levels of uncertainty in the ways in which different human actors (individuals and institutions) will react either to opportunities and constraints of water supply, or to economic and other factors in the wider decision-making environment.

4   The problem of the market

The majority of commodities are allocated through markets. Some advocates of the 'property rights' approach [e.g. 5, 6] would have us believe that the only essential requirement is for the state to establish and enforce property rights. Markets will do the rest. The failure of property rights to emerge is a signal that the transactions costs of introducing property rights exceed the potential gains from the operation of a market, implying that intervention is not warranted. While there is some merit in the argument, it falls a long way short of offering a complete analysis. The implication of markets as being an alternative to the involvement of the state is misplaced.

We should recognise the social nature of markets, defined as 'institutions which make available to affected parties the opportunity to negotiate courses of actions' [7]. The complexity of environmental relationships and the large numbers of people involved mean that an extensive system of government involvement and regulation has developed in order to manage the allocation and control of water and water courses and this should be regarded as an integral element of the markets for water. Despite the recent extension of private operations into the supply of water, the state continues, inevitably, to play a central role in laying down and enforcing the rules. The state plays this role in respect of all markets (none are truly 'free'), but state intervention (whether direct or indirect) in water supply is inescapable, because of the natural monopoly and public goods aspects of water supply. The markets that do operate in respect of water are all to a large degree contrived through regulations.

5   The problem of planning

A market model of water allocation has obvious limits. At the same time, the limitations of planning for resource allocation have also become very clear. The model of the benign state allocating resources to their best uses is not one that attracts many adherents. Who is to judge the values of water when applied to specific uses? Economists have expended considerable effort in attempts to determine monetary values for the non-priced demands placed on the water [e.g. 8, 9]. However, there are fundamental limits both methodologically and ethically. All valuation techniques rely either on data on individual behaviour that reveals preferences for particular states of affairs (e.g. for recreation activities or buying certain sorts of property) or else representing individuals' stated preferences. In the first case, there are difficulties in identifying the appropriate circumstances under which individuals can be observed making relevant choices and in interpreting the information that does become available. In the second, there are uncertainties in determining how to interpret the statements that are made. In all cases, it is necessary to make critical value judgements in order to reach a monetary valuation of any sort. These may be obscured by the method adopted, or be unacceptable, or both, but they have important consequences for the results of techniques such as cost benefit analysis.

In the absence of reliable valuations, a planning system becomes vulnerable to political influence. This is not in itself a bad thing. There has to be a source of values underlying the planning process and this is precisely what a political system is for. The problem arises where the motivations of political judgements are deemed to be inappropriate: seeking votes in marginal constituencies at the expense of votes in safe opposition ones or responding excessively favourably to lobbying from groups giving financial support to a particular political party. A planning system may also be vulnerable to inappropriate bureaucratic meddling. This may be that in the absence of a financial imperative, bureaucrats are simply inefficient or worse that they are preoccupied with attempting to maximise the size of their own local bureaucratic empires.

6   An institutional approach

There are, of course, no simple answers to the problems of allocating a resource such as water. What is needed is an approach that tries to deal with the complexity, and yet allows a clear analysis. An institutional approach recognises the limits of markets and planning. The objective is not to prescribe required behaviour but rather to design institutional arrangements that bring individual incentives closer towards social objectives, removing constraints on desirable behaviour and rewarding actions that in some sense promote the public good. Decision-makers have the best information to make decisions for themselves within a context that establishes appropriate incentives. The aim would be to enable those affected to reveal their preferences through market and non-market arrangements. Where this is not possible, the state would introduce constraints over private actions so as to guide the system towards a desirable outcome.

As with the property rights approach, the starting point will often be with an analysis of the actual and potential arrangements for property rights. The desirable allocations of property rights are essentially social judgements and it is recognised that the state has a much more comprehensive role to play. The choice of forms of intervention will be driven by the polluter pays and the provider gets principles. This may involve the use of economic instruments such as charges for environmental impacts, subsidies for beneficial actions, transferable permits, or deposit-refund schemes. There are both legal and informal institutions. There may be ethical or religious influences on behaviour that have equal weight to legislation, although social judgements may be altered by changes in institutional arrangements. For instance, the privatisation of the water supply companies has clearly affected the willingness of consumers to make voluntary adjustments in their consumption patterns in response to exhortation. It may be an unintended effect of water privatisation that we now need to put what had previously been achieved through informal social norms onto a more formal legalistic basis.

The prices set within the system should represent social opportunity costs. Historically, water supplies have been allocated on a sectoral basis. Agricultural, manufacturing and domestic supplies have been regulated independently, and there has been no particular allocation of water needs to the wider environment, or to specific land uses such as conservation. The optimal solution would be for the prices paid in one sector to represent the opportunities foregone in the highest value alternative use, irrespective of sector. Similarly, water should be only allocated to agriculture or urban water supply if the benefits of so-doing out-weigh the costs of reduced availability elsewhere, for example to dilute pollution in river channels or keep wetland nature reserves wet.

There is an inherent tension between security of supply and flexibility. Any allocation system must be responsive to change, especially where the options available are subject to the vagaries of weather and natural environment. Similarly, social change and the legitimacy of claims on the water environment alter as preferences and scarcities change. The UK water industry has experienced problems due to irrevocable licences granted in the past when supplies appeared to be unconstrained [4]. At the same time, insecurity imposes a cost on the operators within the system. Businesses, whether in farming, manufacturing or nature conservation, need to be able to plan investments over reasonable periods of time. Nature conservation demands constancy of supply over particularly long periods. Conservation values take long periods of time to become established (centuries rather than decades), but can be damaged irreversibly very quickly.

7   Collective interests: revealing and co-ordinating decision-making

While the potential gains to individuals from taking legal or market action may be too small to justify the transactions costs involved, it is possible for like-minded individuals with similar aims to reduce the units costs by taking action collectively. This may take the form of lobbying, collecting information, group actions in the courts, or actions in markets. Clearly there are a number of such collective interests that tend to be neglected in conventional market contexts. Most of these already have some form of organisation established to represent their interests. The obvious examples are fishermen, those using rivers for transport or recreation, farmers, conservation groups and local amenity societies.

The conventional planning approach tends to seek opinions from such groups and take account of them in drawing up regulations for water management. This approach may be expressed in terms of markets through economic valuation exercises, but as noted above, such approaches remain controversial. An alternative would seek to establish an institutional environment within which such interests may be revealed through collective decisions [10]. Some organisations embrace a number of different interests and can seek solutions in their common interest. Thus the Avon Weirs Trust [11], with trustees representing the Environment Agency, District Councils, Navigation Trusts and private owners has been successful in raising funds from local councils, private individuals, charitable trusts, navigation authorities and government sources and applying these to refurbishing weirs on the river Avon. Fishing clubs take action to protect the interests of their members from water pollution [12]. Conservation organisations buy and manage land [13]. Institutional arrangements are important in determining the opportunities available to such groups to take action in the interests of a membership or in pursuit of a more general public objective, reflected for example in the rights of collective interests to take action in the courts [12], the powers for organisations to hold covenants over land without owning land [14], or the rights to instream flows separate from the ownership of riparian land [6].

However, the presence of such collective groups does not avoid the responsibility for government to play a major part, for the well known problems of public goods and free-riding remain. Solutions of this kind might be regarded as government action by intermediary, seeking out and implementing opportunities for promoting the wider public interest in collaboration and with the support of government at all levels.

8   Quasi-market mechanisms

If it accepted that low flows represent a significant problem, the objective must be to either increase the supply or reduce demand for water so as to increase the flows available at critical times and places. In respect to many resource allocation problems these are issues that would be resolved through market forces. In this case, as has been illustrated, markets general fail to promote an appropriate allocation of water. However, we can identify desirable objectives by analogy with the way in which an efficient market would operate. A price for water would establish incentives consistent with a number of objectives:

- A price for water would establish an incentive for least cost reductions in the level of demand. The price could be raised until the level of demand was consistent with the environmental constraints on supply.
- A price for water would establish an incentive for available water to be directed towards its highest value use. It would increase incentives for a search for efficient techniques for water conservation and re-use within households and business organisations. The evidence of price would provide the information to enable a judgement as to the relative value of water in alternative uses, both within and between sectors.
- A market could offer a mechanism for making short term adjustments to demand in the face of short term variations in the availability of supply. A major issue here is the distribution of risk associated with the inherent variability in supply.

Many of these incentives could be established through a system of transferable permits. Within a catchment, it would in principle be possible to define the total volume of water that can safely be abstracted with a reasonable degree of certainty from surface and groundwater sources in the long term. Constraints would be established on utilisation in order to protect environmental resources within the catchment. Permits to abstract water from defined sources would be defined and allocated (grandfathered or auctioned) to users within the area. Assuming that demand exceeded available supplies, there would be a positive price on these permits and an incentive for water to be allocated towards higher value uses. There would also be an equivalent incentive for users and permit holders to limit their demands. The higher the price, the greater the incentive to reduce demand through conservation measures and relocation of activities.

One significant problem constraining action to relieve the problem of low flows in the present situation concerns the problem of licences allocated historically by water authorities that cannot now be revoked without the payment of compensation. In some cases, the existence of such licences means that available supplies are not allocated to the uses and users where water is valued most highly [4]. If such licences were to be made transferable, incentives would be established for their transfer towards higher value uses. Similarly, regulatory authorities could acquire them in the market in order to reduce the total levels of water abstraction. In this way the economic impacts of reductions in water availability could be minimised, as would the level of compensation payable and the possible need for compulsory purchase would be avoided. It would also be possible for collective organisations concenred for environmental quality to acquire abstraction rights where they anticipated some gain to be made and for prospective developers within an area to be required to acquire permits to cover the impacts of their development on water availability.

Permits could define a priority amongst users in respect to the security of supply. In the same way that shares in a public company define the claim that the holder has on company assets, a permit would establish the rights that the holder has to water supplies in the face of limited availability, both seasonally and in the longer term. Thus permits would be expected to change hands at different prices, depending on the degree of risk associated with them. This arrangement would establish an incentive to allocate the risk towards users who are most able to respond. For example, users wanting water for watering lawns for amenity purposes may decide that it is unnecessary for them to acquire high security permits, simply accepting brown lawns in times of low supplies. Users facing severe financial risks from interruptions in supply, perhaps farmers producing high value crops, would be willing to pay to secure low risk permits. In practice, users might be expected to hold a portfolio of different types of permit as a hedge against risk. Alternatively, users may acquire high risk permits at a relatively low price but make their own arrangements to spread availability over time by constructing reservoirs. Similarly, permits might also relate to the quality of water required, although the prospects for a market in water of differing qualities might be hard to envisage in most circumstances.

This implies a critical role for a water management agency. This agency would need to be able to model water availability and allocation so as to be able to define the relative status of permits within different localities and at different times. The system would thus be complex. The agency would need to define the current status of supply and specify at any particular time what rights holders of each class of permit have to water supplies. There would need to be some system of zoning and spatial differentiation in permit definition and pricing. Indeed, spatial variations in prices would establish incentives for water movement, but might at the same time increase problems for nature conservation. Thus a permit system would need to be associated with complementary regulation. It is not an alternative to it.

Experience with the introduction of permit systems in other contexts, notably with the control of air pollution control in the United States [e.g. 15, 16] indicates the critical nature of the detailed administrative arrangements. These need to consider:

• definition of the rights associated with permits: uncertainty in permit markets has been a major factor preventing the development of trading;
• arrangement for the initial allocation of permits
• trading arrangements
• monitoring and enforcement
• spatial constraints and relationships with other permit areas
• relationships with other policy areas

There are inevitable trade-offs between the various goals of efficient resource allocation, effective trading and fair outcomes and apparently minor details can have significant impacts on a scheme's operation. But the potential gains make the approach worthy of careful consideration. There are also issues more specific to water, particularly the significance of the location of activities or the quantity and quality of water that is returned to the system after use, that present further complications.

9   Conclusions

This is not to imply that there are no significant problems. How to promote responsible participation by collective organisations, neither overstating their claims in circumstances when not forced to pay, nor under-representing them because free-riders will not contribute when they are? How to take account of the spatial variations in water abstraction and use by different users? The potential for a permit system will be constrained by the amount of information available to a managing agency; thus we should expect a co-evolution of models and markets. One of the objectives of the approach must be to point towards what information would be required and to design institutions that make effective use of the information that can be available. Similarly alterations in opportunities facing water users implies new risks in terms of environmental impacts and a need to re-think regulatory arrangements.

Our purpose has not been to advocate specific solutions. Rather our aim has been to suggest that an institutional analysis of water resource allocation indicates avenues worth pursuing. This appproach has been further developed in countries more traditionally recognised as experiencing problems of water shortage, such as the United States. While the differences in physical environments and in legal systems [e.g. 17] mean that it is not possible simply to transfer that experience to the UK, there are ideas and approaches that may offer innovative ways of tackling water scarcity and environmental concerns. It is clear that much further analysis and debate is warranted.

10   References

1.   Hill, C. and Langford, T.   (1992)   *Dying of Thirst: a response to the problem of our vanishing wetlands*, Royal Society for Nature Conservation, Lincoln.

2.   Drury Hunt, E. (1996) *High and Dry: The impacts of over-abstraction of water on wildlife*. Biodiversity Challenge, Royal Society for the Protection of Birds, Sandy.

3.   English Nature   (1996)   *Impact of Water Abstraction on Wetland SSSIs*, English Nature, Peterborough.

4.   Rees, J. with Williams, S. (1993) *Water for Life: Strategies for Sustainable Water Resource Management*. Council for the Protection of Rural England, London.

5.   Demsetz, H. (1967) Towards a theory of property rights, *American Economic Review* 57 (2) 347-359.

6.   Anderson, T. L. and Leal, D.R. (1991) *Free Market Environmentalism*. Westview Press, Boulder.

7.   Dasgupta, P. (1991) The environment as commodity, Chapter 2, pp25-51 in D. Helm (ed.) *Economic Policy Towards the Environment*. Blackwell Publishers, Oxford (quote p29).

8.   Hanley, N. (1989) Problems in valuing environmental improvements resulting from agricultural changes: the case of nitrate pollution, in A. Dubgaard and H. Nielsen, (eds.) *Economic Aspects of Environmental Regulation in Agriculture*. Kiel: Wissenschaftsverlag Vauk Kiel.

9.   Garrod, G.D. and Willis, K.G. (1996) Estimating the benefits of environmental enhancement: A case study of the River Darent, *Journal of Environmental Planning and Management* 39 (2) 189-203.

10.   Hodge, I.D.   (1997)   The production of biodiversity: institutions and the control of land, Chapter 12 in A. Dragun and K. Jakobsson (eds.) *New Horizons for Environmental Policy*, Edward Elgar, Aldershot (in press).

11   Sinclair, I. (1996) The Avon Wiers Trust - A successful answer to a formidable challenge, *ADA Gazette*, Spring, 30-33.

12.   Grant, M. (1995) A review of environmental liability and compensation systems in the UK and other countries, paper presented at conference on the Economics of Environmental Liability, Robinson College, Cambridge.

13.   Dwyer, J. and Hodge, I.D.   (1996)   *Countryside in Trust: Land Management by Conservation, Amenity and Recreation Organisations*, John Wiley & Sons, Chichester.

14.   Hodge I., Castle, R. & Dwyer J., (1992) Covenants for conservation: widening the options for the control of land, *Ecos* 13 (3) 39-45.

15.   Foster, V. and Hahn, R.W. (1995) Designing more efficient markets: lessons from Los Angeles smog control. *Journal of Law and Economics* 38, (1) 19-48.

16.   Polesetsky, M. (1995) Will a market in air pollution clean the nation's dirtiest

air? A study of the south coast air quality management district's regional clean air incentives market. *Ecology Law Quarterly* 22, (2) 358-411.

17.   Harbison, J.S. (1991) Waist deep in the big muddy: property rights, public values and instream waters. *Land and Water Law Review* 26 (2) 535-570.

SOCIAL BENEFITS AND ENVIRONMENTAL CONSTRAINTS OF IRRIGATION IN AN ERA OF WATER SCARCITY

E. FERERES
Institute of Sustainable Agriculture (CSIC) and University of Cordoba, Spain.
F. CEÑA
Department of Agricultural Economics, University of Cordoba, Spain.

Abstract
The expansion of irrigation in recent decades has coincided with the growing perception that renewable water resources of the world are finite. Irrigated agriculture is under intense scrutiny as one of the largest consumers of water. The social aspects of irrigation are discussed first, with emphasis on the issues most relevant to irrigation in Southern Spain. The public concerns about water quality deterioration and the impact of irrigation on the environment are also covered. The general theme is that, on the one hand, irrigation can no longer overlook the intense competition for water and the disruptive effects it produces on the environment. On the other hand, there should be incentives for water conservation in irrigation to sustain irrigated agriculture as a strategic sector in world food production.
Keywords: irrigation, social issues, environment, water conservation.

1 Introduction and background

Irrigated agriculture is essential to meet world food requirements but does so at the expense of consuming a good proportion of the developed fresh water supplies. The agricultural productivity of irrigated lands is about twice the world average, as only 18% of the cultivated land on earth is irrigated but such surface is responsible for over one third of the total food production [1]. The contribution of irrigation to meeting world food needs is expected to increase in the future; Carruthers, [2] estimated that the irrigated areas will contribute about 50 % of the total needs by the year 2025. At the country level, the importance of irrigation can be much greater in the arid and semi-arid zones; in countries such as India, irrigated area production (one third of the total cropped area) represents over 60% of the total agricultural production [3]. In Spain, the irrigated area (15% of total) is responsible for 60 % of the total agricultural output.

Water: Economics, Management and Demand. Edited by Melvyn Kay, Tom Franks and Laurence Smith. Published in 1997 by E & FN Spon. ISBN 0 419 21840 8.

Irrigation differs from most other uses in that it consumes much of the delivered water via evaporation from plants and soil, while urban or industrial uses are, for the most part, not consumptive. The proportion of the developed water resources devoted to irrigation can be very high regardless of the level of economic development. For instance, both in California and in Morocco, about 80% of the developed supplies are used in irrigation. In the dry areas, irrigation has been practiced since long ago because agriculture is not viable in those areas without supplementing the scarce rainfall. However, the large expansion of world irrigation has taken place in the 20th Century, particularly between 1960 and 1980. At present, there are 250 million irrigated hectares worldwide, which are growing at a rate of 1.2% per year. Most of the irrigated areas (73%) are in the developing countries.

Public investments in irrigation are subject to criticism as the capital investments for new irrigation developments are ever higher and the rates of economic return of many projects are less than those anticipated. At the same time, increased demands for urban and other uses are competing for a resource which is becoming scarce, particularly during drought events. Tensions between water supply agents and, particularly, among users have mounted in recent years, leading to even questioning the regulations and laws of water allocation. Increased urbanization has caused the dissociation of the majority of public opinion from the realities of rural life. It is therefore not surprising that criticism towards irrigation as a lavish water user has become fashionable in recent years among urbanites in the developed countries

At present, the perception of water scarcity is ubiquitous everywhere, particularly in areas of arid climate. Although it is common to focus on the amount of annual rainfall to characterize the water needs of an area, it is the annual evaporation that frequently makes the difference. Cordoba and Copenhagen have about the same amount of annual rainfall but the evaporation at Cordoba is more than twice that of Copenhagen and that fact is the major determinant of the need for irrigation in Southern Europe. The periodic droughts that dry regions often experience add tensions to the competition for the water available. The recent assessment of a possible global climate change also contributes uncertainty about the future water supply in those regions. Despite all past efforts in water development, increases in demand are now the major cause of the situations of water scarcity that arise in many world areas. In addition to water scarcity, an important problem is the degradation of water quality, which is a consequence of the many human activities that involve the use of water.

All water users and society as a whole must respond to the challenges described in ordner to achieve a balance between water supply and demand and to avoid water quality deterioration. In particular, irrigated agriculture which has historically overlooked the ecological problems caused by irrigation, must face a new reality of scarce supplies, questionable water rights and increased regulations to control pollution. Irrigated agriculture must be made permanent under a changing and uncertain socio-economic environment. To make irrigated agriculture sustainable, both economic viability and environmental protection are required. The achievement of such goals is vital for the future of world agriculture.

In areas where irrigation has been practiced, a whole culture of irrigated agriculture has been developed with its technical, economic, social and legal components. That culture is specific to the region where it developed and so are the problems associated with irrigation and the solutions proposed to solve such problems. Following is a discussion of several social and environmental issues which are most relevant to the irrigated agriculture of Southern Spain and which may be useful in the analysis of other irrigated areas.

2   The social environment of irrigation

Presently, irrigation in Spain which was once considered the engine of economic growth, is the subject of increased criticism. There are four issues that stand up among the many observations made about irrigation:
- Irrigation uses too much water (80 % of the total water used)
- Irrigation is inefficient; about 50 % of the delivered water is wasted.
- Farmers hardly pay for the water they use or they do not pay at all.
- Water pollution problems are often caused by irrigation.

The above unqualified statements are creating a very negative public opinion about irrigation which has yet to meet an organized, objective response on the part of the farmers. The primary cause for this change in public opinion must be linked to the events associated with the most recent drought of 1992-95 which was the worst of the century. It affected primarily the East and South of Spain where precipitation was well below normal for those four years. For the Guadalquivir Basin - the largest river in Southern Spain - runoff into its dams was only 27 % of normal during the drought years [4]. When the drought ended in October of 1995, the average reserves in all dams of the basin was about 6 % of normal. Large cities such as Seville were subjected to severe restrictions with urban water available for only a few hours a day. Under that situation, irrigation which has less priority than urban water use by law, suffered drastic cuts. In fact, water supply restrictions to irrigation in Andalusia started earlier, with a previous drought in the eighties. Normal water supply to irrigation in the Guadalquivir basin was limited to four of the last 15 years. For the other 11 years, irrigation water supply oscillated from 65 % of requirements to no supply in two years. Such limitations are evidence of a structural water resources deficit in Andalusia which has been quantified in the regional water plans at about 20 % of present demand [4]. Uncontrolled irrigation developments are, in good part, responsible for the structural deficit; it has been estimated that there are over 100,000 ha of unregistered irrigated lands in Andalusia. It is assumed that the deficit has to be solved by reducing the supply to agriculture. Thus, there is a current emphasis on the modernization of irrigation in the hope of reducing agricultural water use without an economic impact.

The tensions that developed among users and policy makers during the drought, created an environment of conflict which the subsequent rains of 1996 did not wash away. There is a strong push towards giving market-like forces the primary role in resolving the water conflicts among the users and the regions of Spain. If water would be considered strictly an economic good, the consequences for the agricultural sector would be unpredictable but very negative in the short run. The drought of 1992-95 already caused losses to irrigated agriculture estimated at over 4 billion US $ [4].

Water is a natural resource which is vital for present and future life on Earth. Therefore, society must provide equitable access to all individuals so that they can satisfy their vital needs. Water is also an input to many economic activities but its demand as an input may be considered a derived demand. The price of a good under a derived demand depends on the demand of the output that is generated with that water as an input. Thus, the value of agricultural, industrial or other goods would determine the price of water for each activity. It is obvious that, under strict market forces, water would be allocated to the economic activity where it has the highest marginal value. From strict market rules to free access to all; the resolution of the water conflict in our society lies somewhere between those two opposing views.

3   Social benefits

Historically, farmers' water rights were a guarantee of supply under normal conditions. In the majority of cases worldwide, farmers still do not pay for the volume of water used but rather they pay a fixed fee for the right to use the water on a per area basis. In countries with an old irrigation tradition such as Spain, many different situations coexist, from nearly free irrigation water (less than $100 per ha) to about 0.20 $/m^3 in some of the new irrigated areas in the South East. Questions are raised among the general public about the low costs of irrigation water to farmers that often end up producing surpluses which need to be subsidized; such a view is typical of an affluent society which tends to ignore some of the benefits of irrigation presented below.

3.1   Employment and wealth creation

Irrigation increases both agricultural production and the productivity of the land. Additionally, it increases stability providing better control of production and greater crop diversification. These last factors must contribute to the sustainability of irrigated agriculture in areas of excess production and at times of declining prices for certain crops. In the semi-arid areas, the productivity of irrigated lands is several times that of rainfed agriculture. In Andalusia, the average productivity of its irrigated area is about six times higher than that of rainfed agriculture [5]. In the South-East of that region, though, the productivity ratio (irrigated/rainfed) can exceed 30.

An important fraction of the employment still provided by the agricultural sector is generated by irrigated agriculture. In Andalusia, about 60 percent of the employment in the agricultural sector is attributed to irrigation which represents 16% of the cropped area. Within the irrigated area, about half of the employment is provided by the 28 percent of the total area devoted to horticultural crops [5]. The high levels of unemployment that still exist in the agricultural sector of Andalusia would be even higher if irrigated agriculture would decline. In the event of water transfers from the agricultural to other sectors, provisions should be made to include compensations for the direct and indirect losses in employment due to a reduction in the irrigated area.

3.2   Rural development and land use

Irrigation is an important element of rural development policies in the dry regions. It represents the most important option to increase farmers' income, employment and secondary economic activities derived from an increased and a more diversified agricultural production. Alternatives to irrigation for rural development in the dry areas are scant. A reduction in irrigated agriculture would increase the trend of migration from rural areas to urban centers enhancing social conflicts. Such a reduction would also change the land use patterns as rainfed agriculture is unsustainable in many dry areas and would result in land abandonment.

3.3   Food security

Three fourths of the irrigated areas are located in the developing countries. For them, sufficient food production is vital and, in the dry areas, can only be achieved via an efficient irrigated agriculture. In view of the population growth rates of those countries, increasing land and water productivity is essential to feed the poor [2]. Such goals can be achieved, not only by improving the performance of existing irrigated areas, but also by developing new lands under irrigation. It is probable that the current slow rate of irrigation development would have to be accelerated in the next two decades to meet world food needs.

For a developed country, the food security issue is also very important. If located in a dry region, its irrigated agricultural systems are the most productive and, probably, the most sustainable as well. The risk of not maintaining a productive agriculture is a strategic mistake, and a substantial decline in irrigated agriculture would make the dry areas very vulnerable in the long run, regardless of the economic development level. There is no doubt that irrigated agriculture is the best option to attain sufficient food production in those areas if the environmental problems of irrigation can be solved.

3.4   The social benefits of irrigation: The case of Almería, Spain

Among the provinces of Spain, Almería had, in 1969, the lowest per capita income. Located in the South-East, it has a mild, coastal Mediterranean climate with only 200 mm of annual rainfall. Traditionally, its limited irrigated area was devoted to table grapes and some open-air horticultural crops until the first plastic greenhouses were installed in the early seventies. By 1981, there were about 10,000 ha of greenhouse for vegetable production in Almería and its relative position in per-capita income had gone from the last one to the 30th of the 50 Spanish provinces. At present, there are over 25,000 ha of greenhouse in Almería, making it the largest area of intensive vegetable production in Europe. The development of intensive irrigated agriculture in Almería has raised employment in the agricultural sector relative to the other Andalusian provinces (Almería has 16 percent of all workers in that sector and only three percent of the total unemployed [6]. Almería agricultural production represented about 20-25 percent of the total of Andalusia in 1995 [6]. More importantly, while the subsidy levels of Andalusian agriculture were, on the average 28.2% of total farmers income, subsidies for Almería represented 4.6 % only [6].

The spectacular agricultural development of Almería is due to many factors but it would not have happened in the absence of irrigation. The development of water resources (primarily groundwater) has provided until now sufficient irrigation water, although there have been indications of ground water depletion since the early eighties. The shift from surface to drip irrigation and the emphasis on raising off-season vegetables in winter - a low evaporative demand period- slowed down the water consumption as the greenhouse area expanded [7]. The present estimates of supply and demand in the area suggest an overdraft of 40 Hm^3, equivalent to about 30% of current water use. Improved irrigation management, wastewater reuse and, eventually, interbasin transfers are needed to sustain the irrigated greenhouse industry in Almería. The economic importance of that industry and its social structure (mostly family farms associated in marketing cooperatives) emphasize the urgency of finding a solution to the ground water overdraft problem in the coastal plain of western Almeria.

4.   The modernization of irrigation

If irrigation uses too much water and has been practiced since long ago, the assumption is that technical improvements can reduce water use without affecting farmers income. Modernization of irrigation districts can have many different objectives, but it should aim at the following:
- improve the physical infrastructure.
- guarantee water deliveries under the various supply and demand scenarios.
- provide the level of flexibility in water deliveries needed by users.
- improve water management, raise the educational/technological levels of the users.
- Minimize operational losses and non-point pollution.
- Improve the organisation and promote bottom-up management in the district.

In essence, the objectives should include the modernization of irrigation networks, systems and practices. Unfortunately, quite often there is too much emphasis on the engineering improvements alone. It is true that in Spain, more than 70 % of the irrigation networks are more than 20 years old, and that 29 % has been in existence for over 200 years. However, many of the historic irrigation areas of the East and South-East have very sophisticated operational rules based on a long tradition of efficient use of their limited resources and on equitable distribution of the water available.

A recent, preliminary assessment of the efficiency of the Spanish irrigation networks was performed by the public administration for an area equivalent to 20 % of all irrigated area [8]. The conclusions of the study is that losses were minimum under pressurized irrigation, where water charges were by volume, and in recently developed areas. On the contrary, losses which were on the average of 54 % were associated with surface irrigation, low water charges by surface, and old networks [8]. It is easy to come to general conclusions and to recommend the changes to demand delivery and from surface to drip irrigation as the panacea for conserving irrigation water. Those changes are expensive, though, and the question is: who should pay for the costs of modernization?

It is evident that the costs should be levied on all those who would profit from the improvements, not the users alone. There should be reductions in water use and in environmental problems, and those modifications have a value to society. One fact to be considered is that farmers would not pay for investments designed without their input. The proposals for modernization should emerge from site-specific, detailed analysis of the irrigation performance of an area. Such studies should be interdisciplinary and should involve the users from the start. Consideration should be given by the administration to providing cost effective means of improving the technical level of water management. As an example, the recent analysis of the costs and benefits of the California Irrigation Management Information System [9] indicate that a modest investment on providing statewide information on water use by crops can result, over the years, in a substantial reduction in applied water (13%) and in a yield increase of 8 % among a sample of users.

5   Environmental constrains of irrigation

There are many concerns about the environmental impacts of irrigation; salinity, drainage disposal, return flows from irrigated lands, erosion risks, pollution from nutrients, pesticides and trace elements and other disruptive influences on the natural environment.

5.1   Salinity, drainage disposal and return flows from irrigated areas

It has been known since long ago that long-term salt balance must be achieved for irrigated agriculture to be sustainable. The maintenance of salt balance requires that drainage water must carry the excess salts, leaving the irrigated area. The disposal of the drainage water often creates a serious environmental problem, as a number of incidents have shown that such waters can carry toxic elements which are dangerous to wildlife and to humans. The contamination of the Kesterson Wildlife Refuge with water carrying selenium from the San Luis Drain in California in 1982, and the episodes of pollution and subsequent massive deaths of aquatic wildlife in the Doñana National Park in Southern Spain which occurred during the seventies, are well known examples of problems caused by drainage disposal. Less spectacular but equally important are the episodes of ground water contamination from the percolation of drainage water. Intensive agriculture has been practiced for too long in many areas without control of the amounts of fertilizers and pesticides used. It may take years for some pollutants to reach the ground water, but it is almost impossible to reclaim that water once it has been contaminated.

Surface and subsurface flows from irrigated agriculture often are an important part of the water resources of areas downstream, but their water quality must be monitored and controlled to avoid environmental problems. Irrigated agriculture will have increasing difficulties in finding sinks for its return flows, for free drainage disposal is no longer a politically viable option in most areas. On the other hand, it has always been considered that some drainage is needed to provide for salt leaching.

The best solution to cope with this dilemma is source control by: (i) efficient management of applied water and chemicals; (ii) reducing the amount of drainage water leaving the area; and (iii) promoting the removal of salts via precipitation and the uptake of excess nutrients by an appropriate crop rotation. The episodes of contamination from irrigation return flows have generated drastic responses from the rest of the society. After finding the origin of the toxic elements in the Kesterson Reservoir, the drains from the Westlands Water District were ordered to be plugged. Such decision apparently meant the end of irrigated agriculture in the long run. Nevertheless, there are no symptoms of salinization in the district ten years after the drains were plugged. Improvements in irrigation management and a very detailed monitoring programme are providing the tools to control salinity to manageable levels so far. The message here is that with improved technology and management, irrigated agriculture can be less detrimental to the environment.

5.2 Interactions between irrigated areas and surrounding urban and natural ecosystems

Irrigated areas were established historically in the plains, near the population centers. It is therefore natural that the process of urbanization is competing with land primarily with irrigation. Some of the most fertile lands have been taken out of production due to urbanization. This interaction can become a problem when urban planning is absent from the conversion process, and urban dwellers appear at random within an irrigation district. This situation is becoming quite common in some areas in the Mediterranean, where tourism adds pressures to urbanization.

The maintenance of river flows for wildlife preservation is a competing demand for water resources of increasing importance. Here again, irrigation faces another threat when water supply is limited. The control of non-point pollution from irrigated areas that might interact with natural ecosystems has been mentioned above and should not be overlooked.

6   Conclusion

Irrigated agriculture is directly or indirectly subsidized by the rest of the society in many world areas. The time has come to effectively use those subsidies as incentives for water conservation and for prevention of environmental problems. Market forces and regulations can aid in achieving efficient use of water. Knowledge already exists for a much more precise management of irrigation water, but the adoption of good practices will not occur without incentives and motivation on the part of the users.

7   References

1. Jensen, M.E. (1996) Irrigated agriculture at the crossroads, in: *Sustainability of Irrigated Agriculture*, (eds. Pereira, L.S., Feddes, R.A., Gilley, J.R. and Lesaffre, B.), Kluwer Academic Publishers. Netherlands, pp. 19-34.
2. Carruthers, I. (1996) Economics of irrigation, in: *Sustainability of Irrigated Agriculture*, (eds. Pereira, L.S., Feddes, R.A., Gilley, J.R. and Lesaffre, B.),

Kluwer Academic Publishers. Netherlands, pp. 35-46.

3.   Sinha, I. N. (1996) Irrigation policy for realization of high agropotential of Bihar state in India. *J. of Irrig.and Drain.*, ASCE, 122, pp. 31-37.

4.   Saura, J. (1996) R. Superficiales, in: *La Gestion del Agua en Andalucia ante la sequia.* Fundacion El Monte, Camas, Sevilla, pp. 137-150.

5.   Corominas, J. (1996) El regadio en el umbral del Siglo XXI. XIV Congreso Nacional de Riegos, Almeria, Spain, 47 p.

6.   Analistas Economicos de Andalucia. (1996) Informe anual del Sector Agrario en Adalucia. Unicaja, Malaga, 465p.

7.   Fereres, E., López-Gálvez, J., Castilla, N., and Orgaz, F. (1992) Greenhouse irrigation technology transfer in Spain. *Water Forum'92.* ASCE, Baltimore, USA.

8.   Krinner, W., García, A. and Estrada, F. (1994)   Method for estimating efficiency in Spanish irrigation systems. *J. of Irrig.and Drain.*, ASCE, 120, pp. 979-986.

9.   Parker, D., Zilberman, D., Cohen, D., and Osgood, D. (1996) The economic costs and benefits associated with the California irrigation management information system: final report. Dpt.of Agr. and Resour. Econ. Univ. of California, Berkeley.

NEED FOR NEW APPROACHES IN VALUING WATER: LEARNING FROM A LESS SUCCESSFUL CASE

J.C.CALDAS, P.L.SOUSA and L.S.PEREIRA
Instituto Superior de Agronomia, Universidade Técnica de Lisboa, Portugal

Abstract

Most of the southern territory of Portugal, the Alentejo, is under annual and inter-annual irregular availability of water resources. Furthermore, this is a region where agriculture has a determining influence as an economical activity, and also where significant decreases in population have been observed during the past few years.

The different ways of managing the irrigation projects established by the Government have been strengthening a system of practices and attitudes towards water use, in which water is seen essentially as having value in use and with almost no price.

With the present demand for diversification of the economical activities and for the social and environmental valuation of water, one can foresee the competition for its use, which requires institutional decisions in order to regulate potential conflicts and modify old-fashioned practices and deeply ingrained attitudes.

In this scenario, one can discuss the need for a water pricing system that allows a fair allocation to the different sectors and leads to improvements in the efficiency in water use.

Keywords: Water value, water pricing, water resources, irrigation.

1 Introduction

Under situations of annual and inter-annual irregular availability of water resources, the water is faced as the image of life itself. Particularly it compromises people's permanency in a territory when drought takes place and demand largely exceeds water availability. The possibility of constituting water reserves becomes a decisive element for the correction of natural restrictions and imbalances. Inter-annual water reserves became an outstanding vector in development plans and policies, namely at regional and local level. The resulting balanced distribution of water resources for different uses can indeed contribute to fight against unemployment, to correct demographic asymmetries and to support the improvement of welfare of directly affected populations, including their work conditions.

Water: Economics, Management and Demand. Edited by Melvyn Kay, Tom Franks and Laurence Smith. Published in 1997 by E & FN Spon. ISBN 0 419 21840 8.

The management of water resources quantity and quality has, nevertheless, originated a great number of conflicts. These conflicts are revealed with particular evidence in drought situations. Conflicts occur among users or among the representatives of private interests. Conflicts are shown relative to the strategies defined by public institutions concerning the planning and management of water reserves, namely the respective inter-regional and inter-sectorial allocation. In regard to irrigation, it is known that water has a dominant role in building new social relationships and structural arrangements [1]. These institutional solutions are sometimes complex, with strong political, technical and cultural implications. Questions relative to water rights, to partition of available water streams and to management and operation of hydraulic systems become decisive at local level, however problems and solutions are influenced by type of prevailing technology.

The fact that water is a natural resource often in common property or the object of shared rights, makes difficult the establishment and application of the market economy rules. Also it gives rise to the question and the controversy about valuing the water as just a production factor or as a social good. When water is considered as a production factor it should be allocated to the different uses by the market mechanisms. On the contrary, if it represents a value of social use, then such value is essentially symbolic and non-monetary [2] [3]. Water pricing becomes therefore a question of difficult solution.

Most of the southern territory of Portugal, the Alentejo, is under the above-described conditions of irregular water availability, the water being often scarce. A water-pricing system that allows a fair allocation to the different sectors should lead to improvements in the efficiency in water use.

2   Alentejo, a region with deep ingrained attitudes of symbolic value of water

The 27000 km^2 of Alentejo constitute a large geographical area in the South of the country, characterised by the poverty of population and the scarcity of water

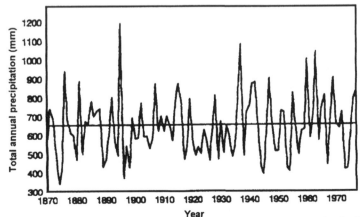

Fig 1. Total annual precipitation (mm) registered in the meteorological station of Évora, Portugal [6].

resources. As for most parts of the Mediterranean area, it is characterised by an irregular climate, with strong spatial and time variability of water availability considering successive years or different periods of the same year. The precipitation mainly occurs in winter, while summers are hot and dry, with a strong water deficit. The total annual precipitation is uneven distributed (Figure 1), which leads to high economic risks in rainfed agriculture [4]. The occurrence of droughts with variable severity and duration is very frequent [5].

The agrarian structure is characterised by a strong concentration of land ownership. Agriculture has been almost the exclusive economic activity in Alentejo, over the last decades. Farming systems are mainly based on extensive winter grains crop (operated by temporary hired labour force). These systems, despite low yields, have been economically attractive when high prices and subsidies policies were adopted.

During the former Portuguese regime (1926-1974), one of the axes of the agricultural policy related to strategies of land consolidation, related to the installation of family farms with the implementation of irrigation areas. Following the example of similar policies in Italy and Spain, it was intended to increase yields and crop diversification, to fight against unemployment, to promote land parcelling, and to improve welfare conditions in the region. Under this framework, in 1957, an improvement plan for the Alentejo region was designed aiming at the irrigation of 170 000 ha. In this context a large number of small and medium size irrigation projects have been built. However, the political influence of the big estate owners made useless the objectives of land reform. The land ownership structure was kept unchanged. Mainly tomato and rice crops have developed, largely practised by small farmers in temporary rented lands. Unemployment decreased because rural migration occurred. Welfare work conditions were consequently improved, and salaries increased because of the exit of a large part of the existing labour force to other areas of the industry. The balance of the irrigation policy until the end of the regime in 1974 [7], points out a substantial increase in the value of land rents, with the reinforcement of the large farm owners' position, without development of small and medium size family farms.

It was foreseen that beneficiaries of the new irrigation systems should pay two kinds of fees. The first should cover the operation and maintenance costs. It was proportionally distributed among the individual beneficiaries of irrigation depending on the irrigated area and taking in to account water deliveries, soil quality, crop socio-economic interests. The second fee aimed at the recovery of investments made by the Government. This charge has always been rejected by the landowners and was never paid. So, the water has never been viewed as an economic good, valued at market prices. The charges for irrigation water remained always under its actual costs [8].

Alentejo remained a region with strong water needs, with a very reduced industrial activity which, in result of continuous population decrease since 1950's, has today only 20 habitants per km^2, strongly concentrated in small to medium size towns, where high unemployment rates are still observed.

3   Requirement of new institutional decisions for management of water resources

The new Common Agricultural Policy (CAP) and the General Agreement on Tariffs and Trade (GATT) represent nowadays the end of a cycle where agriculture modernisation was conducted under strong protectionism by regional policies and public institutions. Agricultural structures are forced to adopt the mechanisms of the international market regulation. This is particularly relevant concerning water pricing. EU directives look to establish charges for the use of water at "full economic costs". These "full economic costs" shall include: operation and maintenance costs; capital costs (recovery of investments plus interest repayment); depreciation reserves for future investments; and a charge representing the environmental and resource costs associated with the relevant abstractions of water. It is also foreseen to prevent cross-subsidies between different sectors of the economic activity and to establish rules for the allocation of shared costs between individual users. Under these conditions, the benefits resulting from increasing yields and from the crop diversification due to irrigation projects in marginal regions, as Alentejo, are dependent upon the capabilities of placing agricultural products on an open market dominated by highly competitive agriculture.

To achieve the competitiveness of irrigated agriculture, among other aspects it is required to guarantee stable water availability, sufficient to satisfy the crop demand. Once the natural precipitation occurs during winter and the irrigation is applied in spring-summer there is a need to store winter water resources for a later use. A large number of new dams have been built in the last two decades, although the majority are of small and medium capacity not allowing an appropriate inter-annual regularisation. The small private dams are normally designed for the average year. Many of the public dams have not been designed to fully respond to farming water demands when successive drought years occur. Therefore the ratio of design to actual irrigated areas vary with time, the range being limited by the full storage of the reservoirs and by the limited availability during droughts. Variations are often larger for small reservoirs, as the case represented for Divor Project in Figure 2 (irrigated area of 488 ha). Rural development objectives in the present context of CAP call for the integration of agriculture with a diversification of activities in the rural environment.

This requires different uses of the territory and uses of the available water resources in order to avoid or to attenuate demographic unbalances and desertification risks in marginal zones.

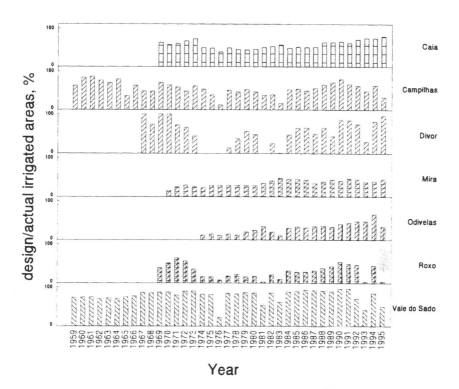

Year

Fig 2 Ratios between design to actual irrigated areas relative to main irrigation projects in Alentejo (1959 to 1995)

Thus irrigation projects should support other rural development activities, the agriculture being forced to share the available water resources. Although agriculture still is the main user of water, the demand from other sectors grows much faster. During the last 20 years the water supplies for domestic use have grown 73.4 % [9].

4   The case of Caia irrigation perimeter

The case of the irrigation perimeter of Caia, in the Alentejo region, is a typical case where old fashioned practices and deeply ingrained attitudes about the value of irrigation water requires institutional decision in order to regulate potential conflicts in the competition for its use. The project was built from 1963 to 1967. It is located along the Caia river and the Guadiana river near the border with Spain, covering about 7200 ha. The water is stored in a reservoir, inundating about 1970 ha and having a total capacity of 203 x 10^6 m^3. The conveyance system is an open channel system with upstream regulation by automatic water level gates. Low-pressure pipes constitute the distribution and delivery system. A scheduled delivery system is adopted. The cost has been entirely supported by Government funds. The beneficiaries did not pay the recovery costs.

The land ownership structure in the area is such that 3 % only of the beneficiaries hold 65% of the area. More than half of the total irrigated area is farmed under rented land by small farmers.

The uneven rainfall regime, which conditions the storage of the Caia reservoir, determines cyclic situations of reduced availability. The last set of dry years, started at 1992 and finished at 1995, imposed drastic reductions on the amount of water delivered, with the reservoir becoming about empty by the summer 1995. (Figure 3).

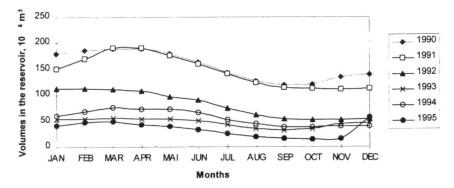

Fig 3   Storage evolution at Caia reservoir during the period 1990-95.

The consumption of water is necessarily affected by availability conditions. When drought takes place priority is to allocate water to the non-agricultural sectors. During the dry year of 1993 it was observed a reduction of 70% on the use of water by irrigation, while the supply of water for domestic uses registered an increase of 35% relatively to the precedent year (see Table I).

Table 1 Use of water (10^3 m^3) by different sectors at Caia, during the period 1990-95.

Year	Irrigation	Industry	Domestic	Total	Irrigation (% of total)
1990	37 129	9	729	37 867	98.1
1991	38 114	7	1 572	39 703	96.0
1992	32 800	14	1 585	34 399	95.4
1993	10 156	11	2 151	12 318	82.4
1994	24 464	12	2 274	26 750	91.5
1995	17 455	17	2 672	20 144	86.4

However, agriculture remained the main user, with 82.4% of the total water withdrawn. For most years the agricultural sector's quota exceeds 90%. This is a characteristic of the arid and semiarid countries, where the agriculture uses up to 80% of the total available resources, although demands will increase significantly in the other sectors over the next years [10]. All the sectors have to fix targets to improve efficiency of water. However, the main efforts to attain efficiency benefits must come from the main user - irrigated agriculture [11].

In Table 2 are given the average water uses by the main crops in Caia for the years 1990 - 1995. They show a great variability but it is also seen that water uses definitely decrease in years when water availability (cf. Fig.3) is reduced. This indicates a potential for water saving when adequate irrigation scheduling and improved water application methods are applied. Appropriate water pricing policies have to be applied to favour the adoption of such programs.

Pressure to reduce crop water uses will also result from increasing the share of water uses to other sectors, namely for domestic uses.

Domestic consumption did not exceed 0.11 % of the total by 1977. It is increasing since then, reaching 17.5 % in 1993 and 13 % in 1995.

The evolution of fees paid by m^3 of consumed water by the agricultural and the domestic sector shows clearly differences in practices and attitudes towards the value of water. For the agricultural sector water is seen essentially as a good for which a right of use exists. At constant prices, the price of a m3 of water for the agricultural sector increased by 155 %, from 1969 to 1994, while the cost for domestic water supply increased 252 %, from 1977 to 1994. This reflects an important unbalance such that the fee paid by industry and domestic water supply is nearly seven times higher than the price paid by the agricultural sector.

Table 2   Average uses of water by crop (mm), for the period 1990-95, in the Caia Perimeter.

Year	Rice	Tomato	Corn	Sunflower	Melon
1990	1 506	528	763	326	295
1991	1 652	714	960	439	522
1992	1 600	700	800	300	560
1993	-	152	472	121	294
1994	2 295	556	780	272	676
1995	1 399	397	680	110	360

5   Conclusions

When agriculture was highly subsidised the irrigated agriculture sector resulting from new irrigation projects has been favoured concerning the fees which should be paid for water use. The water price excluded the recovery of investments and other societal benefits and only covered a part of the operation and maintenance costs. Thus farmers did not develop an attitude favourable to recognise the water as an economic good. They just used the water as a good, with right of use given to them.

Consequently, farmers do not give a value to water saving or to increase the efficiency in irrigation. The situation is changing since CAP and EU environmental policies will progressively impose the need for full payment of water. Also, the competition coming from other uses puts more pressure on the need for efficiency, for water savings and for full cost payment. Thus, it is required to progressively promote a change of farmers' attitude towards water, which also include the adoption of good agricultural practices aiming at water savings, irrigation efficiency and economic competitiveness.

6   References

1.   Caldas, J.C. (1995) *L'opinion publique et l'eau*. Communication presented at the Seminar "Aspects économiques de la gestion de l'eau dans le bassin méditerranéen", CIHEAM, Marrakech.
2.   Aguilera Klink, F. (1991) Algunas cuestiones sobre economia del agua. *Agricultura y Sociedad* 59 : 197-221.
3.   Milliman, J. (1992) "La propriedad común, el mercado y el suministro de agua". In: F. Aguilera Klink (coord). *Economia del agua*, Ministerio de Agricultura, Pesca y Alimentación, Madrid.
4.   Carvalho, M. (1996) *Uso eficiente da água em ambiente mediterrânico. Aspectos agronómicos.* Communication presented at the III Seminário Luso-Espanhol "A Àgua e a Agricultura Mediterrânica", Vilamoura, Portugal, 23-26 Jan. 1996.
5.   Espirito Santo, F. (1993) *Vigilancia e caracterização das secas. A teoria do caos e a previsão a longo prazo*. Simpósio sobre Catastrofes Naturais, Ordem dos Engenheiros, Lisboa, pp.V. 13-16.
6.   Azevedo, A.L. and M.J.Carvalho. (1981) A irregularidade desconjuntante do nosso clima. *Semente*, n° 5: 11-12.
7.   Baptista, F. O. (1993) A política agrária do Estado Novo. *Afrontamento,* Porto.
8.   Frazão, F.F. and L.S.Pereira. (1993) Evaluation of performance indicator applied to several irrigation systems in Portugal. In: S.Manor and J.Chambouleyron (eds). *Performance Measurement in Farmer-changed Irrigation Systems.* IIMI, Colombo, pp.137-145.
9.   Frade, V. and A. Alves. (1991) *O mercado da água em Portugal*, Volume I and II, DGRN, Lisboa.
10.   Hill, H. and L.Tollefson. (1996) Institutional questions and social challenges. In: L.S.Pereira et al. (eds.), *Sustainability of Irrigated Agriculture*, 47-59. Kluwer Academic Publishers. Printed in the Netherlands, pp 47-59.
11.   Hennessy, J. (1993) Water management in the 21st Century. *Transactions of the 15 th Congress on Irrigation and Drainage*, Vol 1-J. Keynote Address, pp 1-30.

APPROCHE ECONOMIQUE POUR CONCILIER IRRIGATION ET ENVIRONNEMENT DANS LE BASSIN VERSANT DE LA CHARENTE

Concilier irrigation et environnement en Charente

M. MONTGINOUL et T. RIEU
Division Irrigation, Cemagref, Montpellier, France
J.P. ARRONDEAU
Institution Interdépartementale du fleuve Charente, Angoulême, France

Résumé

La Charente est un fleuve de la façade atlantique française, drainant un bassin versant d'une superficie voisine de 10 000 km^2.

Du fait de l'apparition de conflits d'usage, lorsque son débit devient trop faible, les agriculteurs sont rationnés et utilisent des pratiques d'irrigation économes en eau. La restauration du milieu aquatique a conduit à la création d'un nouveau barrage.

Ce nouveau barrage permet d'accroître la ressource en eau et pose la question du maintien d'un mode de gestion économe de la ressource en eau par les agriculteurs. Outre les instruments réglementaires déjà utilisés dans la situation antérieure, la collectivité souhaite mettre en place un dispositif incitant à l'économie d'eau.

Deux instruments sont envisageables : les quotas (dont les restrictions périodiques) et la tarification. Ils sont analysés en tenant compte des contraintes à la fois sur la demande et sur l'offre en eau (élasticité de la demande en eau d'irrigation, revenu minimal des agriculteurs, débit minimum pour respecter la « contrainte environnementale »).

Suite aux simulations réalisées, la tarification binôme mise en place et acceptée par les différents acteurs au moment de la décision de construction du barrage n'incite pas à économiser d'eau. Une tarification binôme non linéaire, par paliers croissants, semble constituer une réponse plus adaptée.

Mots clés : Charente, eau d'irrigation, environnement, quota, tarification

Water: Economics, Management and Demand. Edited by Melvyn Kay, Tom Franks and Laurence Smith. Published in 1997 by E & FN Spon. ISBN 0 419 21840 8.

1   Introduction

La Charente est un fleuve de la façade atlantique française, drainant un bassin versant d'une superficie voisine de 10 000 km² et débouchant dans la baie de Marennes-Oleron.

Ce fleuve connaît un régime hydraulique très variable selon les saisons. Le développement des activités agricoles et industrielles, l'augmentation des besoins domestiques (assainissement et eau potable) et la nécessité de sauvegarder la qualité des milieux naturels expliquent l'apparition de "conflits d'usage" qui portent sur la répartition de la quantité d'eau disponible en période d'étiage (juin à novembre).

L'observation des débits du fleuve Charente montre un non respect fréquent du débit de crise (soit 2.5 m³/s en amont d'Angoulême) : une année sur deux et pendant au moins vingt jours consécutifs, la Charente connaît un débit inférieur au débit de crise, alors que le SDAGE ne tolère le dépassement que durant cinq jours consécutifs, au plus. Le débit respecté en période d'étiage est pour le moment compris entre 1.2 et 1.8 m³/s, limites très inférieures au débit de crise, mais qui correspondent à des valeurs transitoires valables jusqu'à la mise en service du barrage de Mas Chaban.

De plus, en l'état actuel des ressources en eau, le respect du débit de crise impose des restrictions fortes aux agriculteurs irrigants. Ceux-ci ont adapté leurs pratiques à l'utilisation d'une ressource limitée et aléatoire, de manière à économiser de l'eau.

Un mode de gestion, basé en particulier sur une tarification binôme simple, est proposé par les acteurs intervenant sur le fleuve de la Charente. Nous montrerons, au cours d'une première partie que celle-ci ne convient pas en regard des objectifs fixés (en particulier en terme de revenu minimum des agriculteurs) avant de proposer deux solutions alternatives, et d'en voir leurs avantages et leurs limites.

2   La Charente : état des lieux. Une tarification linéaire ne convient pas

L'objectif est de mettre en place une tarification qui incite les agriculteurs à conserver leur comportement économe en eau. Une tarification est proposée, qui résulte d'un débat entre les différents acteurs intervenant dans la gestion de l'eau en Charente. Nous les présentons avant de détailler la gestion de l'eau qu'ils envisagent. Enfin, nous montrons que ce type de tarification n'est pas adapté, étant données les contraintes.

2.1   Les acteurs de la gestion de l'eau en Charente

De multiples acteurs interviennent dans la gestion de l'eau du fleuve Charente. Nous nous limiterons ici aux principaux. Analysons ici les objectifs des différents acteurs.

2.1.1   L'Etat

Dans le cas du barrage de Mas-Chaban, nous pouvons établir deux objectifs de l'Etat, représentant la Collectivité :

- le soutien d'étiage (par des aides financières à la mobilisation de nouvelles ressources), et

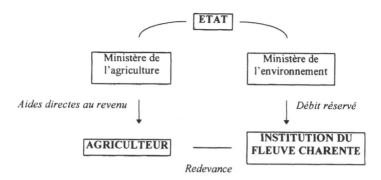

Fig. 1. Les différents acteurs de la gestion de l'eau en Charente

- le soutien du revenu des agriculteurs (ce qui se traduit, pour les irrigants, par le maintien des surfaces irriguées et l'instauration de primes spécifiques).

En Charente, les ministères concernés sont principalement le ministère chargé de l'Environnement et le ministère chargé de l'Agriculture. Tous deux disposent de services déconcentrés en Charente pour mettre en application leurs conceptions de la gestion de l'eau. Lors de la négociation du barrage de Mas-Chaban, le ministère chargé de l'Environnement a mis l'accent sur la préservation et la restauration des hydro-systèmes parmi les richesses issues du milieu aquatique. Le ministère chargé de l'Agriculture a cherché à préserver un revenu minimal des agriculteurs.

2.1.2 L'Institution

L'Institution Interdépartementale, émanation des trois conseils généraux intervenant sur le bassin versant, est maître d'ouvrage du barrage de Lavaud (en fonctionnement) et a été l'initiatrice du projet de création de nouveaux ouvrages, dont celui de Mas Chaban. Son action est subordonnée au principe de gestion équilibrée de la ressource en eau que souhaite faire respecter le Ministère de l'Environnement. Dépendant pour son fonctionnement des dotations financières du département, l'Institution n'est pas soumise à l'obligation d'équilibre budgétaire.

Les Conseils Généraux ont un triple objectif :

1. un maintien des surfaces irriguées, pour soutenir le revenu des agriculteurs et accroître la compétitivité des exploitations agricoles;

2. un respect de l'environnement : l'objectif est d'atteindre un débit d'étiage minimal de 3.0 m^3/s (« débit objectif ») en amont d'Angoulême;

3. un objectif économique vis-à-vis des externalités dues au fleuve : tourisme, qualité de l'environnement dans la région des producteurs de cognac.

Le respect de la consigne est accepté par tous les acteurs. Toutefois, un conflit d'usage persiste sur l'allocation du volume au delà du respect de la consigne. En effet, la maximisation de l'effet sur l'environnement conduirait à allouer ce volume au soutien d'étiage, l'optimisation du revenu des agriculteurs à l'allouer à l'agriculture.

2.1.3 Les Agriculteurs

Le groupe des agriculteurs concernés réunit les agriculteurs irrigants, représentés par le Groupement des irrigants de Nord-Charente. Leur objectif principal de maximisation du revenu se décline en trois sous-objectifs :

1. maintenir la surface irriguée,
2. obtenir une ressource en eau à bas prix ct
3. sécuriser la ressource en eau (de manière à stabiliser leur revenu).

2.2 Le mode de gestion envisagé par les acteurs

2.2.1 Un barrage

Le barrage est construit pour l'environnement :

le but premier est le soutien des étiages du fleuve et aucun volume n'est affecté à l'irrigation.

C'est pourquoi les agriculteurs ne paient pas la construction du barrage. Celle-ci est financée par les Conseils Généraux (15 %), l'agence de l'eau Adour-Garonne (30%), la Région (20%) et l'Etat (35%)..

Toutefois, le barrage, par son rôle de soutien d'étiage,

- régularise le débit de la Charente et donc l'offre en eau et
- limite les interdictions de prélèvement aux sécheresses d'une fréquence décennale.

C'est pourquoi les frais de fonctionnement des deux barrages existant sur la Charente (Lavaud, Mas-Chaban), dont le montant s'élève à 1 600 000 francs sont un élément de comparaison dans la négociation de la redevance.

2.2.2 Une tarification

L'Institution Interdépartementale est chargée de recevoir la redevance établie à la suite de la construction du barrage de Mas-Chaban.

Une tarification binôme est adoptée. Elle ne concerne pas l'ensemble du bassin versant de la Charente mais uniquement les parcelles qui peuvent être irriguées à partir du fleuve et à l'amont d'Angoulême. Elle est structurée de la manière suivante :

a- *une part fixe* de 75 F par hectare irrigué,

b- *une part variable* proportionnelle au volume consommé sur la base d'un tarif de référence de 0.0375 F/m^3,

c- *une pénalité* par multiplication de ce tarif de référence par 10 pour les volumes utilisés au-delà des seuils de référence.

Cette redevance a été établie à partir de négociations au sein du Comité de gestion. En réalité, il s'agissait de mettre en place une redevance "supportable psychologiquement et économiquement par les agriculteurs". Le montant ainsi calculé aboutit à une charge financière de 150F/ha pour un agriculteur qui consomme 2000 m3/ha, ce qui est la quantité apportée en moyenne pour répondre au besoin en eau de la culture de maïs irrigué.[1]

Deux autres dispositions protègent les agriculteurs des aléas liés aux marchés agricoles et au remplissage des retenues :

d- *une actualisation du tarif de référence* fondé sur l'évolution de la somme de la valeur marchande du produit d'un hectare (composé de 90% de maïs et de 10% de pois protéagineux) et de la prime compensatoire "PAC" (Politique Agricole Commune),

e- *une indexation* de cet ensemble tarifaire *fonction du taux de remplissage du barrage* actuel de Lavaud et du barrage futur de Mas-Chaban. Ainsi, si les barrages ne sont remplis qu'à 80% par rapport au remplissage normal, les agriculteurs ne devront payer que 80% du prix de l'eau.

Outre cette tarification, une redevance prélèvement est due à l'agence de l'eau. Elle est actuellement forfaitaire et s'élève à 39 F/m3 pour 1200 m3 par hectare. En l'absence de dispositifs de comptage, les agriculteurs du bassin paient cette somme quelle que soit leur consommation.

2.2.3 Un engagement des agriculteurs

Les agriculteurs s'engagent à limiter leur consommation en eau et à n'irriguer le maïs qu'à hauteur de 85% de ses besoins en eau théoriques, ce qui correspond, en année moyenne, à un volume de 2000 m^3/ha irrigué.

Tableau 1. Volume disponible par hectare et contrainte de débit minimal

Période de retour	Volume disponible par hectare	
	Débit minimal 2.5 m^3/s	Débit minimal 3.0 m^3/s
½ an	7 021 m^3/ha	4 755 m^3/ha
1/5 an	2 681 m^3/ha	490 m^3/ha
1/10 an	1 805 m^3/ha	0 m^3/ha

Or, comme le montre ce tableau, cet engagement ne permet pas de faire face aux sécheresses décennales, en sachant que ces années-là, on cherche uniquement à respecter le débit de crise. Des mesures complémentaires doivent être alors envisagées, comme une tarification exceptionnelle, la mise en place d'un marché de l'eau ou un quota-temps (qui correspond, par exemple, à des restrictions de prélèvement en eau d'irrigation). Le choix est d'autant plus difficile que les années de faible disponibilité en eau correspondent aux années à fort besoin. Une autre mesure peut être de mettre en place un instrument de gestion très restrictif, à savoir un instrument qui incite les agriculteurs à consommer moins de la disponibilité décennale.

De plus, ici nous avons considéré la période estivale comme une période homogène. Or, souvent, seuls certains mois, en particulier le mois d'août, sont critiques en terme d'économie d'eau. Il est donc nécessaire d'avoir une approche plus fine (en mettant en place une tarification de pointe, un quota décadaire, etc.) pour en tenir compte.

Ici, nous ferons l'hypothèse que l'engagement correspondant à un volume de 2000 m^3/ha correspond bien à ce que les acteurs cherchent à atteindre.

2.2.4 Une gestion des crises

En cas de sécheresse exceptionnelle ou de non respect de la consigne de débit d'étiage à Angoulême, un plan de gestion de l'étiage (caractérisé par un système d'interdictions de prélèvements) est toujours de vigueur.

Cette situation ne devrait apparaître que conjoncturellement lorsque la ressource est défaillante, qui ne doit être utilisée que si les autres mesures évoquées ci-dessus n'ont pas été suffisantes pour laisser suffisamment d'eau à l'environnement.

2.3   Une tarification linéaire ne convient pas

Dans cette section, nous analysons les conséquences d'une tarification linéaire sur la consommation en eau des agriculteurs.

Le type de modèle utilisé (modèle de programmation linéaire) permet de construire une fonction de demande. Les modèles correspondants ont été validés mais la précision est de l'ordre de 20%.

La fig. 1 montre que la fonction de demande en grandes cultures irriguées est sensible au prix. Pour être incitative, on constate que la tarification doit se situer dans une plage où la demande est sensible au prix. Ici, cela correspond à un prix supérieur à 0.61 F/m^3.[2]

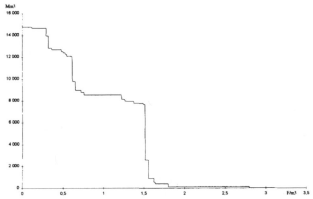

Fig. 2. Demande en eau d'irrigation en fonction du prix

Les modèles nous permettent alors de comparer 4 situations différentes :
- la première, « Lavaud 1.8 », correspond à la situation actuelle, à savoir sans le barrage de Mas-Chaban et sans tarification;
- la deuxième décrit les résultats issus de l'engagement des agriculteurs à ne pas dépasser une consommation 2000 m^3/ha, situation avec barrage et avec tarification (75F/ha et 0.0375 F/m^3);
- la troisième montre ce qu'il se passerait si les agriculteurs ne respectaient pas leur engagement et cherchaient à maximiser leur revenu lors de la construction du barrage et avec une tarification de 75F/ha et 0.0375 F/m^3;
- la dernière décrit le cas d'une tarification incitative (s'élevant à 0.478 F/m^3 et 75 F/ha) en cas de barrage de Mas-Chaban.

	Lavaud 1.8	Engagement	Tarification envisagée	Tarification incitative
Demande en eau totale (m^3)	11 883 826	12 186 261	14 780 789	12 507 903
ommation (m^3/ha)	1882	1930	2341	1981
Revenu des agriculteurs (F)				
- moyen	128 279	126 193	129 677	102 500
- plus faible	90 571	92 264	92 264	82 627
- plus fort	207 189	191 227	205 214	137 593
Recettes du gestionnaire	0	939 715	1 062 402	6 463 600

Tableau 2. La tarification linéaire dans le bassin versant de la Charente

L'engagement à consommer au plus 2000 m^3/ha irrigué est contraignant, en particulier pour les gros consommateurs d'eau aux revenus élevés : leur revenu sera pénalisé par rapport à la situation actuelle, contrairement aux petits irrigants qui appliquent des doses inférieures par hectare. Pour les gros consommateurs, l'unique avantage retiré de la construction du barrage est donc une sécurisation plus importante de l'eau d'irrigation, vu que l'offre en eau sera assurée 9 années sur 10.

Le comportement spontané des agriculteurs suite à la construction du barrage est tout à fait différent de celui auquel ils se sont engagés. Il conduit à une augmentation significative de la consommation en eau d'irrigation (+24%) alors que le respect de l'engagement n'autorise un accroissement de consommation de 2.5%. Si les agriculteurs ne respectent pas l'engagement en matière de consommation, leur revenu moyen augmente.

Pour inciter les agriculteurs à respecter la consigne en appliquant une tarification linéaire, la partie proportionnelle doit s'élever à 0.478 F/m^3. Cette situation est difficilement acceptable pour les agriculteurs, en particulier ceux qui ont tendance à consommer beaucoup d'eau, dont le revenu baisse significativement. Par contre, elle avantage le gestionnaire, dont les recettes couvrent largement l'objectif initial.

La tarification linéaire n'est donc pas un instrument envisageable, compte tenu des contraintes de revenu en particulier, sauf à prévoir une redistribution de la redevance non liée aux prélèvements.

3   Comparaison d'une tarification non linéaire et d'un quota

C'est pourquoi on s'intéresse maintenant à la comparaison entre deux instruments de gestion qui paraissent plus adaptés : un quota et une tarification par paliers.

Le quota à 2000 m^3/ha correspond à la situation où les agriculteurs s'engagent à consommer 2000 m^3/ha; la tarification par paliers, non linéaire, comporte une partie fixe (75 F/ha) et une partie variable qui change en fonction de la quantité d'eau consommée. Pour reprendre l'objectif en matière de consommation en eau, le premier palier de tarification (qui est facturé à 0.0375 F/m^3) est limité à 2000 m^3/ha. Au-delà, le gestionnaire cherche un tarif incitatif pour respecter l'engagement.

Avant de nous intéresser au cas de la Charente, reprenons quelques éléments de comparaison théorique entre un quota-volume et une tarification par paliers.

3.1   Comparaison théorique entre quota et tarification par paliers

Au niveau des conditions nécessaires de mise en place de ces instruments, il est indispensable dans les deux cas d'avoir un système de comptage de l'eau consommée. De plus, la tarification par paliers exige que la demande en eau d'irrigation soit élastique par rapport au prix de l'eau, tandis qu'un quota par volume nécessite la possibilité d'une fermeture des bornes en cas de dépassement du quota.

En matière de définition par rapport aux objectifs que l'on se fixe, le tableau 3 résume les caractéristiques des deux instruments.

Tableau 3 . Tarification par paliers et quota en fonction des objectifs recherchés

Objectifs	Tarification par paliers	Quota par volume
Efficience[3]	Oui	Non
Equité	Dans une certaine mesure	Oui
Recouvrement des charges	Possible	Non

Enfin, le tableau 4 reprend les avantages et les inconvénients de ces deux modes de gestion du point de vue des différents acteurs participant à la gestion de l'eau d'irrigation.

Tableau 4. Intérêts et limites des instruments

	Tarification par paliers	Quota par volume
Avantages		
- *Gestionnaire*	1. Recouvrement des charges 2. Prise en compte du coût d'opportunité	Evaluation aisée de la demande en eau d'irrigation maximale
- *Agriculteur*	Souplesse, en cas de besoin exceptionnel	Volume garanti par le gestionnaire
Inconvénients		
- *Collectivité*		N'incite pas les agriculteurs à économiser de l'eau en deçà du quota
- *Gestionnaire*	Difficulté pour fixer le palier	1. Aucune recette 2. Nécessite un suivi des quantités consommées
- *Agriculteur*	Coûteuse en cas de dépassement	Rigidité de l'offre

Les deux systèmes amènent à prendre en compte le coût du comptage. Au-delà de ce coût, le système de quota apparaît plus coûteux : pour qu'il fonctionne, on doit fermer les compteurs dès que celui-ci est atteint. Par contre, la tarification par paliers permet de n'effectuer qu'un seul relevé par an.

3.1.1 Tarification non linéaire ou quota. Quel instrument le mieux adapté à la gestion de l'eau dans le bassin versant de la Charente?
Pour déterminer l'instrument le mieux adapté à la gestion de l'eau dans le bassin versant de la Charente, analysons l'impact d'une tarification binôme et d'un quota en terme de consommation, de revenu des agriculteurs et de recettes pour le gestionnaire.
En terme de consommation
Le même niveau de consommation peut être atteint. Toutefois, la tarification par paliers permet aux agriculteurs ayant un grand besoin en eau d'irriguer davantage que la limite fixée par le gestionnaire. Le jeu des compensations entre agriculteurs, dû à l'hétérogénéité des systèmes en terme de demande en eau, permet de respecter la contrainte globale de 2000 m^3/ha : si l'objectif est de consommer en moyenne de 2000 m^3/ha sur l'ensemble du bassin versant, qui est atteint avec un prix du second palier de 0.5 F/m^3, certains types consommant moins de 2000 m^3/ha permettent à d'autres de consommer davantage :

Type d'exploitation	Moyenne	La plus faible	La plus élevée
Consommation (m^3/ha)	1998	1679	2232

Tableau 5 : Hétérogénéité des consommations en eau par type d'exploitation avec un second palier de 0.5 F/m^3

En terme de revenu des agriculteurs

Pour l'agriculteur, le quota est l'instrument le moins pénalisant. Toutefois, la tarification par palier permet une baisse de revenu supportable par rapport à la tarification binôme simple. Elle n'a aucun effet sur le revenu des agriculteurs qui consommaient déjà moins de 2 000 m³/ha.

Tableau 6. Revenu et instruments de gestion de l'eau

(en francs par an)	Revenu moyen	Revenu le plus faible	Revenu le plus élevé
Quota	130 329	93 943	202 005
Tarification	126 311	92 776	192 955

En terme de recettes du gestionnaire

Seule la tarification permet au gestionnaire d'avoir des recettes. Analysons le montant de ces recettes en fonction du prix du second palier.

Globalement, la figure 2 montre que l'optimum pour le gestionnaire est atteint quand le prix du second palier est de 0.2892 F/m³ (1 552 473 F). Les problèmes de l'élasticité de la courbe de demande induisent une allure « non standard » de la courbe des recettes. De plus, le premier palier à 0.0375 F/m³ garantit au gestionnaire une recette minimale.

Lorsque la consommation pour chaque agriculteur ne dépasse pas 2000 m³/ha, la recette s'élève à 932 357 francs. En considérant une consommation moyenne de 2000 m³/ha sur l'ensemble de la zone, avec un prix du second palier de 0.5 F/m³, il perçoit 1 156 560 francs.

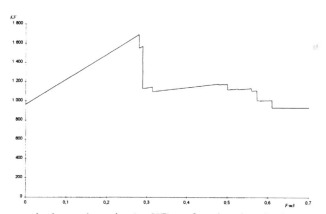

Fig. 3. Recette totale du gestionnaire (en KF) en fonction du prix du second palier

4   Conclusion

La question traitée est la pertinence des instruments économiques incitatifs pour améliorer la gestion de l'eau et éviter le gaspillage dans le bassin versant de la Charente. Pour tout système envisagé, un préalable est l'instauration de dispositifs de comptage volumétrique, actuellement en cours de réalisation.

Les résultats des simulations montrent que, dans le bassin versant de la Charente, une tarification linéaire ne peut à la fois être incitative et assurer un maintien du revenu des agriculteurs. Les solutions alternatives sont le recours à une tarification non linéaire par palier ou un système de quota. Dans le premier cas, le premier palier correspond à la tarification actuellement envisagée et le second est dimensionné pour être dissuasif (0.5 F/m^3) au delà d'une consommation en eau dépassant 85% des besoins en eau théoriques du maïs. Le quota, quant à lui, limite réglementairement la consommation à ce niveau.

En terme d'économie d'eau, les deux instruments sont équivalents. La mise en oeuvre du quota ne dispense pas d'une tarification ayant pour unique objectif le recouvrement d'une partie des charges de gestion du barrage. La tarification non linéaire domine le système de quota en permettant de décentraliser la décision d'économie d'eau au niveau individuel alors que le quota est vécu comme une mesure administrative. Elle apporte aussi plus de souplesse dans l'ajustement de l'offre à la demande.

Néanmoins, l'efficacité de la tarification peut être remise en cause lorsque la demande d'une catégorie d'usagers est inélastique ou par suite à une modification importante des prix des produits agricoles ou des intrants.

Même si la nature des cultures et les valeurs proposées pour la tarification non linéaire garantissent l'efficacité du mécanisme dans le bassin versant de la Charente, le planificateur a maintenu une sûreté en conservant le système antérieur d'interdiction d'arrosage lorsque la consigne de débit dans le fleuve n'est pas respectée à Angoulême.

5   Références

1.   Institut interdépartemental pour l'aménagement du fleuve "La Charente" et de ses affluents (1992) *Protocole relatif à la gestion des eaux du bassin de la Charente et de ses affluents*, Angoulême, 26 mai.

2.   RIEU, T.and PALACIO, V. (1994) Equipements hydrauliques collectifs et réforme de la PAC : des conséquences conflictuelles? Le cas d'une projet de barrage en Charente. *Actes et communications de l'AIP*, séminaire d'Economie et Sociologie Rurale, n°28, pp. 186-203.

3.   Cemagref, Ministère de l'environnement (1996) *Compte rendu d'exécution de la convention 'Tarification de l'eau agricole'*, décembre, 33 pages.

1. Sur la totalité de la zone, cela correspond à une superficie irriguée directement par prélèvement dans le fleuve de 7500 ha. Le montant perçu est alors de 1 200 000 francs. Le modèle n'étant bâti que sur le département de la Charente, le montant que l'Institution doit percevoir est d'environ un million de francs, la surface irriguée directement n'étant que de 6314 ha.

2. Ceci correspond à la partie proportionnelle du tarif de l'Institution. La partie fixe (75 F/ha) ainsi que les coûts de pompage à la charge de l'agriculteur sont intégrés dans la modélisation mais n'apparaissent pas explicitement sur la courbe.

3. Au sens de maximisation du bien-être collectif

THE PRODUCTIVITY OF WATER IN PUBLIC IRRIGATION SYSTEMS OF THE GUADALQUIVIR RIVER BASIN

Water productivity in irrigation systems

M. MARTÍN R. and N. RODRÍGUEZ
Economics Dpt., Granada University, Spain
J. LÓPEZ MARTOS
Hydrological Planning Dpt.-Confederación Hidrográfica del Guadalquivir, Granada, Spain

Abstract

The Guadalquivir River Basin, located in the south of the Iberian Peninsula, presents water resources conditions similar to dry climate countries. Agriculture and, more particularly, irrigation are an important part of the economy in this River Basin.

The share of water demand for irrigation represented in 1992, 79.7 % of total demand and the percentage expected for the year 2012 is 78%. In both cases, we are faced with important deficits in the balance of hydraulic resources, which provoke social problems and economic losses. The traditional solutions applied so far consisted in merely resorting to hydraulic accounts. In other words, demands and resources were balanced with little or no consideration for economic costs.

In this paper, we therefore intend to analyze the real costs of water consumed in irrigation, the participation of farmers in assuming those costs, as well as look at the productivity obtained by irrigation in the areas created to that effect by public initiative in the Guadalquivir River Basin. In other words, we want to project an economic vision on the balance of water resources where, not only the problems of quantity of resources are taken into account, but costs of exploitation of new regulated resources are looked upon with equal concern.

Keywords: Cost of irrigation water, irrigation rate, public irrigation systems, regulation fee, unitary water allowances, water productivity.

1 Introduction

The Hydrographic Drainage Basin of the Guadalquivir river is located in the south of the Iberian Peninsula, and stretches over 57,527 Km^2, 11.4% of the whole spanish

Water: Economics, Management and Demand. Edited by Melvyn Kay, Tom Franks and Laurence Smith. Published in 1997 by E & FN Spon. ISBN 0 419 21840 8.

territory. According to the last official census realized in 1991, the area is populated with 3,817,319 inhabitants, 9.7% of total national population.

The mean annual runoff is 7,230 Hm^3: figure calculated for the Hydrological Plan proposal of 1992 and based on hydrological data corresponding to 1942/43 - 1987/88. This average value might reduce on entering the latest data which will soon be available, for the extremely dry period of 1992/93 - 1994/95. These natural resources come from inland surface and underground waters.

The resources obtained through construction of regulating dams and exploitation of aquifers were 2,710 Hm^3/year in 1992, a figure that has now increased to 3,080 Hm^3/year as several new dams were since put into service, an investment of 82,777 millions pesetas, as Table 1 shows.

Table 1. Dams recently put into service

Name of Dam	Capacity (Hm^3)	Annual regulated volume (Hm^3)	Investment (10^6 ptas) [1]	Investment per regulated m^3 (pesetas)
Giribaile	475	128	25,938	203
Vadomojón	165	92	8,807	96
La Fernandina	245	58	12,435	214
José Torán	101	28	11,286	403
San Clemente	120	22	9,258	420
Francisco Abellán	58	22	8,496	386
Colomera	42	20	6,557	328
Total	1,206	370	82,777	224

Source: Confederación Hidrográfica del Guadalquivir
[1] Includes investments to be made in 1997. The peseta is the currency used at 1995 exchange rate, estimating for 1997a 3% inflation rate.

The agricultural sector is the main water consumer, with a demand for 1992 of 2,870 Hm^3, meaning 80% of total demands. In that same period, the land under irrigation was 443,000 Ha; at present, this figure has increased to near 600,000Ha. The most outstanding increase (nearly 100,000 Ha.) took place in irrigated areas specialized in olive tree growing, due to an important rise in production achieved in changing these from dry farming, the usual farming system for this crop type in the basin, to an irrigated system. Also continuity in production was ensured in this crop type, since the well known "variability" of olive trees under dry farming could be controlled. Another decisive factor has been the subsidies granted by the European Union, evaluated according to production. The periods of drought suffered since 1980 made farmers more aware of the necessity of irrigation, the best system to ensure good production.

Water balance sheets calculated for the years 1992 and 2012 show overall deficits of 490 and 330 Hm^3/year, respectively. These figures represent 14% for 1992 and 8%, as estimated for 2012, of total demands.

The traditional answers of increasing prime necessity water volumes, whether drawn from the catchment area itself or coming from other drainage basins, to balance out water deficits find opposition from scientific, technical, socio-economic and informative spheres, mainly because they assume heavy investments and arduous negotiations, especially on the environment.

Facing this situation and conscious of increasing supply demands, it seems pertinent to wonder if there are any other alternative ways of complementing or substituting the increase in prime necessity water resources to respond to increased demand for water

The purpose of this paper is to assess the water productivity problem in the agricultural sector (first water consumer) of the Guadalquivir basin.

To this end, we selected an irrigated area of 178,103 Ha that represents 40% of the total irrigated area[1]. This area is divided into 22 irrigated zones located as Table 1 indicates. Most of these zones were promoted by Public Administrative Authorities, since the 1920s. The selection of these zones rested on criteria of quality management by the basin agency, in this case the Confederación Hidrográfica del Guadalquivir. It would be equally interesting to carry out a similar study in privately owned irrigation zones. The time period analyzed for each irrigation season stretches from 1983 to 1993.

2   Unitary water allowances

Table 2 exposes unitary allowances in m³/Ha supplied to each irrigated zone from the regulating reservoirs; measurement has been made at headwaters of the water delivery system each zone possesses.

In order to render a correct evaluation, we ought to add a few comments to the data of Table 2.

The first comment refers to the available means of supply for each zone. The first eight zones in the list dispose of only one source of supply, generally a reservoir.

An exception is the Albolote Canal which is fed from a spring and, when supplied, diversions from the Cubillas river during the irrigation season. Also the Bembezar Canal that receives its supply from two reservoirs, the Bembezar and Retortillo dams. Among all these zones, the Guadalmellato is the only area that uses other sources of supply to those the Confederación Hidrográfica usually resorts to.

The other fourteen zones are supplied through the so-called General Regulating System that, at the time of our study consisted of seven great dams, with an average regulation of 911 Hm³/year, and five lesser reservoirs regulating an additional 36 Hm³/year. Superficial runoffs too are to be taken into account. Within the same irrigation system, the quantity of irrigation water for each zone will depend on the quantity of reservoirs located upstream. Thus, the conjunctive management of the twelve reservoirs is of utmost importance to balance out these geographic circumstances. Moreover, the provision of alternative sources of supply play a crucial part in case of drought. Some of these sources are not fully controlled[2] by the basin's agency, as is the case for the Genil M.D., Genil M.I., Valle Inferior and Bajo Guadalquivir. The two zones of the Genil River and Valle Inferior are notable for their location over alluvial aquifers of the same name.

Table 2. Unitary allowances (m^3/Ha)

Zone	Years										
	83	84	85	86	87	88	89	90	91	92	93
Salado de Morón	8764	5623	8619	6656	7280	7027	8537	4756	6932	6676	2946
Cacín	5947	4966	6254	4425	5435	5501	3231	6169	6303	6322	3709
Canal de Albolote	----	5042	6786	4668	4975	6248	4686	2423	3007	3099	2068
Guadalentín	----	7576	10415	7641	5739	8106	4431	4667	5519	3533	2543
Rumblar	4344	5320	9711	9941	8107	7262	7301	7387	6865	4438	2446
Guadalmellato	----	8368	10244	9607	7002	9125	2207	8434	8795	1998	----
Bembezar	----	7042	9201	8907	4792	8200	8044	8518	9223	5837	302
Viar	2384	6016	7844	7859	7189	8211	8524	9321	8673	6313	289
Guadalmena	----	----	4136	1477	4550	6207	2914	3622	3842	2208	378
Guadalen	4075	3396	6029	5820	6223	5431	2720	4431	4082	2033	836
Jandulilla	2797	3634	4234	4173	3658	4488	3252	3776	2779	2600	1287
Vegas Altas	4305	5366	9173	8818	9809	8039	3189	8147	5905	3934	2394
Vegas Medias	3738	5466	7944	8237	8645	6684	3725	7055	6297	3825	1702
Vegas Bajas	3529	5442	9284	8542	7389	6490	3251	5476	5112	2564	873
Genil Cabra	----	----	----	----	----	----	----	----	3634	2023	108
Fuente Palmera	----	----	4275	3855	3987	4590	2202	4116	4221	1692	26
Genil M. D.	3264	5335	7522	7226	7082	7386	2058	6136	4725	2087	374
Genil M. I.	2941	4853	7549	7756	7298	7257	2342	6172	4389	2033	207
Valle inferior	----	4947	8636	7429	7557	7729	2199	6089	4986	2179	252
Bajo Guadalqui.	----	4854	7566	7219	7045	7673	2497	6358	5102	2243	52
B XI	----	----	5993	6332	3308	4688	2202	4947	4718	2082	-----
B XII	3498	5098	6475	5951	5944	7097	2362	6009	5055	2314	3

Source: Report on Irrigated Zones Exploitation-C.H.G., and thesis of N. Rodríguez

If we observe the data of Table 2, three dry periods stand out: 1983-84, 1989 and 1992-93. This last dry period lasted till 1994 and 95, and was so drastic that the Confederación came to forbid irrigation in all the drainage basin, in 1995. The data collected give us the opportunity to observe how little impact the experience of a dry period has on the following years, probably because not enough importance has been given to the carry-over storage of the dams that supply water to the various irrigation systems and zones.

Finally, we should add that the allowances have been calculated at the zones headwaters, taking into account the area actually irrigated each year, although that covered by the irrigation system is different. In the dry years, and especially in 1993, certain zones show very scarce duties. This reveals that the supply of water was concentrated to maintain trees, or crops of high value, extended over a limited area.

As far as water productivity in irrigation systems is concerned, it is important to look at the unitary allowance of water served to irrigated lands, in proportion with net crop requirements.

3   The cost of irrigation water

As stipulated by the Spanish Law in force, water is a free-of-charge resource. Therefore, the State, manager of this good considered public property ("dominio público"), is only authorized to allocate fees to water users and irrigators for investments made to regulate the natural resource and provide water channelling and distribution structures.

The contribution based on the regulation works is called a regulation fee ("canon de regulación"), while when water transportation and distribution equipment is used, and that is generally done in the irrigation zones financed by the State, the contribution is called irrigation rate ("tarifa de riego").

Although the users of all the works built to that end are legally compelled to pay their fee, neither non consumptive recreational uses, nor works for flood abatement are their responsibility.   An approximation is made of the cost generated by unidentified users, which will be assumed by the General Administrative Body, and thus deducted from the fee paid by water users for consumptive and energetic purposes.

The total cost of the regulation fee is divided into three headings: costs of operation and maintenance of the works carried out, costs borne by the Administrative body of the drainage basin attributable to the aforementioned works, and repayment of the investment made by the State. As for the irrigation rates, their financial structure are similar to those of the regulation fee.

The distribution of these fees among all those who benefit from the works will be done with respect to the Water Act provision, on the basis of the following criteria: reasonable use of water, equitable sharing obligation and self-financing of the service.

Apart from the aforementioned criteria, the absence of individualized volumetric meters for each property within the irrigated area, have impulsed the Drainage Basin Authorities to charge the regulation fee as a function of the extension of land irrigated. Beforehand, the total value of the regulation fee is divided into several uses of water for irrigation, supply, and electrical energy production, according to an estimate of the general benefits generated by water uses, in each one of the concerned sectors.

As far as irrigated areas are concerned, which really are the object of this study, the regulation fee is still perceived per Ha, even if a few meters have already been installed in certain areas to start up a distribution based on a binomial formula that would include in its components the volume of water really consumed.

In Table 3, the effect of the regulation fee and irrigation rate are reflected for the period of time analyzed and in pesetas/m^3, for each of the irrigable areas. The values observed were calculated taking into account, on one hand, the fees charged per Ha every year by the Drainage Basin Authorities and, on the other hand, the quantities supplied from the reservoirs during the corresponding year. The values henceforth obtained were converted into 1995 pesetas so as to make the analysis easier to carry out.

If we analyze Tables 2 and 3 together, we can see that the variations of this "legal price", or effect of the regulation fee per Ha over the volume really consumed, are mostly affected by the unitary allowance, since the price value of the fee only varies drastically when it reflects for the first time a new investment in regulation works.

Coming back to our analysis, we observe mainly in the second part of the table, where 14 zones are served by one unique operating system (the General Regulating System), that in dry years when allowances of water are obviously lesser, the "legal price" increases (see 1983, 1989 and specially 1993). In 1993, there are great variations between zones of the same system. For example, let us compare the zones of Bajo Guadalquivir, with 264 pesetas/m^3 and BXII, with 4187,7 pesetas/m^3, and the Vegas Medias and Vegas Bajas, with respectively 14,48 and 15,35 pesetas/m^3.

Table 3. Regulation fees and irrigation rates (pesetas/m^3)

Zone											
	Years										
	83	84	85	86	87	88	89	90	91	92	93
Sala. Morón	1,09	1,03	1,182	1,313	1,449	1,551	1,694	1,86	1,806	1,83	4,672
Cacín	1,55	1,81	1,294	2,503	2,166	2,677	4,655	1,9	1,692	3,1	4,781
Canal Albolote	----	1,44	0,974	1,391	1,714	1,916	2,534	0,41	2,943	2,62	5,314
Guadalentin	----	1,2	0,783	1,143	1,636	1,582	2,637	1,96	1,553	1,79	3,517
Rumblar	1,79	1,31	0,703	0,973	1,387	1,217	2,21	2,06	2,4	3,83	8,059
Guadalmellato	----	1,04	0,894	1,159	1,87	1,886	6,894	1,86	1,806	4,19	(1)
Bembezar	----	1,67	1,134	1,329	3,023	1,704	1,429	1,62	2,223	3,14	47,51
Viar	4,82	1,62	1,182	1,313	1,574	1,399	1,399	1,45	1,339	1,88	46,85
Guadalmena	----	----	----	1,468	0,639	1,019	2,946	2,01	2,905	5,68	28,75
Guadalen	4,42	4,05	1,996	3,615	2,649	3,498	7,218	3,25	3,031	7,19	18,4
Jandulilla	4,99	3,55	3,274	3,708	6,232	4,563	6,629	4,25	7,199	8,58	9,692
Vegas Altas	8,76	6,7	3,641	3,708	2,96	3,955	9,133	3,01	3,031	4,18	8,494
Vegas Medias	6,75	4,38	2,715	2,781	2,727	4,015	7,807	2,6	3,789	7,54	14,48
Vegas Bajas	6,87	4,37	2,427	3,214	3,054	4,715	8,249	3,7	3,41	6,26	15,35
Genil Cabra	----	----	----	----	----	----	----	----	7,78	7,92	60,3
Fte. Palmera	----	----	----	12,65	----	----	3,358	1,7	11,43	20,7	249
Genil M. D.	1,95	0,5	0,703	1,051	1,059	1,232	4,714	1,45	1,819	4,33	29,04
Genil M. I.	2,27	0,57	0,719	0,942	0,982	1,217	4,036	1,41	1,92	4,35	51,73
Valle inferior	----	0,38	0,431	0,603	0,763	0,684	2,548	1,15	1,314	2,32	25,65
B.Guadalqui	----	0,75	1,022	1,174	1,558	1,46	5,73	1,92	2,223	5,36	264
B XI	----	----	1,469	1,344	3,35	2,434	5,804	2,41	2,387	5,77	----
B XII	1,47	0,47	0,799	0,912	1,028	1,247	4,493	1,55	1,793	4,22	4187,

Source: Report on Irrigated Zones Exploitation-C.H.G., and thesis of N. Rodríguez
[1] No water was consumed; however, the regulation fee was charged.

4   Water productivity

To illustrate the question of water productivity in these irrigated zones, we firstly calculated the incremental gross production in pesetas/m^3, on the basis of real production obtained in the different zones, compared to the fictitious production, if these zones were using a dry farming system. This fictitious situation was elaborated respecting the surface distribution of irrigated zones between municipalities, and observing existing dry farming systems of the evaluated zones. The productivity of crops and prices of products were calculated with agrarian statistics available.

The results obtained appear in Table 4. Data from the years in which the water allowance moves away from the quantity necessary for the crops, and given the level of efficiency of the system itself, are not very representative. Moreover, we should account for the years 1983, 1989 and 1993, and in some zones 1984 and 1992, as years of transition from or to a dry period which then might be little relevant.

As the costs of irrigated crops are higher than dry farming crops, we prepared Table 5 to evaluate the incremental net benefits by taking away irrigation systems cost increases. As it has not been possible to calculate incremental costs of all the zones and years, we used the data of the following zones: Cacín, Vegas Altas and Viar, as reference data which we applied to the other zones, in relation to crop similarities.

Table 4. Incremental gross production (pesetas (1995 exchange rate) /m^3)

Zone	Years										
	83	84	85	86	87	88	89	90	91	92	93
Sala. Morón	19,41	28,51	20,77	45,97	39,17	34,23	49,73	51,51	58,97	49,43	34,08
Cacín	63,44	54,10	66,21	103,51	83,11	78.82	114	89,08	86,39	86,42	125,66
C. Albolote	----	74,37	50,36	89,14	83,95	74,61	24,69	137,03	85,17	136,38	190,15
Guadalentín	----	6,77	15,16	17,66	27,46	18,89	13,54	28,79	22,34	21,83	56,96
Rumblar	52,05	47,94	18,68	21,84	25,09	42,64	33,16	40,98	48,50	65,02	83,19
Guadalmell	----	33,20	30,56	23,70	43,37	28,94	91,32	43,14	35,,6	79,05	------
Bembezar	----	37,99	30,85	27,17	39,06	30,33	35,51	34,03	39,41	58,51	817,84
Viar	356,32	64,81	34,50	35,33	41,03	43,86	37,87	37,57	42,17	46,58	719,03
Guadalmen	----	-----	29,73	6,76	6,80	31,61	44,59	33,53	48,90	59,37	277,94
Guadalen	33,63	33,69	18,99	26,05	7,45	46,77	39,04	19,71	47,78	71,90	30,99
Jandulilla	45,12	26,29	15,83	33,48	28,61	46,14	47,03	41,09	64,58	59,14	56,61
Vegas Altas	69,68	45,08	19,59	28,70	15,17	33,57	49,74	37,36	42,59	99,39	34,71
V. Medias	51,57	39,51	23,09	31,89	22,60	43,83	61,38	36,38	42,22	67,28	11,97
Vegas Bajas	44,74	47,76	28,18	33,88	30,39	45,40	59,13	97,21	60,34	66,70	32,98
Genil Cabra	----	-----	-----	----	----	----	----	111,76	115,02	363,19	2514,28
Fte. Palmera	----	-----	47,36	71,57	49,68	41,48	104,83	63,13	77,86	112,21	822
Genil M. D.	114,62	68,42	42,69	48,70	51,58	37,89	165,54	79,33	99,71	189,19	531.47
Genil M. I.	63,40	49,08	36,23	36,79	38,39	36,38	85,80	50,06	82,81	80,09	527,67
V. inferior	----	66,53	39,69	54,03	51,51	48,02	155,02	78,92	87,43	186,52	1112,64
B.Guadalqui	----	43,33	27,85	36,82	38,28	25,86	79,08	32,86	51,82	85,81	1852,48
B XI	----	-----	27,62	39,55	61,51	40,46	87,52	41,45	55,96	95,37	------

Source: Report on Irrigated Zones Exploitation-C.H.G., and thesis of N. Rodríguez

Basing ourselves on the information provided in Tables 4 and 5, as well as Tables 1 and 2, we could make several observations on water productivity in public irrigation systems of the Guadalquivir Valley that are worth considering:

1.   Taking into account that the average investment per regulated m^3 in reservoirs built recently is 224 pesetas (table 1) and estimating an average return of public investments of 5%, or 11,2 pesetas/m^3, the incomes of most irrigable zones are superior to that figure (table 5). This means that the investments in hydraulic works made in the Guadalquivir river basin are economically highly profitable. Moreover, we should mention the potential productivity and employment

stimulation generated by irrigation systems (table 4), which means everything in a depressed area, with high unemployement and low opportunity cost of labour.

2. Let us point out that the "legal price" of water is extraordinarily low, not only in relation to the public investment necessary to fix this price (table 1 and 3), but also in comparison with its own productivity. That is why, adapting the "legal price" of water to its real productivity will permit improved economic efficiency in the sector that for the time being prevails in the Guadalquivir Valley.

Table 5. Incremental net benefits (pesetas (1995 exchange rate)/m^3)

Zone	Years										
	83	84	85	86	87	88	89	90	91	92	93
Sala. Morón	9,03	6,05	6,081	25,88	26,19	16,75	37,43	16,1	37,93	29,3	-40,3846
Cacín	80,6	51,9	76,61	85,86	87,75	78,07	126,6	98,4	93,51	105	139,423
C. Albolote	---	85,7	53,4	70,68	85,93	75,97	16,17	140	79,04	161	205,821
Guadalentín	---	-13	6,323	1,971	8,826	0,736	0,724	20,2	13,72	6,88	42,1538
Rumblar	62,6	54,9	16,69	15,01	21,22	42,78	34,64	42,2	49,72	73,2	78,8462
Guadalmell	---	27,1	26,21	8,802	31,06	16,47	11,84	26,5	16,21	-20,2	-----
Bembezar	---	29,8	23,9	11,3	10,14	15,47	17	15,5	22,51	36,2	228,321
Viar	540	68,2	25,5	17,88	28,49	34,56	21,78	22,1	24,26	23,2	-5,87179
Guadalmen	---	---	16,42	-56,1	-17,1	24,67	36,14	24,3	42,79	49,4	176,372
Guadalen	29	18,1	9	13,02	-8,94	-43	27,24	10,3	42,41	63,8	-41,641
Jandulilla	35,6	8,36	-5,27	14,97	7,899	37,49	41,7	34,3	55,48	53,9	19,7179
Vegas Altas	92,7	50,4	17,39	21,48	9,464	24,89	44,86	38,5	40,98	118	16,0897
V. Medias	57,3	41,7	20,82	24,3	18,55	43,01	63,13	36,1	41,21	73,7	-24,6282
Vegas Bajas	43,7	55,1	31,4	26,94	27,42	44,67	57,61	109	60,93	63,9	-35,641
Genil Cabra	---	---	---	---	---	---	---	84,2	75,43	371	929,936
Fte. Palmera	---	---	21,08	35,95	16,16	9,833	29,37	23	39,13	2,01	-8473,03
Genil M. D.	132	69,3	37,42	30,95	43,32	23,32	101,6	61,1	71,73	136	19,0897
Genil M. I.	36,5	33,3	27,11	19,23	25,13	20,71	10,82	25,2	47,3	-16,5	-520,115
V. inferior	---	61,9	36,65	37,57	45,19	38,69	95,25	60,3	59,37	138	-851,346
B.Guadalqui	---	23,9	13,66	17,63	23,87	7,708	8,316	5,02	17,01	3,12	-2388,46
B XI	---	---	5,097	17,38	21,84	9,528	7,25	5,49	18,3	7,36	-----
B XII	101	32,2	7,597	19,37	15,57	-0,72	111	-9,51	26,73	32,4	-33716,6

Source: Report on Irrigated Zones Exploitation-C.H.G., and thesis of N. Rodríguez

5   References

1. Aguilera Klink, F. (coord) (1992) *Economia del Agua*, MAPA, Madrid.
2. Azqueta, D. and Ferreiro, A. (eds) (1994) *Análisis económico y gestión de recursos naturales*, Alianza economía, Madrid.
3. Calatrava Requena, J., Ceña Delgado, F. and others (1986) *Evaluación del Impacto Socioeconómico de Grandes Proyectos de Regadío: Aplicación a la Zona Regable de Fuente Palmera (Córdoba)*, Junta de Andalucía, Consejería de Agricultura y Pesca.

4.   Cantero Desmartines, P. (1996)   *El Análisis Coste Beneficio en el Sector Agrario: Especial aplicación a una Trasformación en Regadío en Andalucía*, Junta de Andalucía, Consejería de Agricultura y Pesca.

5.   Ceña Delgado, F. (1994)   *Flujos Económicos del Regadío. Presente y Pasado de los Regadíos*, MOPTMA, Madrid.

6.   Garrido Colmenero, A.   *La economía del agua: análisis de la asignación de recursos mediante el establecimiento de mercados de derechos del agua en el Valle del Guadalquivir*, Doctoral Thesis, E.T.S.I. Agrónomos, Madrid.

7.   Gibbons, D. C. (1986)   *The economic value of water. Resources for the future*, Johns Hopkins University Press, Baltimore.

8.   Gil Ocina, A. and Morales Gil, A. (coords) (1992)   *Hitos históricos de los regadíos españoles*, MAPA, Madrid.

9.   Gittinger, P.   *Análisis económico de proyectos agrícolas*, Tecnos, Madrid.

10. Kulshreshatha, S. N. and Tewari, D. D. (1991)   *Value of water in irrigated crop production using derived demand functions: A case study of south Saskatchewan River irrigation district*, Water Resources Bulletin, 27, pp 227-236.

11. Martín Rodríguez, M. and Rodríguez Ferrero, N. (1992)   *Los Regadíos de la Presa de la Bolera*, Confederación Hidrográfica del Guadalquivir, Sevilla.

12. Annual Report on *Exploitation of Irrigable Zones of the Guadalquivir*.

13. Orgaz, L. and Pérez Blanco, J. M. (1986)   *La rentabilidad social de la inversión pública en regadío*, Investigaciones Económicas, Vol. X, n°3.

14. *Plan Hidrológico de la Cuenca del Guadalquivir* (1995), Confederación Hidrográfica del Guadalquivir, Sevilla.

1. The area was 443,024 Ha in 1992. At present, the area irrigated has increased by some 110,000 Ha, of which 80,000 Ha are olive groves equipped with water saving localized irrigation system.

2. They are cases of ground water extraction which have a private jurisdiction, as agreed in the Water Act of 1879 and in the transitory order of the Water Act of 1985, presently in force.

THE WISE USE OF MANGROVE SYSTEMS: THE SOCIAL AND ENVIRONMENTAL VALUE OF WATER

TOM FRANKS
Development and Project Planning Centre, University of Bradford
MOHD SHAHWAHID
Faculty of Economics and Management, Universiti Pertanian Malaysia
HO LIM HIN FUI
Forest Research Institute, Malaysia

Abstract

The paper reviews changing approaches to water resource management, by considering the case of mangrove ecosystems. Three broad options for mangroves are identified: conservation, exploitation and conversion. Major alternative uses following drainage and conversion include irrigated agriculture and aquaculture. The paper draws on recent research in the Merbok estuary of Malaysia to illustrate the discussion. The social structure of the population in the vicinity of the Merbok mangroves is described, and their dependence on mangrove processes in analysed. In the next section the paper reviews the benefits and impacts of the different production systems and updates estimates of their value. The final part of the paper looks at the implications of this analysis for the planning and management of water systems such as mangroves.
Keywords: Mangroves, social, economic, environmental, value.

1   Introduction

In the latter part of the twentieth century approaches to water management are changing in response to changing perceptions of the value of water. We can see such changes taking place over recorded history, with the first large-scale attempts at water management being directed towards increasing crop production through irrigation along the Nile, the Indus and in Mesopotamia. This was as a result of the perception that the highest value of water at that time lay in increasing food production to assist in human settlement. This was closely followed by the realisation that flood control and drainage could make similar contributions towards achieving food security: the three aims of irrigation, drainage and flood control then remained the key objectives

Water: Economics, Management and Demand. Edited by Melvyn Kay, Tom Franks and Laurence Smith. Published in 1997 by E & FN Spon. ISBN 0 419 21840 8.

of water management from the early hydraulic civilisations until about the last two hundred years. From that time, industrialisation and the accompanying phenomenon of urbanisation and trade increased the value of water for domestic use, for production processes, and for navigation. In the twentieth century, as the pace of industrialisation quickened and living standards rose, power generation became a high priority, and the value of water for this application rose correspondingly. At the present time, we see available water becoming an increasingly scarce resource and competing demands from all these sectors for the available supplies.

Along with these competing demands for water for economic purposes there have arisen new uses and demands from different sectors. A particular example of the changing value of water in different sectors, and resulting changes in approaches to water management, is the growing emphasis on the value of water in creating and maintaining wetlands. This arises in a general way from the trend towards "environmentalism" in the 80s and 90s, but more specifically and perhaps more logically from a growing understanding of the importance of the hydrological cycle in many natural and human processes, and the realisation that water is involved in complex and intricate linkages which are often of vital significance to human welfare. This situation is well illustrated by the case of mangroves. Until recently mangrove wetlands have been considered as wastelands: it is now becoming apparent that they have social, economic and environmental value not hitherto appreciated.

2   The importance of mangroves

Mangrove systems occur in tropical regions in the inter-tidal zone. They correspond to salt marshes in temperate zones, but are characterised by a variety of salt-tolerant trees, many of which have the distinctive spreading root system which allows them to survive in this niche. They are widespread in East and South East Asia, parts of East and West Africa and along both coasts of the Americas, though their form and the dominant tree species vary from region to region. An estimate of the mid-eighties [1] put the total area of mangroves then at around 165,000 sq. km., about 0.1% of total land area, but it is known that there has been rapid destruction or conversion since then and the present area is put at much less than this.

Mangroves form complex ecosystems, dominated by the mangrove trees themselves, but including a large number of other plants and animal species which are dependent on them. The source of primary productivity in the system is leaf fall from the mangrove trees themselves: estimates put this at around 10t dry matter per hectare per year [2]. The leaf matter is decomposed by micro-organisms and then becomes the basis of a food chain through detritivores, small carnivores to large carnivores. Of particular importance to-day is the place of fish in this food chain: mangroves are known to be of importance in the life cycle of many pelagic fish, as they provide habitat and food, especially for young fish.

At the present time, three broad options for mangrove management exist: preservation, exploitation and conversion. Preservation involves retaining the systems in their existing state, in which they have both environmental and social values (distinguished by some writers such as Ruitenbeek [3] as carrier and information functions). Carrier functions include their importance in the maintenance of biodiversity and in their provision of breeding grounds and migration habitats, especially for water birds. Preservation does not necessarily preclude human activity in their vicinity, because the possibility exists for educational and recreational activities (information functions), such as in the Everglades Swamps of Florida and in the Sundarbans of Bengal. These may also be a source of revenue for conservation. A variety of ways of exploiting mangrove products exist. In order of increasing ecological disturbance, these include: honey production; fodder provision for livestock; collection of products such as medicines, tannins and genetic material from mangrove species (particularly the Nipa palm in South-Asia); and finally charcoal and timber production, which can be practised on a sustainable yield basis with a replanting programme.

Conversion, which results in the destruction of the mangroves and their replacement with other forms of activity can be undertaken for a number of reasons. Mangroves close to centres of population are often cut down for human settlements or infrastructure. In other areas they may be drained and converted to irrigated agriculture, usually for rice or coconut cultivation. More recently there has been a rapid increase in drainage of mangroves and conversion to aquaculture ponds, particularly as a result of the growing emphasis on prawn and shrimp production. Finally mangroves are also used for salt production, in regions where other sources of salt are not readily available. This may sometimes be combined with the need to use mangrove timber as a fuel in the production process, resulting in an increased rate of destruction.

The broad options described above take account only of the on-site values and functions of the mangrove areas themselves. There are also important off-site values and functions, including flood and storm protection provided when the trees dissipate wind and wave energies, nutrient supply and regeneration, and provision of navigation routes, whose benefits may be both on-site and off-site. Another important off-site function is pollution control provided by the extensive root systems which trap and hold sediments and other pollutants. Ironically, this useful service can hasten the destruction of the mangroves since they become perceived as waste land of low value except for refuse disposal.

3   The social, economic and environmental value of water in the Merbok Mangrove, Malaysia

The complexity of the management of mangrove systems is being investigated as one part of a research project carried out by partners from India, Malaysia and UK [5]. This research has three main objectives: improved modelling of the hydrodynamic processes in the mangrove systems, particularly related to the effect of vegetation on flow patterns: improved analysis of water quality parameters and nutrient fluxes: related to these, an improved understanding of management processes, with a special emphasis on the part which can be played by mathematical models.

Field work on socio-economic values and management processes concentrated on the Merbok estuary, which is a medium-sized mangrove in a rural location on the NW coast of peninsular Malaysia. In the 1950s the Merbok mangrove covered an area of around 6000 ha, of which 4037 ha of forest and 1000 ha of waterway was gazetted as a mangrove forest reserve in 1951. In the 1960s, a total of 1075 ha of stateland mangrove forest were converted to agricultural crop cultivation. In the last 10 years about 931 ha of the mangrove forest reserve were excised for aquaculture development. The area remaining under the mangrove system to-day is therefore about 3100 ha.

The population in the vicinity of the mangroves is estimated to be of the order of 191000 [5]. Of this some 158000 (80%) live in three sub-districts close to the district capital of Sungai Petani, which lies at the upstream end of the estuary. These sub-districts show a much higher population growth rate than the remainder of the district, due to the rapid pace of industrialisation and urbanisation. The population of Sungai Petani itself, for instance, grew at a rate of 7.4% over the period 1981-1990, whereas populations in the more rural sub-districts grew at rates of 1% or less.

Local dependence and utilisation of the mangrove resources near and around the Merbok mangrove can be seen from the four main types of economic activities, namely forestry, padi farming, fishing and aquaculture development. The local employment created by these activities is given in Table 1. This shows that directly productive or consumptive uses of the mangrove system provide employment for some 10% of the rural population living in their vicinity, (the remainder finding employment in petty trading, local industries and urban sector).

Table 1   Economic activities in or near the Merbok Mangrove

Activity		No. in employment	% of Total
Forestry	Kiln operators	15	N
	Timber harvesters	60	2
	Kiln workers	280	9
Padi farmers		635	20
Fishing	Sea fishermen	1290	41
	River fishermen	270	9
	Processing workers	207	7
Aquaculture	Cage operators	100	3
	Pond operators	70	2
	Aquaculture workers	200	6
	Total	3127	100

Irrigated agriculture in the converted mangrove area began in the 1960s. The government developed the area by constructing a 35-km bund to exclude tidal water, with nine gates to allow drainage at the period of low tide. Apart from the physical structures, there were no other facilities provided. The converted land was allocated to local farmers, each family being allocated an average of 1.4 ha. Irrigated rice production in the converted areas faces problems of lack of water and acid sulphate soils, and the farmers were not able to achieve the high productivity found on other schemes such as the Muda River. Some land was abandoned and consequently, an Integrated Agricultural Development Project was introduced in 1984 with the intention of providing additional infrastructural facilities such as roads and machinery, as well as introducing a programme of planting fruit crops, such as mangoes and coconuts, as supplements to paddy cultivation. Simultaneously, some of the abandoned land was replanted with oil palm by the Federal Land Reclamation Authority, under its group replanting programme. Farmers were provided free lime (to reduce soil acidity), free fertiliser and padi subsidy. The use of the converted mangrove for agriculture is continuing but needs capital and operating subsidies to maintain its viability.

Fishing is a traditional occupation for the local population living near to the Merbok mangrove. This includes both sea fishing and river fishing, the relationship with sea fishing lying in the fact that the mangrove ecosystem provides a breeding ground and nursery for various species of sea fish. As indicated in Table 1, this sector accounts for half of the employment created near the mangrove forest area and is an important source of income for local population.

Large scale fishermen, using purse-seine nets for harvesting, provides the greater part of the total employment and income generated in the fishery sector. It is estimated that there are 50 boats using this method of fishing. Such boats are equipped with modern instruments such as radar to detect the availability of fish. Trawl nets and drift nets are used by relatively poorer fishermen. There are about 25 boats using trawl nets and 100 boats using drift nets. Bagnet fishing along the Merbok River and its main tributaries is practised by small-scale fishermen living near the water way. There are about 100 boats practising this type of fishing which involves 200 fishermen. Other methods of fishing such as trap, hooks and lines and shellfish harvesting are also practised.

Aquaculture development within and near the Merbok mangrove began in the early 1980s. There are two categories of aquaculture activities. Various species of fish and shell fish are cultured in cages anchored along the waterway. Cage culture is practised by individuals, Fishermen's and Farmers' Associations, and the Fisheries Department, and is supported by the government. The second category of aquaculture is pond culture which utilises cleared mangrove land for the culture of prawns. There are now three types of pond culture practitioners within and around the Merbok mangrove, government agencies (which initiated pond culture projects in the early 1980s), local villagers and small entrepreneurs operating on a small scale, and large companies operating on land officially excised from the Merbok Forest reserve in 1985. At the present time there is estimated to be some 70ha of small-scale ponds and 620 ha of ponds operated by large companies. In general, aquaculture is becoming more widespread around the Merbok, reflecting its increased financial and economic importance.

4   Financial returns to mangrove production systems

Towards the end of 1996, a survey of households living in the vicinity of the Merbok mangroves was conducted with the objectives of understanding: the economic activities related to mangrove ecosystem utilisation; the structure of financial remuneration and costs incurred by the local communities when involved in these activities; the attitudes and management processes of the various stakeholders. The survey covered a stratified sample of 5% of the operators of the various economic activities, and the results of the financial analysis are shown in Table 2. (Values converted at the rate 1 US$ = 2.5 M$ Source: 1996 field survey)

Table 2 . Cost and Earning Structure of Economic Activities (US$/annum)

	Gross Income	Variable Cost	Fixed Cost	Total Cost	Net Income
Fishermen (sea)	9470	4890	860	5750	3720
Fishermen (river)	2070	1050	190	1240	830
Cage culture	11020	5370	1010	6380	4640
Pond culture	91910	52550	6600	59150	32760
Charcoal kiln	35170	27070	3720	30790	4380

Table 2 shows that, at the present time, pond culture of shrimps is by far the most remunerative activity in the mangroves, giving a net income seven times higher than the next most profitable activity, cage aquaculture. Other types of activity are broadly similar in their returns, except for river fishing, where the returns are much lower. The low fixed cost of river fishing reflects its basis on access to a public good in which the fishermen have no ownership or security of tenure. Together these limit the incentive to invest in capital assets such as modern boats.

Profitability measures for the various activities have also been calculated and are shown in Table 3. As far as profitability is concerned, cage culture, pond culture, and fishing in the river are efficient investment opportunities around the Merbok system. Fishing in the river is efficient with respect to its low investment and operational costs incurred, but does not generate a high net income. Charcoal production is the least efficient investment opportunity, both in terms of returns to investment and on sales. Paddy farming shows a high return on sales because inputs are heavily subsidised by Government.

Table 3. Profitability Measures for Economic Activities

Economic Activity	Return to Investment[1] (%)	Profit Over Sales[2] (%)
Fishing (sea)	31	39
Fishing (river)	74	39
Cage culture	91	42
Pond culture	76	35
Charcoal kiln	16	13
Paddy farmers	10	42

1. Annual net income/total investment
2. Annual net income/annual gross income
Source: 1996 survey

Applying the average net incomes observed to the number of operators of the various production activities allows an estimate to be made of the total returns to these various activities in the mangrove system. The calculation is shown in Table 4. The total returns obtained from these amounted to US $9.3 million per annum.

Average annual household income was estimated at US $5,395 in the Sungai Petani district [6]. The same number of households on the average income would have earned a higher income of US$ 12.8 million. However, this higher average income is skewed and influenced by economic activities, such as manufacturing and commerce, in the urban district centre where the majority of the population are located. The total net returns obtained from the various activities in the mangrove system provide reasonable returns for the rural population in the district.

Table 4. Total Returns (US$/annum)

Activity	Nr. of Operators	Net Income	Total Income US$/annum
Charcoal production	15	4380	66,000
Sea fishing	1290	3720	4,799,000
River fishing	270	830	224,000
Cage culture	100	4640	464,000
Pond culture	70	32760	2,293,000
Padi farming	635	2273	1,443,000
		Total	9,289,000

The figures in table 4 indicate that the fishing and aquaculture activities in and around the Merbok which do not require conversion of the mangrove are important economic activities in their own right, bringing in a net income of over US$ 5 million annually. Of this total, the sea fishing is by far the largest share, though not all of this income can be attributed to the presence of the mangroves. Pond culture is also a highly significant economic activity, with a net return of over US$ 2 million annually, from a pond area of less than 1000 hectare. Padi farming brings in net returns of about US $ 1.4 million, from a similar area, though this of course is heavily subsidised and is, in economic terms, considerably less. Both pond culture and padi farming require drainage and conversion of the mangrove and therefore reduce returns to the non-consumptive uses of fishing and timber production. The charcoal industry is not a significant part of the economic activity in the mangrove, and indeed is reported to be on the decline. It should also be noted that there is intense pressure on the upstream end of the mangrove from urban development in Sungai Petani: in this location land values are very high.

The figures of Table 4 above give an order-of-magnitude estimate of the value of the various productive activities of the Merbok mangrove but need considerable refining to give a true indicator of the economic returns of the different activities and, in particular, a comparison between those which require drainage and conversion (padi farming and pond culture) and those which are non-consumptive (mainly fishing).

Firstly, the figures are given in financial prices and need converting to economic values to allow for the effects of subsidies, and for prices to reflect the true opportunity cost or marginal willingness to pay.

Secondly, and more importantly, the analysis covers only traded economic activities. Households rely on the mangroves in addition for various non-traded goods and derived services, which are not incorporated into the net financial income estimates. Non-traded economic activities can contribute quite significantly to the economic well-being of local communities surrounding mangroves. Ruitenbeek [2] estimated that 85% of the economic value of the Bintuni Bay mangrove area of Irian Jaya, Indonesia are from non-traded portions. The valuation by Mohd Shahwahid and Nik Mustapha [7] of the Semelai people's economic activities in the peat swamp and lake area of Tasek Bera of Malaysia suggests that 24% are derived from non-traded raw and processed goods collected from the area.

Finally, no attempt has been made at the present time to allow for the economic values of recreation and leisure along the Merbok, though there are increasing signs of a leisure industry developing, nor for the specific environmental benefits of biodiversity conservation and other carrier functions.

5   Conclusions and implications

The Ramsar Convention advocates the "wise use" of wetlands, defining this as "their sustainable utilisation for the benefit of humankind in a way compatible with the natural properties of the ecosystem" [8]. Sustainable utilisation is defined as "human use of a wetland so that it may yield the greatest continuous benefit to present generations while maintaining its potential to meet the needs and aspirations of future generations", a concept which clearly has a close linkage with the often-quoted definition of sustainability from the Brundtland Commission. The work done so far during this research programme and by others on the Merbok and other mangrove systems world-wide clearly indicates the multiple values and functions that such wetlands have, the complex linkages between them, and the need to take fully into account the social, economic and environmental value of water in such circumstances. Whilst it is difficult to forecast how future generations will value different functions and products from the mangroves, the experience of the last thirty years should lead us to caution in taking irreversible steps with regard to the mangroves. For instance, the economic imperatives which lead to the drainage of part of the Merbok system and its conversion to irrigated rice production in the 1960s no longer apply: such production may remain financially profitable to farmers to-day but it is heavily subsidised, and its economic profitability is much more questionable. Similarly to-day's high returns to aquaculture, particularly prawns and shrimps, are also likely to change as economic circumstances change, and it is necessary to be continuously aware of these changes.

The management of wetland systems such as mangroves in the future will depend on a greater understanding of the biophysical processes and their linkage with the social, economic and environmental values and functions of the system. This in turn requires increased appreciation of the different values, and a realisation that decisions can not be based on single-value systems such as irrigated agriculture alone. Management of these systems implies a careful choice between preservation, exploitation or, in appropriate situations, conversion to other uses.

6   References

1.   Saenger, P., Hegerl, E.J., and Davie, J.D.S., (eds) (1983) *Global Status of Mangrove Ecosystems* Ecology Paper no. 3, IUCN, Gland

2.   Chan, H.T., Ong, J.E., Gong, W.K. and Sasekumar, A. (1993) The socio-economic, ecological and environmental values of mangrove ecosystems in Malaysia and their present state of conservation. Pp. 41-81 in Cough, B.F. (Coordinator), The Economic and Environmental Values of Mangrove Forests and Their Present State of Conservation in the Southeast Asia/Pacific Region. *Mangrove Ecosystems Technical Reports*, Vol. 1, International Society for Mangrove Ecosystems.

3.   Ruitenbeek, H.J. (1992). *Mangrove management: an economic analysis of management options with a focus on Bintuni Bay, Irian Jaya*. Environmental Management Development in Indonesia Project (EMDI) Environmental Report No. 8. Jakarta, Indonesia.

4.   Franks, T.R.(1997) *Managing natural resource systems: the study of mangroves*. New Series Discussion Paper No. 72. Development and Project Planning Centre, University of Bradford. ISBN 1 898828 80 6 pp 1-14

5.   Lim, H. F. (1996) *Socio-economic impacts of Merbok mangrove ecosystem utilisation: some preliminary findings*. Seminar on sustainable utilisation of coastal ecosystems for agriculture, forestry and fisheries in developing regions, organised by the Japan International Research Centre for Agricultural Sciences, Penang, Malaysia.

6.   Sungai Petani City Council, (1996) *Structural Plan of Sungai Petani 1990-2010*.

7.   Mohd Shahwahid H. O. and Nik Mustapha, R. A. (1991) *Economic valuation of wetland plant, animal and fish species of Tasek Bera and residents' perceptions on development and conservation*. Asian Wetlands Bureau Publication No. 77. Kuala Lumpur.

8.   Ramsar Convention on Wetlands (1990) *Guidelines for the Implementation of the Wise Use Concept*

WATER, CONFLICT AND THE ENVIRONMENT: A CASE STUDY FROM TANZANIA

Conflict over water in Tanzania

M.A. BURTON

Director, Institute of Irrigation and Development Studies, University of Southampton, Southampton, United Kingdom.

C.K. CHIZA

Formerly Team Leader, Rehabilitation of Traditional Irrigation Projects, Moshi, Tanzania.

Abstract

The paper summarises issues related to water resources management and use in northern Tanzania, focusing on the conflicts arising from development of irrigation schemes. Increasing population is placing severe pressure on available water resources in the region. There is conflict over water, the incidence of which is expected to increase unless effective measures are taken to manage the resource. Development of the water resource for irrigation, though apparently beneficial to support populations in arid environments, is seen to be potentially high risk for the environment unless such developments are adequately planned and managed.

Keywords: Conflict, environment, irrigation development, northern Tanzania, waterlogging and salinity, water resources planning, water scarcity.

1   Introduction

Arusha, Kilimanjaro and Tanga regions in northern Tanzania (Fig. 1) are predominantly semi-arid. Exceptions are the lands on the slopes of Mount Meru, Mount Kilimanjaro and the Usumbara and Pare mountain ranges. The annual average rainfall in the area falls from 2000 mm on the highest parts of Mount Kilimanjaro to below 400 mm in the surrounding plains.

Traditionally irrigation has been practised in the mountain areas, with irrigated agriculture extending to the lowlands during the colonial period for estate production, and more recently for smallholder cultivation arising from land shortages on the more

Water: Economics, Management and Demand. Edited by Melvyn Kay, Tom Franks and Laurence Smith. Published in 1997 by E & FN Spon. ISBN 0 419 21840 8.

fertile and well water mountain slopes. The lowland areas have traditionally been used for grazing and cultivation of rainfed maize and beans. There is increasing interest in irrigation of maize and rice in the lowlands to support the growing number of lowland settlements.

The demand for rural and urban water supplies is increasing, as is the need for electricity from the three hydroelectric power stations on the Pangani river (Fig. 1).

Water supplies, arising predominantly from the well watered mountain ranges, are limited in quantity and quality. The demand for water is leading to competition and conflict between water users. Historically, management of the water resource has been by village committees, though their jurisdiction has been limited to the village irrigation systems. In the upland areas this management has been robust and effective, in the lowlands less so. As the pressures on water supplies increase there becomes a need for management of the whole catchment, a task that has recently been taken on by the Pangani Basin Water Office, who have formalised water rights and introduced a water user fee.

This paper describes some of the conflicts that have taken place in the region to illustrate and emphasise the pressing need for an effective integrated catchment management strategy in an area of increasing water scarcity.

2   Background

The main catchments in the area are those of the Pangani and the Mkomazi rivers (Fig. 1). The Pangani rises in Arusha and Kilimanjaro Regions and flows through Tanga Region on its way to the Indian Ocean. The Mkomazi catchment is shared by Kilimanjaro and Tanga Regions. Numerous streams flow off the mountain ranges, which have, and are still being, developed for cultivation by traditional village societies. Many of these traditional irrigation systems are reportedly over 200 years old.

In contrast large scale development of the water resource for irrigation has been relatively recent. During the colonial period settler farmers developed the resource for estate cultivation of crops such as coffee and sugarcane. Successful cultivation of coffee was not restricted to the settler population however, the number of African coffee growers increased from some 3,300 in 1923/4 to over 87,000 by the late 1960s, involving 90% of rural households [1],[2]. Government became involved in assisting furrow construction in the 1930s, and subsequently with larger government initiated schemes such as at Mombo (220 ha,1966), Mto wa Mbu (2500 ha,1986), Lower Moshi (2270 ha,1987), Ndungu Rice Scheme (940 ha,1989) and Kitivo (640 ha,1990).

Irrigation in the regions can be divided into estate and smallholder schemes, with smallholder schemes totalling some 58,000 ha and estates 8,500 ha [3]. Cropping patterns vary with location, estates tend to grow sugarcane, coffee or rice whilst smallholder schemes have mixed cropping patterns; at the higher elevations coffee, bananas and vegetables predominate, in the lowlands maize, rice, beans and cotton are grown.

The majority of smallholder schemes are traditional irrigation schemes developed and managed by farmers. These schemes have rudimentary control systems but, in the upland areas, sophisticated procedures overseen by the village elders. In the more recently developed lowland areas the management structures and procedures are less well defined and adhered to.

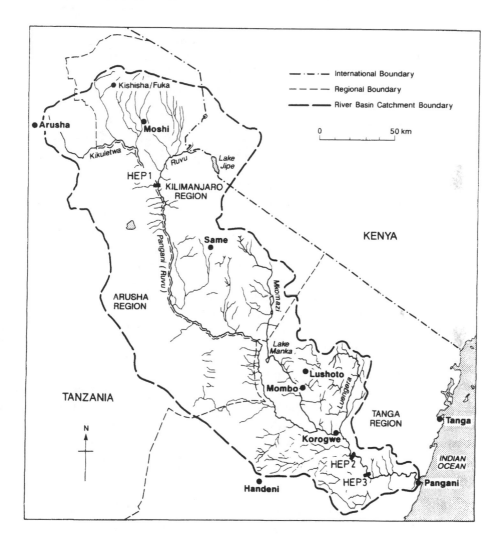

Fig. 1. Pangani River Basin and the boundaries of Arusha, Kilimanjaro and Tanga Regions.

Irrigation development in the Regions is controlled by the Regional Agricultural Development Offices of the Ministry of Agriculture and Livestock Development. Separate Water Offices of the Ministry of Water, Energy and Minerals are responsible for all matters pertaining to the use of water, i.e. granting of water rights, generation of hydroelectric power, etc. This division has marginalised the Irrigation Offices in matters related to irrigation water management.

There are three hydroelectric power stations on the Pangani river, one at Nyumba ya Mungu (HEP1, Fig.1) just south of Moshi, and two further ones at Hale (HEP2) and Pangani Falls (HEP3) just south of Korogwe. These power stations compete with irrigation for water resources, as they did in 1956 when the General Manager, Dar Es Salaam and District Electric Supply Company wrote to the Moshi District Commissioner requesting that there should be no further irrigation development on the Pangani since at times of low flow water levels at Pangani Falls were below critical (Tanzania National Archives, File W.1/13). The current interest in improving basin wide water management in the Pangani catchment is as a result of the recent rehabilitation of the Pangani Falls HEP by the Nordic Power Group funded by NORAD. A precondition for the project was the establishment of the Pangani Basin Water Office to ensure adequate supplies are available for the power station.

3   Water resources development

Within the three regions there are limited opportunities for further irrigation development, certainly on a medium to large scale [4]. Recent government-supported initiatives have been characterised by a failure to achieve target irrigated areas, constraints on the water resource, and salinity hazards. The planned 2500 ha rice scheme at Mto Wa Mbu to the west of Arusha, despite 12 years support from the International Labour Office, has succeeded in cultivating just over 1000 ha and is having serious problems with maintenance of the infrastructure; the planned 940 ha Ndungu Rice Scheme has had to be cut back to 640 ha due to limited water resources and reportedly has a salinity problem; the 1,510 ha NAFCO Kahe Estate just south of Moshi and the 6,480 ha TPC Sugar Estate (Fig. 2) are both suffering from reduced area of cultivation due to waterlogging and salinity [3].

There are a variety of issues facing water resources development which include:

• conflict over water: between water users themselves, between water users and government agencies, and between technical personnel and politicians
• lack of knowledge of water resources availability and projected demand
• inadequate financial and human resources for effective management by government agencies
• development of irrigation schemes with inadequate feasibility studies
• problems with salinity and waterlogging.

These issues are illustrated and discussed below in four case studies.

3.1   Case 1 - Lower Moshi irrigation scheme

The Lower Moshi Irrigation Scheme has been operational since 1987. It was developed for a total area of 2270 ha, 1100 ha of which was for rice cultivation, though by 1990, due to a shortage of water, the maximum area cultivated to rice was only 518 ha.

The scheme was designed by international consultants and constructed by international contractors, with relatively little active participation by the farmers. Land was appropriated, consolidated and handed back to the farmers once the irrigation and drainage system had been laid out and the plots levelled. Agricultural activities have been tightly controlled and supported by management, resulting in a rigid and productive cropping pattern, with rice yields of 6 - 6.5 tonnes/ha.

Paradoxically one of the causes of the water shortage in the River Rau has been due to the success of the project itself. Having seen the level of production on the scheme farmers upstream developed an area of approximately 300 ha, without assistance or sanction from government. This development was not welcomed by the project authorities as it reduced the water supply available to the Lower Moshi Scheme. A proposal was made in 1989 that the offtakes for this "unauthorised" development be gated in order to regulate water abstractions from the river. In 1994 the Pangani Basin Water Office commissioned contractors to construct gates, farmers chased them away and broke the gates. Only after protracted discussions between PBWO and the farmers were the gates finally constructed.

This action by the farmers upstream of the Lower Moshi Scheme was, however, no different to the behaviour exhibited by the Scheme itself towards downstream traditional schemes, whose water supply was cut off by the construction of the scheme. In 1989, two years after the Lower Moshi Scheme opened, farmers downstream of the scheme were looking for financial support to construct a weir on the nearby Mue river (Fig. 2) and divert water via a canal to the Rue river [4].

The disenfranchising of the traditional water users downstream of the project is difficult to accept. In a related feasibility study, in a section discussing water rights, a

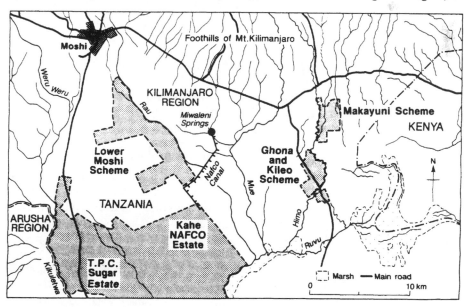

Fig. 2. Location of the Lower Moshi Irrigation Scheme

statement is made that "Other irrigation systems, including all traditional furrows, have no legal right" [4]. There does not appear to be any associated statement discussing the rights of traditional users and the consequences of ignoring such rights.

There is now pressure on government to further develop the water resources to make up for the shortfall created by this relatively large scale development. The first shortfall is to the downstream traditional water users, the second to the scheme itself to fully utilise the developed irrigation infrastructure. There is the danger that additional interventions may cause further problems for other irrigation communities in the vicinity.

3.2   Case 2 - Kishisha/Fuka furrow

The Kishisha/Fuka furrow on the western slopes of Mount Kilimanjaro (Fig. 1) irrigates approximately 200 ha. Water is diverted from the Fuka river to the irrigation area via a 7 kilometre contour canal (furrow[1]). In the mid-1980's the Water Office constructed a potable water supply intake on the furrow to divert a sizeable portion of the discharge for domestic water use in the plains below.

This action incensed the farmers as it reduced the water available for irrigation. They maintained a running argument with the Water Office, closing the 30cm diameter pipe in times of water shortage with a plastic cover. To overcome the problem in the longer term the farmers sought out another stream higher up the mountain and constructed a supplementary furrow to join the Kishisha/Fuka. Whilst the village committee were insistent that the new source was an unused spring, it is clear from ordinance survey maps that it was the headwaters of the Liwati river. The Irrigation Office appeared powerless to sanction or prohibit this unilateral action by the Kishisha/Fuka farmers to abstract water from another catchment.

3.3   Case 3 - Mkomazi valley development

Development of the Mkomazi valley for irrigation has been a subject of protracted discussion between politicians and technical personnel. At issue is the suitability of the valley for irrigation development.

In 1976 the Tanga Water Master Plan [6] identified a portion of the Mkomazi Valley in Tanga Region as having a potential irrigable area of some 10,000 ha, with supplementary irrigation water being diverted from the Pangani river. A 15 month feasibility study was mounted between June 1980 and September 1981 with the express intention of finalising the area for development and producing outline plans and cost estimates. Unfortunately, as a consequence of their detailed studies, the consultants found the valley to be unsuitable for irrigation development at the present time due to a combination of poorly drained soils and salinity hazards with both soils and water [7]. An additional hazard was the plan to use Lake Manka as a storage point for water from the Pangani on route to the Mkomazi. This "lake" is a salt marsh [4].

However, despite this technical recommendation local politicians were insistent that irrigation development be pushed ahead, arguing that they could see no problems as farmers were successfully cultivating in the area on a small scale. In 1984 they allocated funds and commenced excavation of a large channel to divert water from the

Pangani river through Lake Manka and on to the Mkomazi Valley. Only prompt action by the Zonal and Regional Irrigation Offices halted this hazardous venture.

3.4   Case 4 - Selela Scheme

The Selela development is an example of the conflict between irrigation and the environment. The village of Selela is located at the foot of the Rift Escarpment some 20 km north of the town of Mto wa Mbu, to the west of Arusha. At this location several streams discharge from the Escarpment face, supporting a riverine forest and marshland at its base. The forest and marshland support wildlife and provide all-year round grazing for Masai cattle in the otherwise arid Rift Valley.

The villagers of Selela have established a small irrigation system of some 40 ha near the village for bananas, maize, bean and vegetables. They want to develop the area along the lines of that for the Mto wa Mbu irrigation scheme, drain the land and divert the streams to provide irrigation for rice. Such a development would deprive the riverine forest and the marshland of their water supplies, thus putting their existence at risk and creating conflict between the villagers, the pasturalists and the natural environment.

In 1993, at the request of the District Irrigation Office, a proposal was submitted to DANIDA for a small budget to carry out a topographic survey of the area. Though the area requires a full environmental impact assessment [4] this was not included in the request as it was considered too expensive to fund. Thus development of the area may proceed with inadequate studies; issues of environmental impact, water quantity and quality, will all be omitted due to a lack of sufficient funds and adequately trained personnel. Experience in the vicinity (Mto wa Mbu with saline/sodic soils developing and Engaruka where yields are reported to be dropping due to "tired soils" [8]) is such that it would appear shortsighted to develop Selela without adequate investigation.

4   Discussion and conclusions

The above examples of irrigation development in the area have been used to try to give an indication of the situation in northern Tanzania with regard to water resources development and irrigation. The picture that emerges is one where such water resources development is potentially hazardous and where careful management is required.

In the case of the Lower Moshi Scheme there is an incomplete picture of water supply and demand, and a failure to develop a strategy for the Rau river catchment. Traditional water rights appear to have been ignored in the calculation of water availability, adequate records have not been kept of river flows and abstractions, and development appears to have proceeded in a piecemeal manner, creating further problems which then need further investment to resolve.

As exemplified by the Kishisha/Fuka furrow case there is conflict between irrigation and water supply for domestic and industrial use. Again there has been little apparent strategy for water resource development between the two sectors. Certainly in the future the pressure for domestic water supply can be expected to increase, and it will be at the expense of the irrigation sector. The latter will have to increase its efficiency of water utilization if it is to maintain its levels of agricultural production and farmer livelihood.

The conflict between different interest groups is shown in the proposed development of the Mkomazi valley and the Selela scheme. The issue raises the important point of technical versus social and political assessment of development need. In the case of the Mkomazi valley the proposed development was large enough to warrant a full technical study to assess the suitability for development. The conditions were found to be potentially hazardous in the long term and the project has not gone ahead, mainly because funds for large scale development could not be secured. However, in the case of Selela irrigation development is more likely to proceed as the technical input and investment required are of a lower order. In these circumstances the limited nature of the investigations at the planning stage could prove hazardous in the long term.

As a consequence of population pressure the natural resources are being placed under severe pressure in Northern Tanzania. Land and water resources are being committed to developments which may not prove sustainable in the long term. The pressure is such that measures will have to be taken to manage the resources if a degree of chaos and anarchy is to be avoided.

5   References

1.   Maro, P.S. (1974) Population and land resources in northern Tanzania: The dynamics of change 1920-1970. *Unpublished PhD Thesis*, University of Minnesota.

2.   Groves, A. (1993) Water use by the Chagga on Kilimanjaro. *African Affairs*, 92, pp.431-448.

3.   FAO. (1987) Mission Report, Institutional Support for Irrigation Development Project (Preparatory Phase), Food and Agricultural Organisation of the United Nations, Rome.

4.   Burton, M.A. (1989) *Irrigation development potential for northern Tanzania for the period 1990-2000.* Report submitted to the Government of Tanzania and the Food and Agricultural Organisation of the United Nations, Dar Es Salaam, Tanzania.

5.   JICA. (1983) Feasibility study of the Mkomazi Valley Area Irrigation Development Project, Japanese International Cooperation Agency, Tokyo.

6.   Agrar und Hydrotechnik. (1976) *Tanga Water Master Plan*, Agrar und Hydrotechnik Consulting Engineers, Essen, Germany.

7.   IHRC. (1982) *Lower Mkomazi Irrigation Project, Tanzania - Feasibility study.* Instrupa Hydroplan Rodeco Consortium, Bonn, Federal Republic of Germany.

8.   Bertelsen, P. (1995) *Farmer managed irrigation systems, Manyara Division, Rift Valley, Tanzania.* Working Paper, Institute of Development and Planning, Aalborg University, Denmark.

1. A furrow is the local terminology for a supply canal from the river to the irrigation command area. Due to the nature of the topography it is often a contour canal along the hillside.

SECTION D

PAYING FOR SERVICES

COSTS, BENEFITS AND RATES OF IRRIGATION WATER IN ITALY

L. NOLA
National Association for land reclamation and irrigation, Rome, Italy

Abstract

In Europe, Italy has the greatest amount of land equipped for irrigation, amounting to 3.6 million hectares. Irrigation in Italy is carried out in districts of drainage and irrigation by boards called Land Reclamation Consortia. The system adopted by the Consortia to share the costs of irrigation between farmers is that of fixing irrigation rates directly, depending on the benefits the farmer receives. Irrigation benefits and different forms of irrigation rates are explained in this paper.
Keywords: irrigation rates, Land reclamation Consortia, monomial and binomial rates, pressurised or flow irrigation, supply of water, water use optimisation.

1   Introduction

Farmers in Italy have long understood the economic importance of water for irrigation and the need to pay for it. In fact, as early as the eleventh century, spontaneous associations of land owners existed in the Po Valley, building and maintaining drainage and irrigation works and sharing the work and the costs between them. This system continued for a long time, until gradually the spontaneous associations were transformed into "Land Reclamation Consortia" (Consorzi di Bonifica) in which every farmer living within the district of the Consortia was called upon to share the cost of irrigation, depending on the benefits he received.

Before considering the central subject of this paper, regarding irrigation costs, benefits and rates, it is important to explain how the tradition of irrigation has evolved over the centuries and how it functions through associations of owners joining together in Consortia.

It is also interesting to note the findings of the World Bank, which states that in developing countries around 55% of channelled water is lost before reaching the fields to be irrigated. The cause of this problem, the World Bank maintains, is that in most developing countries, the management of irrigation networks is in the hands of the Pubic sector. The World Bank suggests that these countries adopt the model

Water: Economics, Management and Demand. Edited by Melvyn Kay, Tom Franks and Laurence Smith. Published in 1997 by E & FN Spon. ISBN 0 419 21840 8.

(established in Italy with Law n. 215/1933) of Land Reclamation Consortia: public bodies which represent the interests of the farmers who also have to manage and use the irrigation systems. A correct mix of public and private sectors may, in this way, enable greater utilisation of the ever decreasing water resource for irrigation.

The following pages are therefore dedicated to explaining the Italian organisational model of the Land Reclamation Consortia.

2   Land Reclamation Consortia and their strategic role for irrigation

The modern Italian landscape is not purely the result of nature but that of a patient transformation achieved over a long period of time. Indeed, historically, essential land reclamation, drainage and irrigation in Italy have been central features marking the rural landscape.

The Italian irrigation network is an absolute necessity and today comprises of about 3.6 million hectares of irrigable land (land equipped for irrigation), representing an indispensable contribution to Italian agriculture. Italy has 30 million hectares of land, of which 35% is mountainous, 42% hilly and 23% plains. A large part of the few existing plains (about 6 million hectares) needs to be supplied with a drainage system. This is because in the past, extensive land transformation took place, with the development of drainage and irrigation networks that required large-scale works and enormous capital investment [1].

Water drainage began in Italy during the Roman Empire and this activity received a further boost during the Middle Ages when numerous monasteries, especially those built in Northern and Central Italy, continued this work, completing the reclamation of the marshlands with irrigation.

Near Milan, in Lombardy, the use of the River Vettabia's water by the Cistercian monks of the Chiaravalle Abbey can be traced back to the eleventh century (as Prof. Medici reminded us at the ICID conference in Morocco, 1987). It is possible that those lands were the first to create *marcite* (water-meadows): stable irrigous meadows which produced forage during the rigours of winter, thanks to the use of spring waters whose temperature was considerably higher than the outside temperature, thus enabling vegetation to grow and fresh grass to be cut, when the usual way to feed cattle was with dry forage.

In the same period (XI-XII centuries) free associations of land owners sprang up, with the aim of draining the marshlands and irrigating them. These associations, which were later called Consortia, were based on the principle of sharing expenses in proportion to the benefits gained; detailed documents regarding these Consortia can be found in many towns in the regions of Lombardy, Veneto and Emilia.

Over the centuries drainage and irrigation works were also carried out in other areas with voluntary contributions from landowners joined together in Consortia. Finally, in 1865 the Civil Code gave formal recognition to the Consortia as private bodies freely formed by land owners interested in exploiting and conserving the water resources. Subsequent legislation reflected a growing interest in the water sector and as a result the State became increasingly involved in financing works carried out by the Consortia [2].

At the same time, the concept of the public nature of water was established in 1993 (R.D: n. 1775/1933) with the constitution of a basic priniciple, according to which all water which is, or could be made, available for uses of general public interest is to be considered public (it must be stressed that in 1994, law n. 36 stated that all water, whether surface or ground, is public).

In 1933, law n. 215 was passed regarding multi-purpose land reclamation; it stated that land reclamation was of both public and private interest. The Land Reclamation Consortia are those organisations where public and private interests are brought together to provide the synergy necessary to achieve multi-purpose land reclamation. A Land Reclamation Consortium is a public body with an associative base.

Land owners, tenants are real estate owners within the districts of Land Reclamation Consortia are obliged to belong to the organisation. Every member of the Consortium is called upon to bear the costs of managing public land reclamation work and the cost of the services provided by the Consortium, in proportion to the benefits received.

In Italy the results achieved by Land Reclamation Consortia are as follows: 32,400 km of roads, 5,139 km of water supply systems, 38,000 km of drainage canals, 1,300 pumps for raising water, 168 dams and weirs. An area of 3.6 million hectares has been made irrigable.

Having completed the drainage works and created an infrastructure in their districts, the Consortia have gradually abandoned the field of civil works (roads, and water and power supply systems). They are now mainly focusing their attention on the planning, execution, operation and maintenance of works to safeguard the countryside from flooding and to utilise waters for irrigation [3].

In Italy there are now 186 Consortia, covering a total area of some 14.8 million hectares (about 49% of total area in Italy). They still retain their identity as public bodies, whose directors are elected by all the landowners who benefit from land reclamation activities in Consortia districts, and who contribute directly (700 billion lire in 1995) to the payment of Consortia expenditure for the maintenance and efficient operation of land reclamation and irrigation works.

3   Irrigation rates and costs

Let us now turn our attention to irrigation works in Italy and the costs and rates of irrigation water.

As mentioned above, there are 3.6 million hectares of irrigable land in Italy. Of these, 2.7 million hectares (75%) are served by irrigation systems built by Land Reclamation Consortia or other public bodies. The remaining 25% (equivalent to 900,000 hectares) are served by irrigation systems built privately, which use water raised from wells or acquired by means of small deviations from streams and rivers. The water distributed has various origins: it is estimated that about 70% comes from surface water sources (canals, rivers, etc.), 4% from lakes and 26% from spring waters and wells.

It has been calculated that in Italy more than 40% of the gross value of agricultural production comes from produce cultivated on land equipped for irrigation; the remaining 60% is obtained from water sources coming from rain. Furthermore, agricultural products obtained using irrigation waters account for two-thirds of the value of exports.

These facts illustrate the strategic importance of irrigation in Italian agriculture. The technological evolution which has characterised the last century has had a significant effect, leading to the use of more effective irrigation techniques, while at the same time stimulating interventions on an increasingly wide area of land. In the twentieth century, indeed, the irrigable area in Italy has tripled, from around 1,300,000 hectares in 1900 to 3,600,000 hectares today.

This extension in irrigation has been made possible thanks to an investment policy in which the State has taken an increasingly important role, both in direct and indirect measures, to finance the creation of irrigation systems, such that today it covers up to 100% of the costs.

A policy in favour of irrigation must, however, pay constant attention to costs. Indeed, in irrigation there is the problem of the so-called price of water, that is to say, payment by users of the costs of the irrigation service. It should be made clear that for irrigation, users are obliged to pay the related management costs by legislative norms, according to which the expenditure incurred for maintenance and management of irrigation works are the responsibility of the Consortia members who gain benefit from them.

To this end, the Consortia that manage the irrigation are given a specific power to oblige members to pay, and exercising this power allows the Consortia themselves to recover what they have spent for irrigation management. The system of obligatory payments is based on the division of expenditure between users, in proportion to the benefits they receive from irrigation.

The measure of benefit is calculated according to indices, that the individual Consortia must determine with a special act, provided for by law, called the classification plan. A series of factors affect how the benefit is determined and these vary from Consortium to Consortium: such as, for example, the supply of water per hectare, and the way in which the water is distributed, with flow or pressurised systems, with water delivery on demand or in shifts.

A study, aimed at providing the Consortia with a unitary scheme for drawing up classification plans was set up at ANBI (National Land Reclamation Association) in 1989. This research led to the identification of specific benefit indices for dividing up expenditure for irrigation. It also emerged from the study that, though respecting general unitary criteria, the benefit indices could vary significantly according to different situations. However, the fundamental principle on which the system of obligatory contributions for irrigation is based is that whereby the Consortia divide up all expenditure for the irrigation service between the beneficiaries.

The shared expenditure can be broken down into the following elements: the water concession fee; the quota for the shared management of reservoirs, or catchment points and water channels; ordinary maintenance of the irrigation networks; labour and materials (lime, cement, iron, timber, etc.); hiring and use of machinery (tractors, excavators, etc.); water distribution (raising water, if necessary);

labour; use and maintenance of pumps (energy, fuel, lubricants, etc.); administration costs and general costs which can be attributed to irrigation.

Not all elements appear in all Consortia accounting; there are in fact Consortia that do not have to pay concession fees for water, others which do not have costs for managing reservoirs, (in that the water comes from surface sources), others again that do not incur expenditure for raising water.

In the Consortia, the irrigation rates can be one of two types: monomial and binomial [4]. In the monomial rate, expenditure for irrigation is divided up equally between all the Consortium members served by the irrigation system, independently of water consumption.

The binomial rate, on the other hand, has two factors: one for maintenance expenditure, which is divided up between all Consortium members served by the system, and the other for the cost of use, which is divided up between members of the Consortium according to their actual use.

The binomial rate has important implications for water policy, as it could become an instrument for regulating consumption, given that it is possible to divide the burden of costs between the fixed quota and the variable one, so as to stimulate consumption to reach the optimal quantity and to avoid waste, by penalising excessive consumption.

Generally, the binomial rates are considered the most modern method of payment; in particular, given the possibility of graduating the burden of costs, especially at the start of irrigation plans, these rates prevent the cost of the system falling entirely upon those users who start to use the water first. Moreover, the existence of a fixed quota encourages a broader use of irrigation by those who, though not using the water, are still required to share in the cost [5]. In Italy, the binomial rate is used above all in some areas of the South, where the irrigation systems are more recent.

Below are two examples to show how the Consortia apply the two rates: a) monomial and b) binomial [6].

a) Example of the monomial rate

A Northern Italy Consortium serving areas with flow irrigation and areas with pressurised irrigation (rates refer to 1990)

    Area with flow irrigation: fixed rate   Lit.130,000 per hectare

    Area with pressurised irrigation: fixed rate   Lit.145,000 per hectare

b) Example of binomial rate

A Southern Italy Consortium serving areas with flow irrigation and areas with pressurised irrigation (rates refer to 1990)

    Area with flow irrigation   fixed rate   Lit.42,000 per hectare

                 rate for consumption   Lit.70,000 per hectare

    Area with pressurised irrigation: fixed rate   Lit.58,000 per hectare

                 rate for consumption   Lit.100,000 per hectare

4   Conclusions

The law makes the Land Reclamation Consortia responsible for maintaining and using the irrigation and land reclamation systems. Every system that has been built in the past needs to be conserved and updated in the light of new technology and new needs. Currently, the need to manage irrigation in the light of water use efficiency principles is a pressing concern. This means trying to increase the efficiency of the irrigation systems so as better to use the water that, though available today, will be increasingly scarce in the future.

In addition to water shortages water quality is also worsening, as regards salinity and pollution, in some areas. For this reason, Consortia are now increasingly accompanying the traditional activities of land reclamation and irrigation with innovative initiatives aimed at monitoring water quality and the safeguard of surface and ground water. In fact, since the Consortia are present all over Italy, they have for some time now been collecting and compiling a considerable mass of data on water, for both surface and ground water. As regards the levels of underground water, there are a great number of survey sample points systematically used by the Consortia. These surveys have enabled some Consortia to draw up mathematical models of the underground water.

In this way, the Consortia system shows once more how effective it is, based on solidarity and on the fact that irrigation system managers are the very same farmers who have to use the water for irrigation; it is, therefore, in their direct interests to adopt a policy of rational water utilisation and protection and conservation.

In many areas of Italy, the irrigation networks are in need of urgent functional interventions and automation, because of their age, so that water may be used more efficiently and rationally, thus also protecting the resource. Moreover, in southern areas, the possibility of completing systems that are only partially completed (e.g. multiple reservoirs) would make even greater quantities of water available for irrigation.

However, programmes for modernising and completing the irrigation network are greatly hampered by the fact that, at national level, due to the current policy of reduced public spending, funding for such work is increasingly scarce, while, at Community level, investments in the irrigation sector are considered with a certain diffidence. This diffidence stems, above all, from a fear, at Community level, that investments for completing or extending irrigation systems could result in an increase in surplus agricultural production.

The EU attitude, in reality, still reflects a mentality which sees irrigation as an instrument for increasing the quantity of agricultural production, whereas irrigation today is, above all, an instrument for improving the quality of agricultural products and for diversifying and directing production according to market demand.

It should also be pointed out that, while Community funding in Italy is currently limited to modernisation of existing systems, above all aimed at water savings and to prevent increases in irrigable area, in other Mediterranean countries such as Greece, Portugal and Spain, which come under the intervention area of the Cohesion Fund (from which Italy is excluded) the Community finances new work for water storage (storage reservoirs) and the construction of irrigation networks.

Recently the European Parliament has, however, taken a firm stand, and emanated two Resolutions (16th March 1995 and 13th July 1995) aiming to gain recognition of the structural character of drought in some Mediterranean countries and requesting the

Commission of the EU to intervene, not only with emergency aid, but also by formulating plans with the member states involved to guarantee national utilisation of water resources and to modernise the water networks and irrigation systems to optimise the use of such resources.

5   References

1.   Medici, G. (1985) *L'irrigazione in Italia*, Edagricole, Bologna.
2.   Martuccelli, A.M. (1987) Waters and irrigation plants managed in common: the Consortia, *proceedings of XIIIth international ICID Congress at Casablanca*, edited by ITAL-ICID Committee
3.   Nola, L. (1987) Multipurpose Land Reclamation in Italy, *proceedings of 11th Seminar of European Association of Agricultural Economists*, Wissenschaftsverlag Vauk, Kiel
4.   Associazione Nazionale Bonifiche (1981) *Costi e prezzi dell'acqua*, Edagricole, Bologna.
5.   Bergmann, H.K. (1973) *Guide de l'evaluation èconomique des projets d'irrigation*, OCDE, Paris.
6.   Associazione Nazionale Bonifiche (1992) *L'uso irriguo delle acque*, Edagricole, Bologna.

THE VALUE OF WATER IN COMPETING USES
Irrigation and the environment

JOHN J. PIGRAM
Centre for Water Policy Research University of New England, Armidale NSW 2351
Australia

Abstract
The significance of water can readily be appreciated when some of its many diverse functions are considered. However, the role and function of water, and hence its value, have to be reviewed dynamically in keeping with fluctuations over time, as a reaction to changes in economic, social and technological conditions.

In the Australian context, the dynamic character of water and water values can readily be demonstrated by reference to an emerging range of functions identified with a particular stream or waterbody through time. For example, a river, perhaps initially valued merely as a convenient water supply, may subsequently acquire a series of functions as a transport and navigation route, a source of hydropower, a means of waste disposal, a resource for outdoor recreation, tourism, and nature conservation, or a source of water for industrial purposes or irrigation.

The dichotomy between economic and environmental values of water underlines the need to quantify trade-offs involved in allocating water to satisfy societal values associated with non-consumptive functions of the resource. The challenge is to use water for irrigation in a sustainable manner which respects and maintains environmental values in harmony with economically productive forms of resource use.
Keywords: Water resources, water values, irrigation, environment.

1   Introduction

The value of water can be approached from several perspectives and its significance as a resource can readily be appreciated when some of its many diverse functions are

Water: Economics, Management and Demand. Edited by Melvyn Kay, Tom Franks and Laurence Smith. Published in 1997 by E & FN Spon. ISBN 0 419 21840 8.

considered. These include water as:

- A geomorphological agent helping to fashion the face of the earth
- A vital input to biophysical processes
- A means of transport of materials and nutrients, and the disposal of wastes
- A component of many productive processes both in agriculture, industry and energy generation
- An integral part of nature and a focus of environmental interest

Water has been an enduring force throughout history. The earliest civilisations grew up around the great river basins of the earth and the relatively secure existence which that situation made possible. The collapse of those civilisations, too, has been linked to their inability to manage water in an ecologically sustainable manner. In today's world, water maintains its value to society for the attainment of socially valued goals. Yet, conflict over water is a recurring phenomenon reflecting its functional diversity and the dynamics of water values relative to economic potential, sociocultural forces and technological change.

Given this inherent functional diversity and the contrasting attitudes which go with it, it should not be surprising that conflict arises between those who assess the worth of water in ecological and aesthetic terms, and those whose attitudes are shaped by more utilitarian and materialistic motives. The basis for much of the conflict is in the value put on the same resource at the same time for contrasting purposes – on the one hand, economic, on the other sociocultural and ecological.

The potential for conflict can be demonstrated in an Australian context by examining the competing claims on water for irrigation and for the environment. The discussion begins with a comment on economic values and non-market values of water.

2   Economic value of water

From an economic perspective, the value of water is derived from the contribution it makes to production processes, for example, in industry or irrigation. This value is reflected in what users would be willing to pay for the resource in a particular use. The economic value of water is a function of relative scarcity and is strongly affected by quantity, location of use, and timing of demand. Quality is also important because it affects the range of uses to which water can be put.

Scarcity of water impacts on users through shortages of supply or deterioration in quality, usually triggering a regulatory response from governments to manage the dwindling resource. Economic instruments, e.g. prices and market mechanisms, also signal water scarcity and can be helpful in allocating supplies among users. Economic incentives can change the pattern of use affected by scarcity, benefiting those users for whom the water has a higher marginal value and who are prepared to pay the price and put the resource to a higher valued use [1].

A problem with using price as a signal of water scarcity is that society does not place a consistently high or low value on water. Maintaining a constantly high price, for example, does not reflect the reality of changing water supply and demand conditions through time [2]. Information on economic values of water must always be indicative rather than absolute, and determining appropriate flexible prices is virtually impossible.

Despite this imprecision, an approximation of the economic value of water can be obtained with reference to its role in the production process. For example, an indication of the value of water in irrigation can be derived from the impact of losses of production during a period of water scarcity. In the Gwydir Valley of Northwestern New South Wales, drought shortages during the three years, 1991–94, seriously affected irrigated cotton production. Table 1 shows that the total value of output of cotton under irrigation in that period declined by over half a billion dollars. In percentage terms cotton fell from 46.4 per cent of regional output in 1990-91 to 30.6 per cent in 1993-94 [3]. Comparable declines were experienced in levels of value added, income, and employment in the cotton industry in the Gwydir region.

Table 1. Impacts of the Gwydir Cotton Industry on the Gwydir Region

	Direct Growers	Effect Ginning	Prod'n Induced	Cons'n Induced	Total Flow-on	TOTAL EFFECT	TOTAL REGION	% of Region
OUTPUT ($m)								
1990-91	278	319	91		139	735	1.564	47.0
1991-92	253	296	81		124	674	1.452	46.4
1992-93	98	117	72		102	317	1.158	27.4
1993-94	94	108	77		109	311	1.018	30.6
VALUE ADDED ($m)								
1990-91	155	24	61		89	267	635	42.1
1991-92	143	25	57		83	252	632	39.9
1992-93	-4	11	46		63	70	419	16.6
1993-94	-21	9	52		71	59	395	14.9
INCOME ($m)								
1990-91	23	6	27		46	75	248	30.0
1991-92	20	6	26		43	69	246	28.1
1992-93	10	4	23		35	48	246	19.6
1993-94	11	3	25		37	50	231	21.7
EMPLOYMENT (No.)								
1990-91	1,171	249	1,106		1,925	3,345	11,155	30.0
1991-92	1,031	279	1,010		1,725	3,035	10,641	28.5
1992-93	796	205	883		1,370	2,371	10,314	23.0
1993-94	726	142	944		1,449	2,317	9,703	23.9

Similar calculations show the value of water to cotton irrigation (or the value lost) in terms of impact per megalitre of water applied. The relative impact of the shortage of water (and hence its implied value) per hectare of cotton and per hectare of alternative enterprises, also demonstrates the repercussions of drought for the cotton industry in the region.

Whereas these effects arise from naturally occurring water scarcity, it is interesting to speculate on the magnitude of impacts on irrigation of what might be termed "political drought", i.e., water shortages as the result of political or bureaucratic decisions. Such effects underline the need to quantify tradeoffs involved in allocating water to satisfy non-market values.

3   Non-market values

When it comes to estimating the worth of non-consumptive functions of water, dollar amounts are not easily assigned to environmental values. Some would assert that, in societal and ecological terms, water is a priceless element, inherently too essential to be valued in economic terms or managed by market mechanisms [4].

From this perspective, not only is water valued as an integral component of the present environment, but this assessment can be reinforced by an "existence value". This represents the satisfaction derived from the assurance that the resource is being maintained intact, even though there may be no intention of putting it to immediate use. In the same way, the resource may have an "option value" or the value associated with retaining the option of future use. For others, the concern is for future generations and the "bequest value" of the water resource, reflecting collective concern for the continued existence of this valued attribute of the environment.

Taken together, option value, existence value, and bequest value, represent the overall worth which some sections of the community place on particular resources, in this case, water. Awareness of these resource values is important because they help explain concern over development of water resources sometimes expressed by those with no obvious direct connection with the specific issue. The strong opposition from mainland Australia in the 1980s to the hydropower proposals in southwest Tasmania is an example. The concern is heightened when proposals are seen to involve irreversible decisions which pre-empt alternative forms of resource use.

Accepting that such divergent values of water exist is one thing; quantifying the magnitude of the values involved is another. A number of approaches have been put forward to capture the monetary values placed on environmental amenities such as water and water quality [5]. However, there is a considerable degree of uncertainty and mistrust associated with estimates generated using these techniques.

A widely used survey technique, for example, is the contingent valuation method. This seeks the personal estimates of values by respondents for changes in environmental attributes, e.g. water quality, contingent upon a range of choices offered. The technique is based on participants' willingness to pay, or to accept compensation, for different environmental outcomes. However, there are potentially serious biases arising from the use of contingent valuation and the estimates derived should be viewed as broadly indicative, rather than knowledge-based.

Despite these qualifications, Izmir [5] believes that environmental valuation has a number of policy applications, primarily in the appraisal of development proposals with potential environmental impacts. Environmental valuation can also assist in developing and setting priorities for strategies focusing on protection of environmental values. It is important to note that the techniques should be seen as useful aids in informing the decision making process and narrowing the range of choices of alternative courses of action, rather than as ends in the themselves.

4   Reconciling market and non-market values for water

It is easy to dismiss environmental concerns over resource use as economically irrational and based on misconceptions or fear of change. Yet, despite non-market values being inherently intangible and not readily amenable to precise measurement, they are receiving serious attention in resource management decision making [6]. In Australia, environmental issues, and the societal values which underpin them, can no longer be ignored in decisions regarding the development of the nation's water endowment [7]. The challenge is to use water in such a way as to respect and maintain those values in harmony with productive forms of resource use. Indeed, this is the rationale behind calls for the sustainable use of resources including water. Sustainability seeks to avoid forms of development which carry the risk of irreversible outcomes, or impose unacceptable costs on future generations.

In conceptual terms, sustainable use of water resources is based on the notion of "a safe minimum standard", i.e. boundaries are set to resource use which should not be exceeded if irreversibility is to be avoided. However, this decision should not be determined entirely by ecological criteria any more than economic criteria alone. A balance should be sought between potential gains and losses from imposing limitations and conditions on the quantity and type of use of the water resource.

5   Water policy issues

The foregoing discussion underlines some of the complexities involved in developing water policy to satisfy competing objectives. Market mechanisms can go some way towards achieving desired outcomes in circumstances where enforceable property rights are defined for specified uses of water. However, in the real world, not all the functions of water which are of value to the community are subject to such well defined property rights. Therefore, policy makers are hesitant in endorsing a totally market-driven water sector, in the belief that some government intervention will always be necessary to correct for market failure and the delivery of trade-offs acceptable to the community. Environmental groups, too, question the role of market mechanisms in resolving conflicts over water [8]. Water markets may promote efficiency, but significant social and economic factors, along with equity considerations and political realities, constrain their unfettered operation.

On some issues, such as the "preservation-related values" noted earlier, sections of the community might look to the government for policy initiatives not subject to the disciplines of the market place. In this context, it might be expected that water policy decisions would be guided by stakeholder preferences as reflected in an efficiently functioning market, but constrained by community values which emerge through the political process (David Campbell; pers. comm.). This pragmatic approach recognises the existence of societal goals and values, and external obligations and constraints, as well as the achievement of economically efficient resource allocation, in the framing of public policy for water. The overall objective for water policy becomes the maximising of net economic and social benefit. Adoption of such a policy objective is entirely compatible with the use of market mechanisms for water management and allocation, where appropriate safeguards to accommodate non-market values can be incorporated.

That said, careful consideration must be given to the most effective means of implementing this policy objective. Rigorous, consistent and transparent processes are required if sound principles are to be put into practice and water management agencies held accountable for their performance. This approach implies:

- identification of primary stakeholders with equal weight given to benefits and costs imposed on all interests
- equal weighting to economic benefits and to alternative manifestations of value
- application of market mechanisms, suitably constrained, as appropriate
- consistency of purpose between water management agencies and related resource management bodies (David Campbell; pers. comm.).

6   Consultation not confrontation

Water management agencies in all states of Australia, along with the Murray-Darling Basin Commission, are now engaged in a range of measures to achieve efficient, sustainable use of water and more equitable sharing of the resource. Care will be needed in the process to ensure that impediments do not emerge at the implementation stage and derail the reform agenda.

Typically, there is a more effective and a less effective way of implementing reform; it is worth noting that the difference between "confrontation", "consultation" and "cooperation" is only a few letters. Consultation is a key feature of the Council of Australian Governments (COAG) principles to underpin implementation of reform and policy change, and the COAG Water Policy Agreement stresses the role of community involvement and public education in meeting the challenge of proper water management. If these principles and guidelines are observed, a smooth transition to a more efficient and equitable basis for sharing Australia's scarce water resources should be achieved, acceptable to all stakeholders and reflecting the values placed on the resource [9].

Through consultation, the knowledge, experience and commitment of water users and interest groups can be harnessed on a regional or even catchment basis. In this way, the water using community – urban, rural, and environmental – comes to "own" the problems and can identify the constraints to be overcome, and the incentives and sanctions which may be needed for their resolution. The community frameworks and consultative mechanisms already exist in the form of local and regional Landcare Groups and Catchment Management Committees. These organisations, in tandem with Waterwise and Rivercare initiatives, offer a constructive pathway to water reform, and to securing community support for community-based solutions [10].

7   Implications for the irrigation industry

In the process, Australia's irrigation sector will be called upon to adapt to a more complex water management regime. Increasing demand for water, along with drought-induced scarcity, and the reforms signalled under the COAG Water Policy

Agreement and the National Competition Policy, are bringing about significant changes in the allocation and use of water. In this evolving policy context, the irrigation industry can expect to:

- pay more for water
- allocate water to environmental uses
- explore further opportunities to trade water entitlements
- achieve even higher water use efficiencies and "do better with less" through adoption of best management practices in irrigation and drainage
- conform to more demanding environmental regulations

    In meeting these challenges, the industry should take the lead in adopting a professional approach to water use and management, documenting its claims on the resource in a credible fashion, and pursuing excellence in the practice of irrigation and drainage. This approach should ensure that the full range of values – economic and otherwise – placed on water in competing uses is observed. Moreover, the industry will demonstrate a positive and proactive commitment to rationalising conflicting claims on the resource, and to achieving consensus on the redistribution of water demand in time and space between existing and emerging uses and values.

8   References

1.   Birch, A. and Kulshreshtha, S. (1994) The value of water *Newsletter of the Canadian Water Resources Association*, Vol. 13, No. 1. Supplement 1–4.

2.   Dudley, N. (1995) Microeconomic reform and water use, presented to DPI Rural Outlook Conference, Twin Waters, November.

3   Powell, R. and Chalmers, L. (1995) Regional economy impacts of drought and the value of water, presented to the 7th Ministerial Water Forum, Sydney, October.

4.   Day, D (1996) Water as a social good, *Australian Journal of Environmental Management*, Vol. 3, No.1, pp. 26–40.

5.   Izmir, G. (1995) Valuing the environment, *Search*, Vol. 26, No. 10, pp. 304–8.

6.   Fisher, T (1995a) Water for the environment – what value? *Newsletter of the Water Research Foundation of Australia*, No. 325, pp. 1–2.

7.   Pigram, J. (1986) *Issues in the Management of Australia's Water Resources*, Longman, Melbourne.

8.   Fisher, T. (1995b) Environmental perspectives on managing the Murray-Darling Basin, presented to 7th Ministerial Water Forum, Sydney, October.

9.   Pigram, J. (1996) Water reform – principles and practice – the New South Wales experience, presented to I.I.R. Conference, Sydney, February.

10.   Millington, P. (1996) The Murray-Darling Basin, Australia, Keynote paper presented to Basin Management Context, Regional Consultation Workshop, Asian Development Bank, Manila.

PROBLEMS OF PAYMENT FOR WATER SERVICES IN RUSSIA

Ê. ARENT
The Moscow State Land and Nature Management University, Moscow, Russia

Abstract

Water as a natural resource does not always have a cost and is not a commodity as it is not sold freely in the competitive market. With the growth in water use the necessity for water economy aimed at increasing water stocks is becoming more urgent. This means that water acquires the character of commodity. The experience of introducing payment for water in Russia has shown the existence of economic and technical problems. To solve these problems it is necessary to introduce measures such as the installation of water meters as well as measures such as training. At present it is planned to introduce a water tax and tariffs to compensate for the state expenditures on maintenance of state water resources utilization systems. However the size of these payments does not correspond to the financial ability of the majority of water users to pay and they will need help from the state which will play a significant role in forming the water market in Russia. In this sphere it is also necessary to use the experience of countries with market economy.

Key words: payment problems, state role, transition to market, water payment, water user.

Still Adam Smith wondered about the low cost of water without which life would be impossible and the high cost of diamonds bringing no practical use. In contrast to using many other natural resources up to recently water use has been regulated mainly with consideration for political reasons and moral traditions. Too much importance was attached to water and so it could not become a simple object of sale and purchase.

It has long been known that people paid water-carriers for drinking water. At present, payment is collected both for supply of water to its consumers (potable, technical or irrigation water) and for its removal (drainage, sewage water), as well as for waste water treatment or for preparation of special water (bottled drinking, mineral water) and for using rivers and other water bodies for transport, fishery, recreation and other purposes.

Various kinds of payments are effected: taxes, tariffs, rent and insurance payments. They are to ensure rehabilitation, protection and rational use of water

Water: Economics, Management and Demand. Edited by Melvyn Kay, Tom Franks and Laurence Smith. Published in 1997 by E & FN Spon. ISBN 0 419 21840 8.

resources, compensation for expenditure on water supply to consumers, and indemnification of losses caused by harmful effect of water, solution of social problems. The payers are water users, those who pollute and consume water, and sometimes payments are voluntary as it is in case of insurance. To solve major economic, social and ecological problems it is not uncommon that payments are made by local and state authorities.

Fast growth in the amount of water consumed at the end of 70's and the beginning of 80's and of the amount of polluted water discharged into water sources necessitated the introduction of payments into practice for water management. At present in Russia payments are charged for water intake from water resources utilization systems by industrial enterprises and enterprises providing to cities, as well as for underground water use and for discharging polluted substances into surface water courses. It is only in some regions of Russia that local authorities collect payments for water intake from all water users.

After rejecting the centralized model of water protection management in Russia specific water use and volumes of discharge of non-purified waste water into potable water sources began to grow. During Russia's economic transition to a market economy the extent of state financing of water economy measures is being greatly reduced and the problem of compensation for expenditure and self-financing of these measures is becoming especially pressing.

The experience of introducing water payment in Russia has shown the existence of a number of problems, the solution of which is necessary for successful functioning of economic mechanisms for water management.

The technical problems are connected with a lack of water control equipment such as water meters and the absence of personnel trained to solve these problems. Some technical and organisational difficulties are expected to arise due to the introduction of water payments for small farms using irrigated lands when administrative and controlling services will be required and which will lead to significant expenses in order to make their activity possible. In this case it may be more reasonable to introduce a land tax as a form of payment.

The economic problems are connected both with the choice of a method of determining water payments and with the forms of compensation for these additional costs to water users or state programs of help to low-profitable farms. In addition to this is the absence of economic interest of the parties concerned in realizing these measures and the economic relations between suppliers of water and its users.

The most convenient forms of payment are such simple forms as a cubic metre of water supplied for a hectare of land irrigated or their combination.

In Russia irrigation norms change annually in all regions where irrigated agriculture is practised, therefore the application of a combined two-rate tariff for water used in irrigation helps to stabilize the economic conditions supporting water management organisations.

For minimization of losses, the optimum distribution of limited water resources and maximization of net profit from irrigation, the rate of the water tariff should be determined with considering the "efficiency" price. That is the contribution of water to the production of output. But this theoretical model is known to find little application in practice.

Unlike other countries, where rate of the water payment tariff is determined by the expenditure, Russia it is planned to use the principle of ability to pay. When creating the system of collecting water payments several schemes could be used. Two of them are most widely practised, the difference being in the state role. The first is characterized by a rigid state regulation of the rate of payment. In this case all payments are accumulated in the state budget and these are used by the state to finance the whole water economy activity in the country. The second system involves a corporate water resources use and water protection management on the part of water users who finance it. Water payments in this case have a form of water users' contributions to maintain and finance the activity of an association of water users located in one river basin. The state withdraws part in the form of taxes and uses them for crediting and subsidizing new water resources construction. Now Russia is guided by the first approach.

According to the new water code, payments are made by all enterprises except for those that take in water for irrigated agriculture. These water users make payments according to the land legislation.

The water tax is collected for water intake from surface water and from underground sources. For the first time water payments introduced for navigation, water-power engineering, for timber floating, for work of excavating machines in water objects and for use of water objects surface. Another kind of payment is intended for rehabilitation and protection of water sources. This means money collected as water taxes are directed to the federal budget (40 %) and the budgets of the Russian Federation subjects - regions, territories and autonomous republics (60 %). There is no target payment. It is based on the income received from water users as a result of using water resources.

The second kind of payment is a compensation for expenses on maintenance, repair and operation of multi-purpose water economy objects. These are also distributed between the federal budget (40 %) and the budgets of the Russian Federation subjects (60 %) and are used on the purposes described above.

It is stipulated that the basic rate of payment is fixed at the centre, for individual river basins or their parts and correction factors are taken into account such regional peculiarities depending on quantity and quality of available water resources. However the rate of water payment does not depend on water quality and efficiency of water users' work.

It is important to review the economic consequences of introducing payments from enterprises using water. The prospective rate of payments does not correspond to financial capabilities of the majority of enterprises and so this can result in inflation growth. There is no instrument for setting the rate of water payments under the influence of external factors such as inflation and others. It will be necessary to determine the conditions, validity terms and the order of revision of the present rates of payments.

It will be necessary to provide a step-by-step introduction of the suggested system of water payments ensuring synchronization of payment growth in accordance with the financial stabilization of Russia's economy. It will also be necessary to provide qualified personnel training in the field of water resources estimation.

It is important to work out suggestions on creation of a new type of water economy enterprise in Russia - joint-stock companies, concessions, trusts and other structures, that could take an active part in implementing programs on rational use and protection of water resources on the basis of market relations in water use.

Taking into account the existing social problems when determining water tariffs and the degree of the state involvement is of great importance. Financing land reclamation work in our country in the near future cannot be solely the task of the existing agricultural enterprises or newly created farms, as their efficiency is too low. The state should play a significant role in formation of the water market - ranging from the market of water resources and environmental protection machinery to the market of quotas (limits) on pollution of water sources and services to protect them. Thus, for example, in the USA the market of discharge permits has also arisen as a result of administrative pressure on industry in the form of technological standards and ecological requirements. In Russia during the economic recession the authorities in charge of the environment could not resort to strict measures for the infringers of the laws. The instability of ecological and economic policies of the state can also play a negative role. We can conclude that the lack of human effort directed at environmental protection has shown that the problems concerning the rational use of water resources and the problems of social development cannot be considered separately. Russia's stable development will need the help of such economic measures that meet the needs of the population and of the national economy in water and will make utilization of water resources effective.

The economic mechanism in itself is not a universal means of management in all aspects of water protection activity. An approach to solving the problem, involving creation and the purposeful activity of a state control system in water management development and realization of economic, administrative and legal methods will be essential. Payment should cover all aspects of water use and all kinds of activity influencing the quantity and quality of water, i.e. in reimbursement of national-economic expenditures. This would need to involve all enterprises bound by water use relations irrespective of their departmental subordination and the type of property. To reduce the negative influence of the fact that natural resources in Russia are free of charge, a number of administrative measures (limiting of water supply, planning of water use and others) are being used. The transition of Russia's economy to a market economy opens additional opportunities in satisfying the requirements of rational use and protection of water resources both with the help of the state regulation and the introduction of paid water use.

The experience of countries with advanced market economies with appropriate amendments will be of great benefit to Russia as its water market develops.

WATER PRICING FOR IRRIGATION DURING THE TRANSITION TO MARKET ECONOMY IN BULGARIA

M. KOUBRATOVA

Institute of Agricultural Economics, Sofia, Bulgaria

Abstract

Six years after the beginning of the radical political and economic changes in Bulgaria the price of water is determined by a centralised manner. For 1995 the irrigated area represents about 10% of the command area which is the lowest level after a progressive decreasing of the irrigated area after 1990. In 1996 the irrigated area augmented due to the very low price of water that had as consequence an intervention of the government subsidising the Irrigation System Company. One of the main reasons for the low level of irrigation use is the land and agrarian reform in Bulgarian agriculture and the general decline in agricultural production. Another important reason is the inflexible manner of determining the price of water for irrigation. The methods of irrigation in Bulgaria vary from gravity to drip. Logically different irrigation systems have different costs of maintenance and operation that suppose a different water price in the related fields. In actual practice different systems in a particular branch of the Irrigation System Company have the same price of water. At the most the mechanism of setting the water price includes within itself not only the cost of maintenance and operation of the irrigation system under consideration but the cost of the maintenance of all the systems in the branch.

The author discuss the changes that should be done in the organisation of the irrigation sector in Bulgaria and in the determination of the price of water for irrigation. The existing and possible constraints for the implementation of a new method of setting the price of water are identified. The speed and essence of the changes in this field depend upon the willingness of the government to improve the policy and principles of management of the irrigation systems and facilities in Bulgaria and adapt them to the changing economic conditions in the country.

Keywords: price of water, water user organisation, economy in transition

Water: Economics, Management and Demand. Edited by Melvyn Kay, Tom Franks and Laurence Smith. Published in 1997 by E & FN Spon. ISBN 0 419 21840 8.

1   Introduction

The Law of Concessions, voted in September 1995, stipulates that the irrigation and drainage facilities are public property. They can be leased for a period of time (not longer than 35 years) after a determined procedure.

The owner of the irrigation funds is the government and especially the Minister of Agriculture. There is a governmental company - Irrigation System Company (ISC) that is managing all the irrigation and drainage facilities on behalf of the Minister of Agriculture. The ISC Board of Directors is nominated by the Minister of Agriculture and the Directors' contracts have a so called business task to follow.

The ISC has 22 branches in the country that differ from each other because of the particularities of the water resources used for irrigation, the difference in the types of systems constructed during the socialist period and different natural (climate, soil) and economic conditions.

The incomes of ISC are formed by:
- subsidies of the government for the maintenance of main dams and facilities of national importance
- incomes from the sale of water for irrigation, for industrial enterprises and for fish production
- incomes from additional complementary activities such as leasing halls and rooms, construction, labouring, growing and selling agricultural production or fish

During the six years of reform the ISC had many difficulties in irrigation because of the changes made and still continuing in force in the country.

The main economic conditions that have an impact on irrigation are the land and economic reforms in Bulgaria. The Law of Use and Property of Agricultural Land voted in 1992 stipulates that the property rights of the agricultural land are restored to the former (before the socialist period i.e. before 1944) owners. In practice this means a fragmentation of the land and a very long procedure of proving the property and receiving land. In December 1996 only 7 % of the land is having its owners with a document. Another 80 % have a temporary use of land. This means that they are not investing money in land and because there is not a register it is quite difficult to understand who is using the land.

The economic and political reforms. Their result is a loss of international markets for Bulgarian agricultural products and a decrease of demand for agricultural products on the national markets. As result there is about 7 thousands of hectares of agricultural land that are not cultivated and not used.

In the same time the agricultural producers' strategy is to produce just enough for the needs of their households. A very small part of the production goes to the market. So, in general the need of augmentation of the production does not exist because of the lack of markets and low prices of agricultural products. Very often this is not clear to people dealing with irrigation. They want to have a huge surface of irrigated land forgetting that irrigation is made for the only purpose of response to the markets as quantity and as period of delivery.

In fact the Bulgarian climate is such that in the South Bulgaria it is not possible to grow any crops without irrigation. In the North part irrigation is not so much needed. The cropping pattern in the different parts of Bulgaria is different as well and in the North wheat and cereals are very popular.

The next section is for the existing procedure and methodology of formulating the water price in Bulgaria and the main problems that the SIC has dealing with them. The third section is about the necessary radical changes in the organisation of the ISC as a national company and in the methodology of determination of the water price. The fourth section is summarising of the reserves that should be search for the decrease of the price of water for irrigation. In the last chapter main conclusions and recommendations are presented.

2 Existing procedure for formulating the price of water

2.1 Types of price of water

There are two types of prices used in irrigation in Bulgaria.

The first one is a fixed rate per irrigable decare (1 decare = 0.1 hectare) defined every year by the Council of Ministers. It is to be paid by every one whom land can be irrigated without any reference to whether he or she is irrigating. Because of the dominant temporary use of agricultural land the ISC could not collect the fixed rate per irrigatable hectare. In the prognosis for 1996 the planned incomes from the collection of fixed rates was 15 % of the total incomes of the ISC (excluding subsidies). The experience from the last six years shows that the ISC collects no more than 20-25 % of the total amount of the charged fixed rate.

The main price of water is defined as a fixed rate per cubic meter of water for irrigation at the delivery point. Every agricultural producer using water pays the water price multiplied by the quantity after every irrigation made. In some cases even the collection of the rates on the volumetric basis is not collectable because of thefts of water.

There is a tendency for the irrigated area to decrease over the last 5 years. 1996 is an exception when the irrigated area was about 107 thousands of hectares is 13 % of the irrigable area. The increase of the irrigated area is due to the sensitive decrease of the water price paid by the agricultural producers and losses for the Irrigation System Company.

Table 1 shows the changes of the price of water during the last three years and the prices of two main irrigated crops - maize and tomatoes. The sensitive decrease of the price of water is obvious when the prices of the both agricultural products are not diminishing so quickly.

Table 1. Price of water, tomatoes and maize in USDollars for 1994-1996

Year	Fixed rate of water, US$/ha	Price of water, US$/m^3	Price of maize, US$/tonne	Price of tomatoes, US$/tonne
1994	7.3	0.025	131	270
1995	5.3	0.018	86	310
1996	1.6	0.009	105	260

The water price is one of the main factors that, as in the case of every one good, has a strong impact on its demand and use in agricultural production. The practice shows that during the last year with the decrease of the price of water there is an increase of the irrigated areas.

2.2   Role of the preliminary assumptions

One important factor in the determination of the price of water is the preliminary calculation of the price on the basis of some general suggestions concerning the macroeconomic environment. These suggestions are about the annual inflation rate, the devaluation of the national currency, the price of electricity and the possible surface to be irrigated by the producers. The importance and the dynamic of all these parameters can be shown on the basis of two examples. In the prognosis for 1996 the planned area to be irrigated was about 134200 ha and at the end of the year the actual area is 107000 ha. The losses only from this wrong assumption costs 200 million levs (15% of the total planned income). The changes in the price of electricity plays an important role as well. In the prognosis for 1996 the suggested electricity price was 2.96 levs per kilowatt but the price of electricity varied every month and during the irrigation period this price was between 2.87 (May) and 7.64 (August). Only this augmentation of the price shows an immediate impact. When the suggested 1996 electricity price (2.96 lev/kilowatt) is replaced by the real price of electricity during the year, the calculated average water price should be 2.28 levs/m^3 instead of the charged 1.6 levs/m^3 price.

In the structure of costs for 1996 that define the basis of the water price the fixed costs had about 60 % of the total costs. The participation of every one item in the structure of the fixed costs for 1996 was as follows: amortisation - 28 %, different materials - 18 %, fuel - 15 %, salaries and related costs - 9 %, services - 11 %, other costs - 13 % and overhead costs - 6 %. In the structure of the variable costs the dominant place is occupied by the electricity - 94 % of the total amount of the variable costs.

2.3   Procedure for formulating the price of water

The applied procedure of formulating the price of water begins in each ISC branch. Every branch is defining its average price of gravity and pumping water on the basis of a prognosis about: the area that should be irrigated, the costs for the delivery of water to this area and the costs for the maintenance of all facilities in the branch. Because every one of the branches has different systems with different costs for operation and maintenance and difference participation of the amortization in these costs the average price of water is such that it augments the price for producers irrigating in systems with low costs and decreases the price for producers irrigating in systems with high costs. In this manner there is a cross subsidy between the agricultural producers within the limits of the ISC Branch - the agricultural producers in the systems with low costs subsidise the producers in systems with higher costs and the ISC Branch itself.

When the branch prepares its own water price the price is presented to the Central Office of the ISC where further changes take place. Thus there is a second subsidy between the branches i.e. between the agricultural producers irrigating in the different branches. The producers in regions with limited rainfalls and irrigation systems with low costs are subsidising the other agricultural producers (with higher price of water) and the ISC branches.

There are some proposals to have an unique price of water for the whole country. The main argument is that no one had made investment or no one had a personal responsibility for the construction of the facilities. This idea is not economically good because:

1.   The prices of the agricultural products - rainfed or irrigated, are different in the different parts of the country.
2.   The unique price does not permit to reduce the costs for operation and maintenance of the facilities.
3.   The unique price should, as a consequence be an important amount and not a reasonable cross subsidy between the agricultural producers.

2.4   Example of formulation of the water price

Let use an example to show the way of formulating the price of water in the irrigation field of Katuniza - a village in Central Bulgaria. The irrigable area is 1500 ha separated in two parts - North and South. For the irrigation of a part of the field (250ha) two intake structures in the River Chaya are used - a permanent one and a temporary one. The average amount of gravity water used during the last four years is 750 thousands m^3. For 1996 the Branch set the gravity water price at 1.20 leva per m^3 (excluding value added tax). In practice the ISC charges the water price with a value added tax that now is 22 % for all the products in Bulgaria.

The calculation of the costs includes all the costs for the delivery of the water to the intake structures: costs for amortisation, for operation and maintenance, salaries for the workers related to the field and the complementary costs to the salaries, costs for the making of the temporary intake structure, percentage of the administrative costs of the Branch. All these costs are the full annual costs and on them a profit of 12% is calculated. The water price calculated on the basis of the real costs for the delivery of the water in this irrigation scheme is 0.97 leva per m^3 instead of the charged (by the ISC) water price of 1.20 leva per m^3. For the total amount of water used for irrigation within the scheme this difference in the water price means a loss for the agricultural producers of 1,725,000 leva.

There is another fact that is popular among the users of water and represents a constraint for the real application of the water price. The producers are paying one amount of water of irrigation but they are using more water because there is no measurement instruments installed. As usual the workers responsible for the delivery of water are people from the village and they are not defending the interest of the company. In case that the producers are using the double amount of water they should pay 1500000 m^3 multiplied by 1.20 leva per m^3 and not the costs that are almost permanent divided by the double amount of water. In the first case the price of the water remains the same - 1.20 leva/m^3, and in the second case the price is diminishing - 0.49 leva/m^3. So, it is evident that the water users have a good motivation to ask for the determination of a specific price of water.

3   Necessary changes to be made

It is necessary that the use of irrigation facilities is defined on the basis of the existing economic conditions and the future expected changes if the ISC intends to manage the governmental property in an effective manner and continue to exist. Changes are needed in:

the methodology of formulating the price of water and its application so that the water price should be specific for every one irrigation system,

the financial and economic assessment of every irrigation system and definition of its future use based on the assessment.

3.1   Classification of irrigation systems

The systems must be classified into three groups as follows:
1.   systems with low costs of operation and maintenance and with an acceptable price for the agricultural producers. It is necessary to organise the water users in associations by helping them and creating groups for consultancy and support. In fact all gravity systems that represent 55 % of the total command area are included in this group;
2.   systems with high costs of operation and maintenance in which the price is so high that it is not acceptable by the agricultural producers. Such type of systems are the systems using pumping water. In this case it is necessary to define the level of water price acceptable by the producers and for the difference over this level it is necessary that the government agrees to subsidise it;
3.   systems with so high costs of operation and maintenance that it is not possible to irrigate and an abandonment of the facilities is needed. This should be subsidised by the government.

The assessment of the hydromeliorative funds must be made in every ISC branch. The branch managers are quite experienced and know from their practice the possibility and perspective of use of every one of the irrigation systems within the branch. The need to have a comparability between the systems in the different branches obliges to have a unified methodology for the assessment of the systems and its application.

3.2   Independence of the branches of Irrigation System Company

The most important factor is to have a specific price of water for every one irrigation system. For this it is better to have 22 or more independent companies in the country instead of one national company. The water is a specific good and it should be used in a better manner at local level.

When every irrigation system will have its own price the existence of Water User Organisations (WOU) will be possible. The specific price of water is a good basis for use of different methods of determining the water charge upon the water volume used, grown crop or time of delivery applied in the world practice.

What could be the changes in the price of water if the facilities were managed by the water users? The costs for the amortisition should be lower because a part of it is used for maintenance. This part would be kept and used in the Organisation. The costs for the operation and maintenance of the systems would diminish because of the better care that the water users should take. The costs for salaries would decrease as well because the salaries in the ISC are bigger than the agricultural producers used to have. As result the related amounts for social security would decrease as well because it is calculated on the basis of the salaries. The administrative costs for the ISC would be suppressed but some money for the organisation of the Water Users Organisaiton itself should be added. There would be some additional costs for the control the ISC should make to be sure for the right use of the facilities.

When the WUO would take the management of the irrigation system or separated facilities they could profit from the decrease of the price of water by the augmentation of the quantity of water used for irrigation. In any other case with a centrally defined price of water the users should have no profit and no motivation to irrigate more and use water in efficient way.

3.3   Changes in the fixed rates per hectare

Two possibilities exist for the future of the collection of the fixed water rates. The first one - to suppress these rates: they were never collected by the ISC. Within the existing conditions it is not possible to collect them and they are a loss for the ISC.

The alternative is to continue to collect the fixed rates. In this case a reconsideration of the existing indexed irrigable hectares is required. The ISC wants that the whole amount of these costs were subsidised by the government. In fact there is no possibility for subsidising because of the financial problems the country has. On the other side this subsidy will be a subsidy for the ISC and not for the farmers because farmers used not to pay them rates. The alternative is to make such type of changes in the laws that should make the payment of the fixed rate obligatory.

The best alternative is to suppress the fixed rates. This will simplify the payment for irrigation and will make more transparent the costs and incomes of every one of the branches. The augmentation of the price of water per cubic meter will be small because the fixed rates cover between 15-20 percentage in the calculations of average price of water at present.

4   Constraints for decreasing the price of water

There are many constraints for lowering the price of the water. The essential precondition for this is the existence of a specific price of water for every irrigation system.

The first constraint is the existence of a percentage of profit, now charged (12 %) on the total costs. It is not logical to calculate a percentage of profit if the irrigation itself is subsidised. So, this pourcentage should be eliminated or decreased.

The second constraint is the amortisation. At present the ISC is calculating only 33% of the real amortisation that should be included. During 1996 and in 1997 the amortisation, calculated in the price of water, is not so important because the rapid devaluation of the assets due to the macroeconomic changes in Bulgaria. In 1997 a re-evaluation of the assets of all enterprises in the country is suggested. In this case the amortisation should increase the water price.

The main purpose of amortisation is to collect money for the construction of new facilities. In fact no one in Bulgaria is doing any construction and the money for amortisiation collected in the ISC are used for other purposes. The amortisation is not centralised in a national fund for further public investment. The amortisation in Bulgaria is divided into two parts - one for annual maintenace and one for construction of a new facility. As there is no intention to collect this money and use it for new construction it is better to eliminate this part and leave only the percentage used for annual maintenance. This requires a change in the laws and concerning documents.

Another constraint is the use of the same water prices for irrigation and for other industries. In the prognosis for 1996 the difference between these two prices was only 0.03 levs which is insignificant. The changes should be made in augmenting the price of water for the industrial and fish production having in mind their bigger profitability in comparison with agriculture.

5   Conclusions and recommendations

The manner of formulating the price of water used in the ISC is not helping the efficient use of the irrigation water. The use of a unified price of water for all the systems which have different costs for the delivery of water and maintenance, has as result a double cross subsidy - producers irrigating in systems with low costs are subsidising the irrigated producers in the same branch and in the other branches with a higher price of water. If there are no producers irrigating in the high costs systems agricultural producers in low costs irrigation systems are subsidising the ISC itself.

The correction of the price of water in the central office has as result the cross subsidy between agricultural producers irrigating in different branches and respectively between the branches.

The result of this policy is not only the decrease of the irrigated area but losses for the ISC and an inefficient use of governmental subsidies.

It is crucial that the Ministry of Agriculture requires that the formulation of the price of water for every one irrigation system is made after a detailed financial analysis of the costs for operation and maintenance and the incomes from the sale of the possible amount of water. On the basis of this assessment the ISC should define the systems in which: Water Users' Organisations should be organised; the price of water should be subsidised and the amount of the needed subsidy should be calculated; the facilities should be abandoned.

The ISC branches have to be independent and responsible for the irrigation in their branches. A unit in the Ministry of Agriculture should be created to control and direct their activities,

The creation of WUO is one of the main conditions for the efficient use of water resources for irrigation in Bulgaria. The existing practice is to transform in an artificial way the existing producer cooperatives into WUOs. This type of WUOs are without future because the financial incentives for their creation and existence are missing. It is very important that the WUOs are organised in such a way that they follow the principle of the hydromeliorative independence and completeness of the irrigation. Only in this case the ISC will have stable and long life partners and will manage in the most efficient way the hydromeliorative funds, diminish the number of the clients, augment the irrigated area and succeed in collecting all the water charges.

PARTICIPATORY IRRIGATION MANAGEMENT: A CASE STUDY OF ABILITY TO PAY

L.E.D. SMITH
Department of Agricultural Economics and Business Management, Wye College
University of London, Wye, UK
A. SOHANI
International Irrigation Management Institute, Hyderabad, Pakistan.

Abstract

Using estimates of the value of irrigation derived from survey data an assessment is made of the ability of landowners in the Lower Indus to meet the recurrent costs of irrigation and drainage. Participatory management of irrigation is financially feasible but new drainage infrastructure must deliver significant increases in output. Improvement in the equity of water distribution and determination of a basis for the sharing of drainage costs are critical issues that need to be resolved.
Keywords: irrigation, drainage, water rates, costs, operation and maintenance, Pakistan

1   Introduction

Irrigation is vital to the economy of Pakistan and as a source of livelihood for the rural population.   Government and donor agencies recognise the need to improve performance in terms of output, farm incomes and returns to investment in improved infrastructure. To this end major institutional changes are being introduced including increased involvement by water users in management of irrigation systems.

In Sindh province, interventions in social organisation at the distributary/minor level are being tested in a pilot project mode within the catchment area of the Left Bank Outfall Drain Project Stage I[1]. Three pilot projects are being implemented by the International Irrigation Management Institute in collaboration with the Government of Sindh, in which Water Users Organisations (WUOs) are being established to operate and maintain irrigation and drainage facilities in distributary canal command areas. The WUOs will consist of associations of all water users on each watercourse, and federations of these associations that will manage irrigation at the distributary level.

Water: Economics, Management and Demand. Edited by Melvyn Kay, Tom Franks and Laurence Smith. Published in 1997 by E & FN Spon. ISBN 0 419 21840 8.

These pilot projects relate closely to wider reforms proposed for the institutional environment for irrigation management. These involve establishment of Provincial Irrigation and Drainage Authorities (PIDAs) and Area Water Boards (AWBs). PIDAs are intended to be financially autonomous authorities with accountability to Government and the people of the area served. AWBs are to be created under PIDAs to manage irrigation and drainage in designated canal commands. They are expected to be financially self-accounting, with landowners and senior professionals represented on the Board of Directors. Below the AWBs, landowners are encouraged to set up WUOs along the lines of the three pilot projects.

Implementation of the pilot projects has been based on a "participatory learning process" for all parties involved. Detailed work plans and organisational structures will emerge as the projects mature and as institutional development of the WUOs takes place. Formation of the WUOs is at an early stage and their responsibilities, activities and relationship with the existing Provincial Irrigation and Power Department (IPD) are still being determined.

The aim is that a WUO will operate and maintain the distributary and drainage infrastructure within its command area, thereby reducing the budgetary requirement of the IPD and improving system performance through the positive effects of participation by water users. Specifically it is anticipated that WUOs can achieve more equitable distribution of water, improved reliability, and through collaboration with the Agricultural Extension Department and On Farm Water Management programme[2] (OFWM) increased adoption of improved irrigation and agricultural practices.

2   Current irrigation operation and maintenance

Prior to the pilot projects the only involvement of water users in the operation and maintenance (O&M) of distributaries has been the provision of labour for de-silting. Landowners (*zamindars*) typically provide two of their sharecropping tenants (*haris*) for de-silting of the section of canal that immediately supplies their watercourse. The work may be initiated by water users themselves or by the IPD[3], but the labour does not receive payment[4]. Water users comment that this arrangement tends to be treated as a formality and the work is disorganised and frequently not done to the required standard. Use of manual labour alone is often inadequate for the scale of work required and for removing major obstructions. However, any de-silting does tend to improve water flow and the practice is more common in tail reach areas. Some head reach *zamindars* may see no need to de-silt the canal and the work on their stretch may be carried out by those from the tail. De-silting of watercourses is similar but generally superior, through more effective collective action by fewer water users.

The IPD regulates the discharge in the distributary and sanctions and monitors the withdrawal of water via the outlets. It should resolve conflicts, enforce the 'warabandi'[5] system where necessary, and respond to demand from users subject to balancing competing demands during periods of scarcity. In practice very little maintenance work is carried on the distributary. Permanent staff are responsible for minor routine repairs to embankments and removal of blockages. The IPD aims to de-silt canals every three years, or less frequently for those with a high flow velocity. Small channels with low flows require silt clearance each season and this is done by the water users themselves as described above. Channel maintenance completed by IPD may include silt clearance and re-sectioning, re-handling of spoil, earthworks and reconditioning of banks. Such work is done when needed, and not according to a preventative schedule. Similarly gates may be lubricated and painted during the closure, but other repairs to structures are only done when urgently needed.

3   Future irrigation and drainage operation and maintenance

Tables 1 and 2 outline the possible transfer of responsibilities to WUOs. Joint management of the drainage system is a current policy aim and the O&M Directorate

Table 1. Irrigation Operation and Maintenance: Feasible WUO Responsibilities

INFRASTRUCTURE	WATER USER RESPONSIBILITIES
Head regulators	• regular inspection and reporting of repair and maintenance needs to IPD, or employment of contractors • greasing of mechanisms
Distributaries	• regulation and monitoring of water distribution, conflict resolution • regular inspection and minor repairs • vegetation control and clearance • annual de-silting and channel and embankment maintenance[6] • maintenance of inspection path • repair of outlets or report to IPD • collection of water user contributions to WUO • collection of *abiana* (water tax) including levy to be paid to IPD/PIDA for main system • improve water management and agricultural practices with assistance of OFWM and Agricultural Extension • improve maintenance practices for irrigation and drainage facilities with assistance of IPD

of the LBOD project is developing modalities for this [1]. They envisage evolution of drainage beneficiary participation through three main stages: provision of pump watchmen, disposal channel maintenance, and sub-drain maintenance. Contractors will be monitored, and established WUOs may be able to employ these directly. A drainage cess will be levied on users, though this has yet to be implemented. Partial exemption from this will provide an incentive for participation in O&M. In Sindh water users have had little experience in O&M of drainage infrastructure. Consultation indicates that they are not ready to assume this role in the immediate future as it is regarded as the responsibility of Government (see also [2]). They are willing to consider this once WUOs are fully established and they have observed the drainage in operation. Clearly much depends on the delivery of sustained benefits by the drainage system.

Table 2. Drainage O&M: Possible Beneficiary Roles and Responsibilities

TECHNOLOGY	POSSIBLE BENEFICIARY RESPONSIBILITIES
Drainage/scavenger well	• pump house security and monitoring of power supply • maintenance of disposal channel • monitor performance of O&M contractor and report problems • for scavenger wells, operate and maintain watercourse from well • collection of beneficiary contributions
Tile drainage	• pump house security • manhole security • monitor power supply • maintenance of disposal channel • monitor performance of O&M contractor and report problems • monitor tile drain performance (waterlogging) • collection of beneficiary contributions
Surface sub-drains	• regular inspection and minor repairs • vegetation control and clearance • de-silting and channel, embankment and inspection path maintenance • collection of contributions and cess to be paid to O&M Directorate

5   The case study distributary

The case study distributary canal off-takes from the Nara main canal. It is 10.6 km long, has a design discharge of 58 cusecs (1.64 m^3/sec) and supplies a cultivable command area (CCA) of 12,336 acres (4992 ha). There are 24 watercourse outlets that serve 435 landowners. A minor canal also off-takes from the distributary: 5.12 km; 10.62 cusecs (0.3 m^3/sec); CCA 3074 acres (1244 ha); 7 outlets; 104 landowners. Within the CCA of the distributary are 8 drainage tubewells (saline) and approximately 50 km of surface drains. The CCA of the minor lies in the seepage zone between the main canal and the distributary. Located within its CCA are 16 scavenger tubewells, 6 drainage tubewells and approximately 11 km of surface drains.

The distribution of land ownership is more equitable than in many other parts of Sindh. The mean farm size is 33 acres (conf. interval at 95%, 6.7, n=32) and only two *zamindars* in the sample had a holding of greater than 50 acres. Nine of the *zamindars* were owner operators; the remainder cultivated their holding with *haris*.

6. The value of irrigation to landowners

Table 3: Net annual farm and household income, Rs per acre CCA

	Mean	C. I.$_1$(95%)	Median	Sample size
Net farm income	**3115**	**1051**	**2536**	32
Other h.hold income	1229	513	684	32
H.hold living expenses	3253	1053	2519	32
Net h.hold income	**1091**	**1061**	**1260**	32

Source: survey data for rabi 1995/96 and kharif 1996.   ₁confidence interval

Table 3 indicates the value of irrigation to *zamindars* in terms of the net farm income generated per acre of CCA owned. This was derived from a formal sample survey of 32 *zamindars* which recorded farm management input-output data. The sample was stratified to ensure equal representation of *zamindars* from the head, middle and tail reaches of both distributary and minor. *Zamindars* surveyed were either owner operators, or landlords cultivating their farms through share tenants. For the latter the net farm income was calculated net of the *haris'* share of output and variable costs (usually 50%). It includes sales of livestock or livestock products and is net of farm fixed costs. Existing taxes on water, land, drainage and agricultural income have not been deducted. Table 3 also shows other non-farm household income, and household living expenses (inclusive of utilities, food, clothing, education and medical expenses).

7   Cost assessment

7.1   Irrigation
Data obtained from the IPD for expenditure on "maintenance and repairs" for the case study distributary shows that maintenance by the IPD has been minimal and infrequent. It is concluded that water users already bear the bulk of actual maintenance expenditure for smaller distributaries by undertaking the de-silting works described in section 2. Given this it is sensible to use water user estimates as the basis for assessing future costs. Minimum estimates in Table 4 include costs of de-silting and channel maintenance for the case study distributary and off-taking watercourses; the tasks already largely undertaken by water users. The estimates are based on work done to the standard required and in a timely fashion[7]. A small administration overhead is also included. Maximum estimates are more comprehensive estimates of the full costs of channel O&M including permanent maintenance staff costs. Costs per acre are higher for the minor because of the lower flow velocity and greater de-silting requirement.

Table 4: Estimates of Irrigation O&M Costs, Rs per Acre CCA

	Distributary: CCA 12336 acres		Minor: CCA 3074 acres		Distributary and minor combined	
	min.	max.	min.	max.	min.	max.
Labour contribution	82	82	95	95	84	84
Cash contribution	21	37	28	47	23	40
Total contribution	103	119	123	142	107	124

Source: survey data and water user estimates based on prices in the rabi 1996/97 season.

Estimates include only recurrent costs of O&M as WUOs were not planning major investments in rehabilitation other than channel improvement[8]. Other capital expenditure, for example a permanent office or plant, is also not envisaged by WUOs in the medium term. Contingency funds to provide for canal breaches and emergencies are also not included. WUOs indicate that, as in the past, they will prefer to respond to such events when they happen with costs shared as determined by the WUO.

At present all WUO office bearers are willing volunteers, though experience elsewhere suggests that such officials are more accountable and effective when paid [3]. As a general principle WUOs are encouraged to keep overheads as low as possible, at least during their development phase. The potential for WUOs to develop into multi-functional community organisations is also recognised by water users. To date activity has been limited to collective purchase of seed in two of the pilot projects.

Although zamindars from the distributary and the minor have agreed to work closely together, those from the minor have emphasised that they want to have their own organisation and manage their own affairs. In addition water users currently envisage allocating the O&M costs of irrigation and drainage to the watercourse where that infrastructure is located. For example, each watercourse association will be responsible for maintaining its reach of the distributary, i.e. the upstream reach from that watercourse to the next (or head regulator). The need for some adjustment is recognised, for example, a watercourse with lower costs contributing to one with higher costs, but the potential for inequities and conflicts is clear. This is particular so in the case of drainage, see examples below.

7.2 Drainage
Estimates of drainage O&M costs were obtained from the O&M Directorate of the LBOD Stage I project, WAPDA. Figures should be regarded as illustrative as the Directorate is in the process of assessing actual costs and O&M procedures. Table 5 presents estimates of the combined O&M costs of both irrigation, and drainage infrastructure being installed. For irrigation the maximum estimates from Table 4 are used. For drainage, estimates include costs of pump house watchmen, pump O&M, electricity supply, and O&M of disposal channels from drainage tubewells and scavenger wells. They do not include O&M of surface drains, nor any contribution to O&M for the main drainage system beyond the case study command area.

Table 5: Estimates of Irrigation and Drainage O&M Costs, Rs per Acre CCA

	Distributary: CCA 12336 acres	Minor: CCA 3074 acres	Distributary and minor combined
Total	214	1181	407

8   Cost recovery mechanisms

It is envisaged that WUOs will recover costs/raise revenue from water users in three ways.

1. A subscription or membership fee paid to the WUO at the watercourse level with a proportion forwarded to the federation of WUOs that will manage the distributary. This is primarily to cover routine administration costs and to create a sense of membership of, and participation in, the organisation[9].

2. Direct recovery of the actual costs of seasonal maintenance operations. Water users will contribute to these costs in proportion to the area of land owned or duration of irrigation turn (in practice these are equivalent). Contributions are made both in kind (labour) and in cash[10].

3. Direct collection of *abiana* (water tax) from water users, with a proportion of this to be forwarded to the IPD (or AWB/PIDA) for O&M of the main system.

Abiana rates for 1996/97 vary from Rs35 per acre for fodder crops to Rs158 per acre for sugar cane. Based on the rates for each crop for 1996/97 and an estimate of the cropping pattern in the case study area from survey data for 1995/96, the total amount to be paid by landowners for 1996/97 is equivalent to Rs 84 per acre of CCA owned. Many *zamindars* pay considerably more than this in bribes to IPD officials to secure supply.

9   Conclusions

Estimates of farm incomes generated from irrigated agriculture and O&M costs of irrigation and drainage infrastructure have been compared by expressing both per acre of CCA owned by *zamindars*. Conclusions drawn from the comparisons must be tentative given that they are based on a single case study distributary and a relatively small sample of farmers. Farm management data, and data on non-farm income and household expenditure are also typically subject to substantial non-sampling errors.

Despite these limitations the results indicate that it should be financially feasible for WUOs to assume responsibility for the irrigation O&M activities envisaged. The contributions in cash required appear modest given the mean levels of farm income generated and existing *abiana* rates. They also appear affordable compared to the mean net household income for households, who while generally not wealthy, have living standards well above subsistence levels.

The labour contribution has been valued at the cost of hired casual labour for manual work in the area. Under the terms of the sharecropping relationship the burden of this contribution falls on the *hari* rather than the *zamindar*, particularly if *haris* do have alternative casual employment opportunities. Though inequitable given *hari* living standards, there will be little change compared to the current situation in which *haris* already perform de-silting, and *haris* will actually benefit if the work is done to a higher standard and system performance improves.

The O&M cost estimates in Table 4 will not be paid in addition to *abiana* if WUOs are empowered to collect and utilise this tax, but they are an under estimate because a proportion of *abiana* to be determined, will be passed on to the IPD or PIDA for main system O&M. WUOs will also have to continue to pay bribes to the IPD, at least until reform in the irrigation sector is fully established. These further burdens also highlight the need for WUOs to improve system performance in terms of value of output.

The mean values in Table 3 conceal considerable variation as indicated by the confidence interval. Only one zamindar showed a negative net farm income, but nine (28% of the sample) generated less than Rs1000 per acre and thirteen households (40%) showed a negative net household income. For participatory irrigation management to be financially sustainable there is thus a clear need to improve the equity of water distribution and the extension of improved agricultural technology and practices to less efficient producers.

Including the costs of drainage greatly emphasises the need for performance improvements. The combined irrigation and drainage O&M cost estimates are approximately ten times greater than for irrigation alone. Most of the increase also requires contributions in cash from *zamindars*, as labour use is mainly restricted to the maintenance of tubewell disposal channels. Figures in Table 5 are again under estimates to the extent that O&M of surface drains both within and beyond the distributary CCA have not been included, though financial responsibility for this may have to remain with government for the reasons given below.

Table 5 clearly raises the issue of how drainage O&M costs should be allocated and shared across the whole LBOD command area. The infrastructure is not evenly distributed, yet it can be argued that all components must be well maintained and operated for all water users to benefit. This particularly applies to the surface disposal network where upstream users depend on downstream maintenance, but also relates to the more diffuse hydrological boundaries for the effects of tubewell drainage compared to irrigation supply. It may be difficult to identify all the beneficiaries of a given drainage facility, while these may not be exactly the same as the water users for the nearest distributary. Water users more remote from a tubewell may be indifferent to its maintenance and unwilling to participate in joint management. *Zamindars* will certainly be very reluctant to share the costs of scavenger wells if they do not receive a share of the irrigation water lifted.

Resolving such issues will be vital to a successful participatory irrigation and drainage management policy in Pakistan. The tentative analysis in this paper must be broadened to other pilot schemes and refined through collection of actual O&M costs incurred by WUOs. Monitoring of system performance and water user's incomes is also required.

10   References

1.   WAPDA, *Five Year Business Plan for Operation and Maintenance*, 1995, Water and Power Development Authority, Government of Pakistan: Hyderabad.
2.   SDSC, Operation and Maintenance and Cost Recovery: With Special Reference to Farmer Participation in Drainage Activities, 1995, Sindh Development Studies Centre: Hyderabad.
3.   Coward, E.W., Irrigation Development: Institutional and Organisational Issues, in Irrigation and Agricultural Development In Asia: Perspectives From Social Science, E.W. Coward, Editor. 1980, Cornell University Press: Ithaka.

---

[1] The LBOD project is a major investment in main system re-modelling and watertable control through vertical drainage. Highly saline groundwater also necessitates an extensive surface disposal network.
[2] OFWM is an on-going Government and donor supported programme that promotes watercourse re-modelling and lining, and improved water management practices.
[3] Often following payment of bribes.

---

[4] Tea and food is provided which does carry some value for *haris*.

[5] Rotational allocation of constant supply in proportion to land ownership.

[6] IIMI are providing technical training and guidance to the pilot projects to achieve "design standards".

[7] Accuracy of the estimates were confirmed by monitoring actual de-silting of the distributary done in the rabi 1996/97 closure period by the WUO. The actual labour contribution was as estimated. The cash contribution was lower because of less expenditure on machinery and food (the work was conducted during Ramadan), but will be higher, as estimated, on future occasions.

[8] The case study WUO did repair the distributary main gate in March 1997 at a cost of Rs5390.

[9] To date the pilot WUOs have been unable to agree membership fees, reinforcing the view that willingess to pay depends on demonstrated benefits and clear expenditure needs.

[10] Water users on the case study distributary contributed the full costs of de-silting the distributary, completed by the WUO during the rabi 1996/97 closure period.

WATER AS AN ECONOMIC GOOD IN AFRICAN SMALLHOLDER FARMS

F CHANCELLOR
Overseas Development Unit, HR Wallingford, Wallingford, UK

Abstract

The cropping patterns and marketing strategies observed in smallholder irrigation schemes in a variety of economic situations in Africa are reviewed. The constraints that farmers face in amending their farming decisions in response to changes in the marginal costs and returns of water are discussed. At farm level, rational decisions are often dominated by scarcity of factors other than water resulting in inefficient use of scarce water.

The impacts of paying for development costs, operation and maintenance or services on farmers' short-term and long- term decisions are considered.

Keywords: Smallholder irrigation, Africa, payments, water-use.

1   Introduction

Fresh water is scarce and is becoming expensive to capture so that development costs for new irrigation are higher than before. Scarcity of funds and unpopularity of large schemes for a combination of social, environmental and cost reasons combine with water shortage to focus new development onto small schemes. In addition, in Africa, suitable sites for large developments are increasingly hard to find. Maintenance of existing systems is costly and developing country governments already find the costs insupportable. For all these reasons there is considerable pressure to devolve costs to the farmers.

Thus irrigation development in sub-Saharan Africa in the foreseeable future is likely to be small-scale, financed through loans raised by the farmers themselves and farmer managed. There are already a number of small schemes which have come into being in this way. In Kenya, recent construction of small schemes for horticultural

Water: Economics, Management and Demand. Edited by Melvyn Kay, Tom Franks and Laurence Smith. Published in 1997 by E & FN Spon. ISBN 0 419 21840 8.

production has gone ahead on this basis through SISDO[1] loans. There are older small schemes throughout sub-Saharan Africa, which have moved from agency, top-down management to farmer-committee management, often through a process of self-financed or grant-aided rehabilitation. In the main, design of these small schemes is simple. Opportunities to control water closely and to trade volumes of water are limited.

2   Schemes

There is wide variation in the size of schemes classified as small. Typically, in Africa, small schemes fall in the range between 50 and 400 irrigated hectares. The age of the scheme, characteristics of the area and institutional conditions surrounding construction all affect design. Pumped schemes relying on diesel or electricity, often delivering water through pipes to be applied by sprinkler, are typical of modern technology. Gravity flows from run of river intakes delivering water through unlined earth channels and simple flood recession schemes are typical of low technology systems. There is no hard and fast link between delivery systems and crop patterns, although broad concepts apply such as rice schemes utilising surface flows and horticultural schemes demanding a high degree of water control for successful production.

On the whole, new system design is influenced by shortage of funds as well as low operation maintenance capabilities. Measurement of water volumes is therefore a luxury seldom included among scheme requirements. However, lack of routine measurements reduces potential for tight management control and renders volumetric pricing of water impossible at farm or even group level.

Small schemes are usually farmed by smallholder farmers who each irrigate an area in the range between 0.5 hectares and 5 hectares. Smallholders are essentially household based, relying on the farm family to provide labour for the most part, but they include a minority of families who have to hire most or all of their labour for a variety of reasons. Usually African smallholder-irrigators have other agricultural interests, such as rainfed crops and livestock. Increasingly, smallholder families are in receipt of income from the paid employment of at least one family member. Farmers contribute to schemes by providing labour for maintenance and time and commitment for management tasks in part payment for water for agricultural production. In some places, charges are levied by the government department, or agency, which controls the intake and/or the primary distribution system, in others a farmer committee or a loan agency levies a payment.

The most common bases for charging are land areas and membership. There may be seasonal charges, to account for differing service areas or there may simply be an annual payment. It is increasingly common for schemes to adopt a review process to refine the payment system by assessing cases where water shortage has significantly reduced crop yield. All these methods, however, are essentially flat-rate or service fees and only loosely relate costs to volumes of water used.

3   Cropping patterns

In smallholder irrigation schemes mixed farming is common and there is normally a high degree of interdependence among the various farm enterprises and between farm enterprises and non-farming activities. Profit maximisation theories fail to explain decisions made by smallholder farmers. Risk avoidance considerations are also important. Interactions between the socio-economic conditions of the community, intra-household complexities and economic objectives all influence decisions relating to irrigated crops. In the long term, cropping patterns emerge from interaction between a large number of parameters. The farmer judges the success of his/her choice by how well the physical, social and cultural needs of the family are met. Although the yield of subsistence crops will be important to the farmer, it will not be the whole story and it is increasingly recognised that benefits must include a substantial flow of cash. Where the available irrigated land is limited in area farmers achieve high cash incomes from high value crops rather than cereals. Cash is essential to buy the increasing range of goods and services needed to satisfy the family. High in the list of priorities is cash for school fees and supporting children in higher education. Cash is also essential to operation, whether for payment of staff salaries and essential materials or, in farmer-managed schemes, only for material costs.

3.1   Crop choice

Sensitivity to price changes of factors of production is evident, as in the case of reduced fertiliser applications resulting from increased costs that followed structural adjustment. Crop choice is sensitive to selling price and input costs. Most inputs, including labour have defined unit prices. Water is the exception.

Observations from a number of countries suggests that sustainable schemes regularly market over 60 % of the crops produced (Chancellor and Hide, 1996). Demand is therefore crucial to success. In the subsistence sector of the farm, crops are likely to be determined by the needs and preferences of the family and will be influenced by factors such as taste, storage life and, if selling a surplus is important, possibly local demand. How much irrigated land will be devoted to subsistence farming will be influenced by the reliability of rainfed agriculture in the surrounding area. Reliability of irrigation water delivery also influences the area devoted to subsistence crops. In schemes where tail-end delivery is unreliable farmers grow a greater proportion of subsistence crops. The high cost of inputs for commercial crops represents a high potential for loss. If water supply is known to be unreliable then the rational response is to minimise risk. In the short term, it is unlikely that farmers will reduce the area devoted to subsistence crops as a consequence of costs associated with water. Other parameters have a strong impact. In the commercial sector of the irrigated farm, demand, unit price, risk, production expertise, costs and marketing conditions, will all influence choice of crop. The importance of crop choice should not be underestimated. A recent study in Kenya found that small irrigation schemes that had consistently good financial performance were characterised by cultivation of crops for which the scheme could develop specialised supply (Chancellor & Hide, 1996). In some cases farmers responded to demand for a range of crops and, over time, built a reputation for quality or timing for specialised market sectors, such as

Christmas demand. In other cases supply was geared to a small specific market demand such as mangetout peas. Expertise in production and attention to the finer points of marketing are strongly linked to good financial results.

Inevitably the range of crops from which choices can be made is determined by physical parameters such as the quality and type of soil and water available to the farmer. The farmer uses knowledge and experience to determine the best option for his/her situation. Determining the impact of individual factors of production is commonly based on a combination of experience and extension advice. In general, despite a lack of farm records among rural smallholders, farmers form sound judgements.

4   Paying for services

Confusion exists about payments for irrigation. In many countries, payment for water is considered politically unacceptable; however, payment for operation and maintenance is encouraged; this sometimes includes payment for services. The level of payment is not always clearly defined and may be partly composed of farmer labour for repair and maintenance. Governments and agencies often levy fees as a cost-recovery exercise in which a flat rate is preferred for simplicity and because it assures the supplier of funds to cover the costs.

4.1   Volumetric payments

Metering of water is costly and impractical for many reasons in irrigation schemes. In theory, volumetric charging is possible in modern piped systems, but it is difficult in practice. Findings from a brief study in Egypt (Chancellor 1992) indicated that a close approximation to volumetric payment, through farm-level financing of pump costs led to intensive cultivation featuring high-value crops, utilising small plots as nurseries. Intensive crop patterns were later abandoned when costs were financed externally by commissioning of pipeline supply and large volumes were again freely available.

The cost of water was clearly a strong influence on crop choice although it was only one of several cost changes. Table 1 illustrates the change in intensity, which occurred in the period from 79/80 till 81/82 when water costs were linked to volumes used. Cropping intensity in these years was approximately 70 % higher than the average rate at other times. Increased control over water during that period may have influenced the rise in intensity of use but local farmers indicated that cost was the major issue.

4.2   The effect of irrigation flat rate charges

It has already been established that cost and practical difficulties militate against volumetric charges in most schemes. The two common forms of charging are based on either the number of farmers or the area of land irrigated. Both strategies have the effect of making the cost of water one of the farmers' fixed costs. In that case, there is little incentive to the farmer to apply less water, unless a situation exists in which less

Table 1. Cropping pattern at El Hammami

Percentage of cropped area

	78/79	79/80	80/81	81/82	92/93
Wheat	8.2	6	5	1.9	5.5
Maize	25.2	10.8	13.5	12.5	14.5
Vegetables	42.2	55.7	59.5	62	53
Peanuts	2.1	2.1	3.6	4.5	3
Other	1	10.6	5.9	2.8	1
Cropping Intensity	160	340	371	335	235

water will improve yield, such as when the soil is poorly drained or waterlogged.

Indeed, a fixed charge may act as an incentive to farmers to apply more water if, in the farmer's opinion, water can be substituted for other factors of production which are part of the variable costs. Using the example of rice, more water applied might provide greater ponding depth; this would, in turn allow less weed growth, and would have the effect of reducing labour costs.

There is some evidence to suggest that when farmers are asked to pay a fee then access to water is regarded as a right which encourages heavier use. A similar effect has been observed in the UK in relation to payment for hosepipe services. As a result of having paid, people became more determined to use hoses in times of shortage than they had in the past. Such behaviour either increases the total volume of water used by the group or results in increased inequity as head-reach irrigators increase their share of water. The negative effects of using water extravagantly are pernicious and are not easily recognised.

Better water delivery is expected when farmers pay. When a good service does not materialise, payment may be withheld. In SW Kano in Kenya, a new system delivered reliable flows of water, for a fee. Previously less reliable flows had been freely available to the farmers in the form of outflows from an upstream system. Farmers were reluctant to pay, as they did not regard the new service as providing sufficiently significant improvement (Chancellor and Hide 1996). In this case, conflict resulted until fees were paid. In the short run the effect was to reduce production and farm incomes. Farmers, to whom irrigation water had not been previously available, paid relatively willingly.

4.3   Short term effect of a service charge

Clearly costs will rise when a charge is levied. The increase will be the same regardless of crop and should have the effect of reducing the amount of irrigated land allocated to crops with narrow margins. However, low value crops are often those grown to meet subsistence needs. It is unlikely that reduced margins will have much impact on the land devoted to these crops. A move to higher value crops in the commercial enterprise is a more rational response but is one which will increase input costs still further and will necessitate support of production loans. It is difficult to accurately predict results.

The effect on the volume of water used will depend on physical characteristics of the delivery system, farmer knowledge and the land. The design of control structures and channels may not allow the system to be flexible.

4.4 Water licences

Generally, a nominal charge to cover the administrative cost of investigating the claim and issuing the licence is incurred to secure licences to abstract water from common sources such as a rivers or aquifers. A scale of initial and annual charges is enforced based on the category of the licence. The licence usually refers to maxima for abstractions in given time periods and is only loosely related to the actual amounts taken from the source. In water-short regions the problem of enforcement is immense and the system is increasingly vulnerable to political manipulation and corruption. As competition between users and water for the ecosystem becomes an increasingly strong political issue it will be necessary to consider incorporating some form of volumetric pricing in the licence system, if it is to serve any function at all in future.

5   Produce Markets

Rural demand is often insufficient to support the commercial output of irrigation. Produce leaves the local markets for urban markets or for export. Market structure and infrastructure are essential to success. Developing countries have relatively cheap labour and thus have a comparative advantage when selling produce in the markets of developed countries. There is substantial international trade in irrigated crops such as cotton, sugar and rice. Air-freight has opened markets for fresh produce in recent years which resulted in rapid growth of international horticultural markets supported to some extent at national level by urban growth and a rise in urban consumption levels.

Smallholders are essentially 'price-takers' and are vulnerable to fluctuating prices. Producers from all over the developing world compete for trade within the North American and European markets. These sophisticated markets impose quality standards and regulations which either the smallholder fulfils at a cost or faces penalties for failure to comply. In some places, the middlemen take part of the risk and in others, risk falls to the smallholder. Marginal revenue is hard to calculate or predict in these circumstances

On the supply side, the lags inherent in agricultural production contribute to fluctuations in price particularly for producers in remote areas where glut conditions often lead to poor levels of sales

Natural disasters such as flood and drought are frequent and unpredictable adding to the difficulties that all smallholders' experience in producing crops to comply with quality assurance standards. Other problems beset smallholders relating to availability of fertilisers, seeds, pesticides, equipment and timely credit. In irrigated production, there are complex interactions between all the factors contributing to the production process. The relationship between fertiliser application and yield is dependent on soil moisture, application of more or less water can change the cost of labour for weeding and so on. These relationships, or multicollinearity, make it impossible to disentangle the effects of the cost of water from other input variables.

In the foreseeable future it is unlikely that smallholder farmers will be able to accurately take account of marginal costs and returns to water for irrigated crops because of the limited capacity to measure and price water. If water is to become an economic good design must focus on increased efficiency, measurement and control. These features are currently too costly to be included in the budgets of farmers developing irrigation schemes through loans

6   Water markets

In general, there is little trading of water between groups of irrigators on different schemes. This is largely because water has been regarded as a communal resource to which access was free, apart from the costs of capture, storage and distribution. The nature of water supply and demand and the problems inherent in transporting and storing water have meant that although transaction costs were high, water itself has a low unit value.

Externalities, arising from using water, either in agriculture or in domestic-use, in developing countries were ignored in the first half of the century. Recently, population pressure and ever increasing demand for water have brought the costs of externalities to notice. It is increasingly recognised that downstream users are stakeholders in all proposals and that negative effects of water-use must be considered against the benefit of the proposed scheme. Environmental costs have received progressively more attention since the Rio summit.

7   Discussion

Technical difficulties and high costs associated with measuring water have constrained volumetric charging for irrigation water in smallholder schemes in sub-Saharan Africa. However, African farmers often pay development and operation and maintenance costs. Water related costs for smallholders in Africa are therefore commonly assimilated into fixed costs of production. It is unlikely that this will lead to farmers treating water as an economic good in the way they do other inputs which have a unit price.

Small-scale abstraction works commonly lack measurement and control facilities sufficient to allow farmers to consider trading water. There are political, social and cultural barriers to trading of water between countries, provinces and even between communities in the same watershed. There are, however, some circumstances where some trading of water rights may be possible, such as where water is supplied to groups from a well-regulated main canal system or through a piped network. At field level, it is sometimes possible for trade in water to take place between individuals who farm adjacent land. In a piped systems there are possibilities for trading access to water for crops with very different water requirements cultivated by different groups. In Punjab, in India, the Warabundi system allows trading of watering turns and is common practice. It is possible for a water price of sorts to emerge from such trading.

In the short term, paying a flat-rate for development loans, operation and maintenance or services is likely to have the effect of increasing water use by individual irrigators and could encourage over-use of water with detrimental consequences.

Payment has some beneficial effects by stimulating farmers to participate in determining cost-effective operation and maintenance policies, providing funds to carry out policies and encouraging capacity building in skill areas associated with irrigation management, operation, maintenance and procurement. Payment is a key issue but the form it takes will produce different impacts. It is necessary to identify objectives to be achieved and consider how payment will fulfil those objectives.

New smallholder schemes are being developed using loan financing and the degree of farmer involvement is impressive. The objectives of these systems, however, do not normally include saving water because the size of the system is constrained by lack of land or lack of finance.

Costs and technical difficulties are likely to militate against volumetric payment for water in smallholder irrigation scheme for some time. Promoting flat-rate fees to meet development, operation and maintenance costs will not automatically lead to water pricing.

8   Conclusions

Volumetric payment for water in smallholder irrigation schemes is impractical in the short term

Paying for services can encourage a constructive attitude to management and maintenance

Flat-rate payment for services does not encourage water saving. Rational farmers will use more water.

Volumetric payment will lead to water being used as an economic good

Practical, political and social constraints work against volumetric pricing

9   References

1.   Chancellor, F. and Hide, J. (1996) Small-holder irrigation : Ways forward. HR Wallingford Ltd, OD 136, Vols. 1 and 2.
2.   Chancellor, F. and Fawzy, G. (1994) A study of socio-economic conditions facing farmers at the El Hammami pipeline project, Egypt. HR Wallingford Ltd, OD/TN 66.

1. SISDO Small-Scale Irrigation Support and Development Organisation is an NGO which has been in operation in Kenya for a number of years. It extends credit to groups of farmers, using a carefully monitored participatory approach to development and a group guarantee repayment system. It specifically targets horticultural developments to ensure both satisfactory payback and healthy farm profit. Farmers are supported with good quality technical support, extension and market advice.

SECTION E

MANAGEMENT SYSTEMS

AN ASSESSMENT OF THE POTENTIAL OF OPTIMISATION IN REAL TIME IRRIGATION MANAGEMENT

R.B. WARDLAW
Department of Civil and Environmental Engineering, University of Edinburgh,UK
D.N. MOORE and J.M. BARNES
Mott MacDonald, Cambridge, UK

Abstract

An optimisation approach has been developed to aid decision making in the real time allocation of water in irrigation systems under stress, and to assist in the assessment of different operational rules and rotational systems and their impact on production. This work has been undertaken as part of the ODA funded TDR project on "Improved Irrigation System Planning and Management". The optimisation approach is based on quadratic programming, and has as its overall objective, to maximise crop production through appropriate water allocation, while maintaining equity between schemes and units within the system. The potential of the approach has been assessed through application of a simulation model of the Lower Ayung irrigation system on the Island of Bali in Indonesia. The indications are that significant improvements in crop production can be achieved through application of the optimisation approach. The paper reviews the basis of the optimisation approach and proposes means by which the model performance can be assessed, using the Bali Scheme as a working example to demonstrate result interpretation and to identify practical considerations.
Keywords: Irrigation planning, optimisation, real time operation, simulation.

1 Introduction

This paper outlines work undertaken as part of the ODA funded TDR project on "Improved Irrigation System Planning and Management", under a subheading of "Optimal Allocation of Irrigation Water Supplies". It is concerned specifically with the optimal allocation of scarce water resources in real time between competing users.

The project builds on earlier work carried out under an EPSRC grant (Wardlaw and Coals, 1994), and is directed primarily at complex distribution systems. The real time irrigation water management problem is one of providing the best distribution of scarce water resources, such that overall crop production is maximised. It is not a

Water: Economics, Management and Demand. Edited by Melvyn Kay, Tom Franks and Laurence Smith. Published in 1997 by E & FN Spon. ISBN 0 419 21840 8.

planning problem, in that crops are considered to be in the ground (although the approach could be used to aid planning also), nor is it a water scheduling problem, in that storage is not considered. The application is, at present, primarily to run of river systems.

2   Definition of the Optimisation Problem

2.1   Objective function
If all inflows to an irrigation system are known and their spatial locations are defined, the requirement is to determine the optimal allocation of the available water resource which will maximise some defined objective, subject to a set of constraints. The objective, as discussed above, is one of maximising crop production through water allocation, and this would be subject to constraints of equity between users, canal capacities, and continuity or water balance constraints throughout the system. Crop yields will be reduced as a result of water stress. Yield response to water is dependant on the stage of crop growth, and can be quantified in a number of ways. The more useful have been summarised by Wardlaw and Barnes, 1996. Yield response functions can form the basis of the objective function, although initially the problem can be considered as one of water allocation, and of maximising supplies. The water allocation problem formed the basis of preliminary work.

A number of alternative objective function formulations is possible for the water allocation problem. The most appropriate function has been found to be of the form (Wardlaw and Barnes, 1996):

Minimise

$$Z = \sum_{i=1}^{n} \frac{(d_i - x_i)^2}{d_i} \tag{1}$$

where,

n = number of irrigation schemes
d_i = irrigation demand for scheme i
x_i = irrigation supply to scheme i

Fig. 1. Nodal water balance

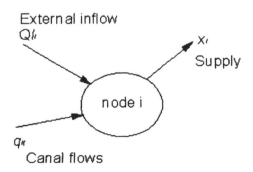

This function preserved equity of supply in the system, ensuring that when water stress does occur, relative supplies are the same to all schemes.

2.2 System constraints

System constraints exist in the form of nodal continuity and reach capacities. Continuity is fundamental to all water resource problems, and for the nodal system shown in Figure 1 may be expressed as follows:

$$QI_i + \sum_{j=1}^{m} q_{ij} - x_i = 0 \tag{2}$$

where,

QI_i =   external inflow to node i (i.e. not from canal system)

q_{ij} =   flow to node j from node i (i.e. flow in canal connecting nodes (schemes) i and j)

The following additional constraints are required:

$$q_{ij} \le qcaq_{ij} \tag{3}$$
$$x_i \le d_i \tag{4}$$
$$x_i \le q_{ij} \tag{5}$$

where, $qcap_{ij}$ =   capacity of the canal between nodes i and j

2.3 Problem solution

A number of approaches to solution of the optimisation problem have been considered, including non-linear programming (NLP) and dynamic programming (DP). Other techniques such as genetic algorithms or simulated annealing may offer advantages in large systems, but their applicability and performance on problems of this nature has yet to be investigated. Under the category of non-linear programming, quadratic programming offers an attractive approach to the problem. There are a number of quadratic programming solvers available commercially, requiring only expression of the problem in the correct format. Earlier preparatory work (Wardlaw and Coals, 1994) had indicated that this was a promising approach. An allocation algorithm based on dynamic programming has been developed (Wardlaw and Coals, 1994), but it is difficult to set up and apply. While the approach certainly has potential, it was considered that quadratic programming offered more immediate advantages in terms of demonstrating what could be achieved.

3   Preliminary evaluation of optimisation on the Tukad Ayung system, Bali

3.1   General

The irrigation systems from the Tukad Ayung in Bali were studied in some detail by Mott MacDonald in 1988. Rice is the principal crop, and over most of the area two rice crops and one dry foot crop are grown annually, under four basic cropping patterns. A simulation model of the system was developed to assist in assessing the potential impacts of potable water abstraction from the Tukad Ayung on the irrigated agriculture, and has been described by Wardlaw and Wells (1996). This simulation model of an existing irrigation system has been used as a control against which the potential of the optimisation approach could be evaluated. A schematic of the irrigation network is shown in Figure 2. The Lower Ayung simulation model runs on a half monthly time step, incorporating a system simulation model, a field water balance model, and a crop production model.

For preliminary evaluation of the optimisation approach, the simulation model was simplified, and the water balance and crop production components of the model removed. This required that drainage returns were incorporated in the model as a function of irrigation supply, representing losses and contributions to downstream

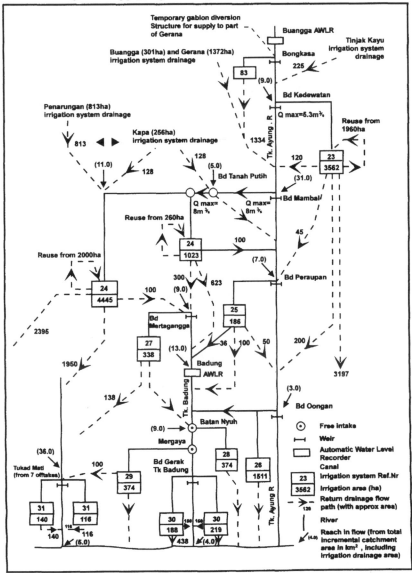

Fig. 2.   Detailed irrigation system schematic for the lower Ayung basin

schemes. Inputs to the system, and gross irrigation demands for all nodes were as defined in the full simulation model. The optimisation model was configured to run with exactly the same inputs as the simulation model, and with the same network description. The models can be run with 34 years of historical inflows.

In setting up the optimisation model for the lower Ayung irrigation system, it became apparent that further constraints were required to permit diffuse drainage to modelled correctly. It will be noted from Figure 2 that for many schemes, estimates have been made of the proportions of scheme area which contribute to particular drainage routes which are shown dotted. These routes are represented as normal reaches in the network diagram (Figure 3), but there should not be free choice over how drainage from a scheme is distributed through these. Fixed proportional distribution factors are specified for these reaches in both the simulation and the optimisation models.

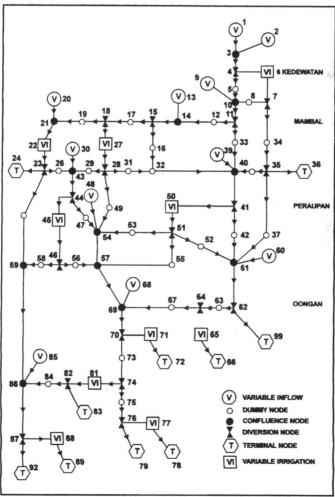

Fig. 3. Network diagram for the lower Ayung system simulation model

3.2 Model performance

Model performance was evaluated initially over the full 34 year simulation period, using drainage returns of 10%, 30% and 50%. Actual drainage returns are of the order of 40-50% for the system, and the range adopted has permitted demonstration of the potential benefits of optimisation under wider extremes of water stress than exist naturally in the system. The simplest indicator of model performance is a comparison of drain flows leaving the system. From Figure 3, it can be seen that outflow from the system occurs from nodes 24, 36, 66, 72, 78, 79, 83, 89, 92. The total outflows from these nodes have been summed over the entire simulation period. Table 1 summarises the results in terms of total outflow volumes over the simulation period, and in terms of average outflow rates. As drainage returns are reduced and water stress in the system increased, particularly in the lower part of the system, then the optimisation model produces progressively higher benefits in terms of water allocation. The indications are that significant benefits are achievable through optimisation, particularly under severe water stress.

Table 1. Evaluation of average water allocation benefits

Drainage return	Optimisation model outflow		Simulation model outflow		Reduction in outflow
(%)	(Mm³)	(m³/s)	(Mm³)	(m³/s)	(%)
50	8440	7.867	8815	8.216	4.3
30	6525	6.082	7142	6.657	8.6
10	4787	4.462	5670	5.285	15.6

Irrigation water supplies are maximised in an equitable manner, and in part this is achieved through reducing supplies to schemes from which there are large drainage flows out of the irrigation system. Figure 4 shows relative irrigation supply deficits, defined as the difference between irrigation demand and actual supply. Scheme 6 generally receives adequate irrigation water supplies, and deficits do not occur on average. As a result, there are significant drainage outflows through node 36. Scheme 22 on the other hand is generally short of water, and as a result, outflows through node 24 are not large. The optimisation model reduces supplies to scheme 6, and attempts to provide the same relative deficit in irrigation supplies in all schemes, subject to supply and capacity constraints. From Figure 4 it is clear that the model does this very well. It could be argued that this could be achieved easily by more appropriate operating rules at the head of the system, but for schemes 71, 77 and 81, where interrelationships with drainage returns are more complex, derivation of more appropriate operating rules would be difficult. As a management and decision aiding tool, the approach clearly has significant potential.

4 Extension of the modelling approach

4.1 General

The results achieved with the simplified model indicated that the approach had sufficient potential to warrant further development towards practical application. Of particular importance for systems such as that in Bali, where there is significant reuse

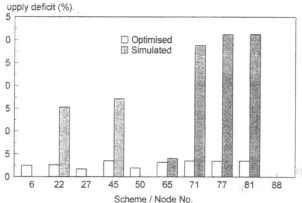

Fig. 4. Evaluation of optimisation model performance

of drainage water, is treatment of drainage and efficiencies in a deterministic manner. It was considered that real time implementation of the approach would require:

- real time monitoring of river flows at inflow points;
- continuous simulation of in scheme soil moisture balances to determine drainage returns;
- continuous simulation of in scheme soil moisture balances from which actual irrigation demands could be computed in real time;
- monitoring of cropping such that the influence of stage of crop growth on damage could be incorporated in the optimisation approach.

Better representation of drainage returns, and evaluation of crop yield response has been achieved through incorporation of the SOILBAL model into the optimisation model. SOILBAL was developed by Mott MacDonald (1982) for use in irrigation and basin planing

studies, and had been incorporated in a modified form in the Lower Ayung Simulation model (Wardlaw and Wells, 1996). It's inclusion in the optimisation approach required significant further modification to permit the model to keep track of soil moisture levels across all schemes in the system.

The model computes water balances continuously on up to six irrigation blocks in each scheme, in response to irrigation supplies, rainfall and the cropping calendar. Outputs from the model include flows at any point in the system, recharge to groundwater, and potential and actual evapotranspiration for each block in each scheme in each time step. These data can be used in a crop production model to assess overall system response. The model actually optimises between blocks within a scheme as well as between schemes. There is a penalty for this in terms of computational effort, although in real time applications this would not be significant.

Operation of the optimisation model for a single time step on a Pentium PC takes about four minutes for the Bali system.

4.2   Results

The fully integrated optimisation and simulation model has been applied to the Tukad Ayung system in Bali, and results compared with those of the simulation model alone. The optimisation model was run with the water allocation objective function given earlier. The optimisation model increased water allocation by 12.5%. Crop losses for the 1957 water year, a representative dry year, were reduced from 618 million rupiah to 197 million rupiah (£1 Rp2500 in 1988), and equity improved throughout the system. The relative crop losses are shown in Figure 5.

Within the present network, there is little freedom or control over drainage routes. By permitting the model to optimise on drainage, further improvements in supply can be achieved. By relaxing drainage constraints and canal capacity constraints, the model gives an indication of the full system potential. Under these circumstances, water allocation is increased by 20% and crop losses reduced by a further 25%.

5   Conclusions

The results obtained indicate that application of a relatively simple optimisation approach for real time water allocation in complex systems can provide significant benefits, both in terms of crop production and equity of supply amongst users. The extended modelling approach permits objective functions based on crop yield response as well as irrigation demand to be incorporated in the model. Ongoing investigations include the assessment of objective functions based on
yield response, and assessment of the benefits of including real time irrigation demand calculations.

Implementation of the approach would not be difficult, and need not rely on sophisticated equipment. The approach could be viewed as a decision support tool for operations staff. Approaches to implementation are being considered under the present ODA project.

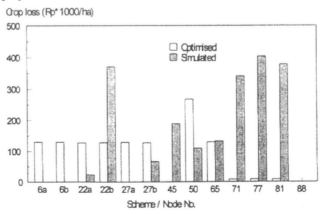

Fig. 5. Comparison of crop losses for 1957

6   References

1.   Wardlaw, R.B., & Coals, A.V. (1994) Assessment of appropriate algorithms for optimal allocation of irrigation water in real time. *Rept. no. cenv0294,* Dept. of Civil and Environmental Engineering, the University of Edinburgh.
2.   Wardlaw, R.B., & Barnes, J.M. (1996) Improved irrigation system planning and management: optimal allocation of irrigation water supplies.   *ODA TDR Research Project No 6261, Phase I report,* April 1996.
3.   Wardlaw, R.B., and Wells, R.J. (1996) Simulating crop production and resource development impacts: the Lower Ayung Simulation Model. *Proc. Instn. Civ. Engrs Wat., Marit. & Energy.* Volume 118, Issue 2, June 1966. ISSN 0965 0903.
4.   Mott MacDonald. *SOILBAL user guide.* WRP internal report, MM Cambridge, 1982.

PROVIDING A WATER DELIVERY SERVICE THROUGH DESIGN MANAGEMENT INTERACTIONS AND SYSTEM MANAGEMENT
Achieving control for water pricing

B.A. LANKFORD
School of Development Studies, University of East Anglia, Norwich, UK
J. GOWING
Centre for Land Use and Water Resources Research, University of Newcastle upon Tyne, UK

Abstract
The premise that water for irrigation should be treated as an economic good, in the same way as other agronomic inputs, holds only if its supply is controllable. The design and management capability of an irrigation system to provide control over water is therefore critical. Three main types of potential water supply control can be identified based on increasing accuracy of allocation under higher standards of performance. The three types, which arise via design management interactions, are water provision, water distribution and water partition. Water control can improve within each type or it can move from one to another by engaging in design management interactions. It is the water distribution and water partition levels which enable higher service standards and a realistic means to charge for water used. Recognition of these concepts is important both to diagnosis of on-going management problems and to the design of new and rehabilitated systems.
Keywords: Canal irrigation, control, delivery, design, management, pricing.

1   Introduction

Much current interest in irrigation management relates to the broader issues of the changing role of the state in natural resource management. The motivating force is the belief that users of natural resources should be responsible for their management to a much greater extent. Accompanying this shift of responsibility is a desire to introduce "market forces" in order to promote increased efficiency.

Irrigation management transfer embraces a range of initiatives aimed at devolving O&M responsibility to users. There is an underlying assumption that farmer managed irrigation schemes (FMIS) are more efficient than those run by public officials and are

Water: Economics, Management and Demand. Edited by Melvyn Kay, Tom Franks and Laurence Smith. Published in 1997 by E & FN Spon. ISBN 0 419 21840 8.

more responsive to the needs of their users. However evidence for this is insubstantial and further attention needs to be given to efficiency and equity assessments [1]. Much of the research into FMIS has concentrated on small-scale traditional schemes and there has been an implicit assumption that lessons from such studies translate effectively to large scale public systems. This assumption is flawed [2] and there is growing recognition of the importance of technology/management interactions [1].

It is argued that the costs of irrigation service provision should be borne by the beneficiaries and there is an extensive literature on cost-recovery strategies [3]. But as Bottrall (quoted in [4]) points out; "despite the desirability of raising water charges, it should be seen as a secondary issue in terms of sequential actions, first because it is a highly politicised issue; and secondly because in most cases farmers will not become better disposed to the idea of higher charges unless other changes are made first - the most important of which is an improved water distribution service."

Analysts point to the dominance of irrigation demand in the face of increasingly scarce water resources and advocate market-based allocation based on an assessment of the economic value of water. It follows that in order to ration water use by price, direct volumetric measurement of demand is inescapable. This has significant technological implications in that the means of making such measurements seldom exists on large-scale public canal systems. The limitations of efficiency gains from introduction of market forces should therefore also be seen in relation to technology management interactions. This paper deals with the issue of the standard of delivery service in large irrigation schemes based upon gravity (i.e. canal) distribution. The importance of control over supply to users is emphasised and the dependence on engineering is discussed in the context of design management interactions.

2   Definitions of design management interactions

Design management interactions have been analysed at different levels which are described below in order of an increasing appreciation of their complexity:
1. *System-technology*: The 3 main irrigation technologies; gravity (canal), sprinkler and drip (trickle), each have specific management requirements. This paper is concerned with design management interactions on gravity irrigation systems.
2. *Size of scheme*: Researchers have identified links between scheme size, complexity and water control [5], with very large and large schemes, typically 1000 to 10,000 ha and over requiring "full water control"; medium-scale schemes, 100 to 1000 ha, needing "full or partial water control" and small scale schemes, 1 to 100 ha, being controlled by farmer groups or single farmers.
3. *Structure-use*: Researchers have investigated the difficulty or "user-friendliness" and degree of manual or automatic operation of structures for water control. These address control of either water levels or discharges or both using single structures or combinations. A good example of this type of analysis is found in Plusquellec et al [4] who discuss use-related interactions under 'Robustness' and 'Ease of Operation', and include a subsection entitled 'Specific structures to avoid.'
4. *Management-aim*: Closely related to the previous analysis, is the discussion of design in the light of how it assists with a water management aim, the common one

being to measure water flow. For example: "Usually water measurements should be planned at all points where it can be reasonably established that information on the flow rate will affect management decisions" [6]. Furthermore, "the flexibility to cope with given situations rests largely upon the cost and type of regulation structures and techniques used in the canal network" [7]. Plusquellec et al [4] examine management aims under 'Functionality of control structures', where they identify the need for sufficient structures to ensure adequate water-level control, via such functions as on-off and flow rate adjustment. The problem with focusing only on use-related and management-related interactions is that while these are critical to water control, they exclude many other factors which enable irrigation schemes to have enhanced self-running capabilities under conditions of required higher performance. For this it is necessary to look at wider design management interactions that explicitly affect performance.

5. *Implicit performance-affecting interactions*:   Prior to discussing the explicit performance-affecting interactions, it is useful to examine how design and management are implicitly presented as factors influencing performance.   For example Uphoff et al [8] describe a 3 dimensional matrix of irrigation management arising from 'water delivered to the crops' (made up from acquisition, allocation, distribution and drainage), 'control structures activities' (design, construction, operation and maintenance) and 'organisational activities' (decision making, resource mobilisation, communication and conflict management). While on the face of it, this matrix hints at interactions, the text does not explain how and fails to address the design-side at all: "Because this study focuses on improving the management of existing systems, we concern ourselves only with the latter. Happily, this reduces the number of management variables to be considered." It is this the lack of focus on interactions with design that hinders management from investigating new designs to observe how they improve allocation and performance.

6. *Explicit performance-affecting interactions*: Researchers are beginning to believe that explicit system-wide design management interactions influence performance. Dargouth [9] highlights the interdependence of system design and water management but acknowledges that research on the subject has been carried out by a few. Ankum [10] states; "flow control systems have developed through a gradual evolution from simple towards high technology, keeping pace with the increasing performance requirements." Bos and Nugteren [11] describe conditions that favour the efficient use of irrigation water and second in the list is the control over water flow to the system and within it: "Only then will it be possible to establish a close match between crop irrigation water requirements and water supply." Turral [1] notes that experience of successful management transfer from USA, Japan, Taiwan and Korea may be misleading in that the physical infrastructure was functioning well and in some cases was improved as a precondition for transfer: "In contrast, many systems in developing countries may suffer from inherent design or construction problems." The remainder of the paper develops the concepts behind these system-wide design management interactions which characterise and affect the delivery of water on irrigation systems.

3   System-wide design management interactions

Design management interactions give rise to three main types of irrigation systems termed water provision, water distribution and water partition. Table 1 presents the 3 types and their main characterising features, which arise from management and engineering choices regarding system infrastructure design, system configuration design and design operational procedures. Further details and discussion on the method of characterisation of irrigation systems is found elsewhere [12]. New and different names are suggested for the main/secondary canals and tertiary canal/field delivery systems. For example, water partition systems consist of division canals supplying water to apportionment canals.

Table 1. Diagnostic features of three main types of design management interactions

Design management type	Main and secondary canals	Secondary canal to tertiary/fiel interface	Distinguishing/diagnostic features
Water provision	Transfer network	Dispersal network	• Lack of flow measurement • Variety of canal system operational schedules • Discharge control mainly at headworks • No or rare head/level control • Simple sluice gates with on/off capability • Proportional distributors are common • Single flow is rotated • Reservoir or canal night storage is rare
Water distribution Sub-type: Level control	Conveyance network	Distribution network	• Emphasis on controlling water levels (cm, m) • Variety of canal system operational schedules • Steady or non steady water level control • Manual-active to automatic level adjustment • Continuous water level adjustment • Flow measurement is rare • Various types of rotation • Occasional reservoir or canal night storage
Sub-type: Discharge control			• Emphasis on controlling flow rates (l/sec) • Variety of canal system operational schedules • Manual-active/passive flow measurement • Variable/active head control structures • Manual-active to automatic adjustment • Occasional reservoir or canal night storage
Water partition	Division network	Apportionment network	• Emphasis on controlling supply hydromodules to match with demand hydromodule (l/sec/ha) • Structured system of canal scheduling • Adjustable headwork discharge control • Automatic/passive water measurement and head control • Stepped/fixed/passive flow adjustment • Matching command areas • Built-in design for strict rotation of water • Congruence between canal levels • Possible night storage via reservoir or canal

3.1   Water provision systems
The key feature of water provision systems is the absence of water flow measurement. Water level and discharge control are sufficient to minimise over-spills.   Good examples of such systems are small-scale farmer operated schemes and some warabundi-based schemes at the larger scale.

3.2   Water distribution systems
These are the most common type of larger scale irrigation systems.   Regarding their main distinguishing features, two sub-types exist either with an emphasis on discharge (l/sec) control or water level control.   The kinds of technologies found on these schemes is wide-ranging from intensive manual methods to automatic control.

3.3   Water partition systems
These systems are the rarest, with an emphasis on the accurate, strict control of the ratio of water supply to area (litres/second/hectare) which is termed the supply hydromodule.   On such systems, water is seen as a scarce resource and is measured and managed in fractions of l/sec/ha, and is carefully matched with crop and system water demands, termed the demand hydromodule.   These systems differ from distribution systems in that water supply control is reliant on a multi-factor, systems approach which includes the use of 'structured' canal scheduling [13], accurate design sizing, carefully chosen gate and water level control technologies, strict rotation patterns, good infield design and a matching of main to tertiary canal operation and design.   During peak demand periods, flexibility of allocation is reduced to increase the frequency of cycling of water between fields.   At present, only a few examples of such systems - in drought-prone areas in Southern Africa - are known to the authors.

4   System management
The above design management interactions do not solely define the standard of water allocation.   There is the possibility of improvement of water management within each of the design management types, and this gives a further dimension to the analysis. Two broad management types have been identified, termed 'normal' and 'actualising'.

4.1   Normal management
This term is borrowed from Chambers' [14] analysis of "normal" irrigation professionalism.   In such systems, the system is operated with much visible daily activity, but with little progression in long-term performance.   Brief examples of activities associated with normal management are outlined in Table 2.

4.2   Actualising management
This term implies the set of skills required to improve system management which results in consistent long-term enhanced performance.   Described by Chambers under the banner "new" professionalism, some examples of these skills are provided in Table 2 under two main activities; diagnostic analysis and practical action.   Both go hand in hand in iterative circles encouraging managers to learn how to improve performance.

Table 2. Approaches to irrigation system management - normal and actualising

	Normal	Actualising	
		Diagnostic analysis	Practical action
Infrastructure and in-field design	Accepts design	Questions design	Introduces trials and new designs, layouts, methods, etc.
Main system management	Accepts current methods or operates incorrectly	Examines effects on allocation	Alters methods and diagnoses new impacts
Monitoring and evaluation	Omitted or by rote	Investigates need and methods of data collection and analysis	Regular summaries and feedback into management. Active use of computers
Leaks/spills/ equipment and canal maintenance	Omitted or by rote	Determines level of cost-effectiveness	Flexible approach. Tries different methods & equipment
Interdisciplinary interaction	Isolated, does not seek interaction with others	Questions source & level of own knowledge	Seeks out and works with other specialists
Management & technical skills	Possibly too narrow	Questions own skills	Seeks new training & motivation
People/farmer management & management transfer	Exclusive, conflict-orientated, partial transference	Diagnoses situation, need, abilities, opportunities	Flexible, participatory, conflict-resolving, communicative

5   Framework of water supply control

The two dimensions of design management interactions and system management provides a 3 x 2 framework of water supply control. This framework of 6 classes is presented in Table 3 along with some water delivery symptoms that might be found within each class. Also included in Table 3 is the proposal that the three levels of design management interactions have associated means of charging users for water. Charging accurately for water on a volumetric basis might be better suited to partition systems where either the time of delivery converts directly to volume applied or, because of the degree of control, even simply counting the number of irrigations gives the amount of water used. Volumetric charging on distribution systems is possible but imposes a greater burden of measurement of flows. Alternatively, fixed and non-volumetric direct charges may be more appropriate on distribution and water provision systems. Moreover, a dependable fair delivery service, necessary for successful collection of water payments, can only occur when irrigation management is in the actualising mode, and this may apply to all three design management types.

6   Conclusion

If water is to be introduced as an economic good on irrigation systems, managers need to be able to treat it as such, and farmers need to have the same view based on an

Table 3. Water supply control via design management interactions and system management

		System management dimension	
		Normal	Actualising
	Water provision (WPV)	• Incorrect settings of structures • Leaks & spills • Excessive water abstraction • Or poor irrigation scheduling • Unable to match delivery service with water charges	• Infield losses low • Settled system with few leaks/spills • Scheduling or organised cycling minimises over-irrigation • Able to deliver and charge by fixed/direct means (payment per irrigation, irrigation services, land & crop taxes, water rights)
Design management dimension	Water distribution (WDB)	• Misuse of existing structures leading to lack of control • No measurement or collection of water flow data • Breakdown of water cycling • Spills and leaks • Unable to match delivery service with water charges	• Correct use of canal structures • Water rotation followed • Collection and analysis of data • Computerisation of scheduling and irrigation management • Able to deliver and charge by fixed/direct means (payment per irrigation, irrigation services, land & crop taxes, water rights)
	Water partition (WPT)	• In-field losses high • Broad design approximations • Leaks and spills • Inequalities of supply/demand match • Delays in irrigation scheduling during peak periods • Unable to accurately determine allocations	• Improvement of infield-irrigation methods • Accurate design of infrastructure • Removal of canal supply bottlenecks and leaks • Appropriate maintenance • Evaluation of equity of supply • Able to deliver water to meet evaporative demand and to charge for water volumetrically

experience of improved delivery services. Both of these depend on accurate control of water. The concepts presented in this paper enable the diagnosis of control problems on irrigation schemes via a framework based on a systems nature of water supply control arising out of many design and management factors and their interactions. It is because of these complex interactions that water control in some cases has not responded to the introduction of new control technology alone and why a distribution system, which has the means to control flows but does not do so, may be similar in performance terms to a provision system which does have good water management.

However, this analysis does not necessarily call for systems to move through the levels of design management interactions from provision to partition. Each system should be evaluated on a case-by-case basis, and may be found to be appropriately designed. Most systems (for example small scale systems and those in humid areas) operate reasonably well with approximate information on volumes suggesting that exact measurements are often not necessary for adequate system management [8].

Even so, in some instances control is insufficient to allow for water to be treated as an economic commodity and this analysis suggests that the introduction of (new) market forces demands an understanding of the systems nature of water supply control on irrigation schemes. Then, for individual schemes, the questions are; what type of system is it, what type of market transaction is most appropriate, and what kinds of technical and institutional interventions are required to enhance water supply control in order to promote the success of those water charges?

7 Acknowledgements

B.A. Lankford gratefully acknowledges the Royal Academy of Engineering (1996 award) and Simunye Estate, (esp. Richard Orr) for their support.

8 References

1. Turral, H. (1995) *Devolution of Management in Public Irrigation Systems: Cost shedding, empowerment and performance.* ODI Working Paper 80, ODI, London.
2. Merrey, D. J. (1994) Institutional design principles for accountability on large irrigation systems, in *Proc. Int. Conf. on Irrigation Management Transfer, Wuhan, China.*
3. Small, L. and Carruthers, I. (1992) *Farmer Financed Irrigation,* Cambridge UP.
4. Underhill, H. W. (1990) *Small-scale irrigation in the context of rural development.* Cranfield Press, UK.
5. Plusquellec, H., Burt, C. and Wolter, H. W. (1994) *Modern Water Control in Irrigation. Concepts, Issues, and Applications.* World Bank Technical Paper No. 246. Irrigation and Drainage Series. World Bank, Washington DC, USA.
6. Clemmens, A. J., Bos, M. G. and Replogle, J. A. (1993) *FLUME: Design and calibration of long-throated measuring flumes,* ILRI Publication 54, Wageningen.
7. Verdier, J. (1987) Computerised control of irrigation water distribution. *ODI/IIMI Irrigation Management Network Paper 87/1d,* 18 pp, ODI, London.
8. Uphoff, N., Ramamurthy, P. and Steiner, R. (1991) *Managing irrigation; analysing and improving the performance of the bureaucracies,* SAGE Pub., New Delhi.
9. Dargouth, M. S. (1992) Institutional Aspects of Irrigation Development in North Africa: Experience with Morocco, in *Developing and Improving Irrigation and Drainage Systems* (eds G. Le Moigne, S. Barghouti, and L. Garbus), World Bank Technical Paper No 178, World Bank, Washington DC.
10. Ankum, P. (1990) Classification of flow control systems for irrigation. In *Int. Workshop on Design for Sustainable Farmer-Managed Irrigation Schemes in Sub-Saharan Africa in Zimbabwe,* Agricultural University, Wageningen.
11. Bos, M. G. and Nugteren, J. (1992) *On Irrigation Efficiencies.* 5th Edition, International Institute for Land Reclamation and Improvement, The Netherlands.
12. Lankford, B.A. and Gowing, J. (1996) Understanding water supply control in canal irrigation systems, in *Water Policy: Allocation and Management in Practice,* (eds P. Howsan and R. Carter), E & FN Spon, London, pp 186-193.

13. Shannan, L. (1992) Planning and management of irrigation systems in developing countries. *Agricultural Water Management*, Vol 22, pp 3-14.
14. Chambers, R. (1988) *Managing Canal Irrigation. Practical analysis from South Asia*, Wye Studies in Agricultural and Rural Development, Cambridge UP.

DEMAND MANAGEMENT IN SUPPLEMENTARY IRRIGATION
The benefit of short-term weather forecasting

J. W. GOWING and C. J. EJIEJI
Department of Agricultural and Environmental Science,
University of Newcastle upon Tyne, Newcastle upon Tyne, UK

Abstract
A model using weather forecast for predicting short-term supplementary irrigation demand is described. The model was applied in simulating potato irrigation under UK conditions at three levels of irrigation costs each for a dry year, wet year and average year. The results were compared with the case of scheduling to maintain acceptable level of soil water deficit without the benefit of weather forecast and with the case of no irrigation. The efficiencies of irrigation water use in terms of profit and yield per ha mm of irrigation were higher for irrigation with weather forecasts than without in the wet year at all the irrigation cost levels (1 ha mm = 0.01Ml). The efficiencies were comparable at each level of irrigation cost for the average year and at the highest irrigation cost for the dry year. It was therefore concluded that short-term weather forecast can be used to improve the management of supplementary irrigation in wet years and in situations of high irrigation cost or water scarcity.
Keywords: Demand management, supplementary irrigation, weather forecasting .

1   Introduction

In humid regions, rainfall is generally adequate to meet seasonal crop water need. Intraseasonal variability of rainfall distribution however often results in dry spells during the growing season. In addition, seasonal differences occur in both total amount and temporal distribution of rainfall. In such situation, supplementary irrigation is necessary to maintain crop quality and to stabilise crop yield and returns.

In the UK, supplementary irrigation is concentrated in the summer months of May to August. A recent survey shows that in the dry years of 1990 and 1995, the irrigated area in England and Wales exceeded 164,400 ha and 156,300 ha respectively. The water application was more than 133 million m^3 and 156 million m^3 respectively. About half of the water was from direct abstraction from rivers and streams while about   one-third came from wells and deep boreholes. Potatoes are the   single   most

Water: Economics, Management and Demand. Edited by Melvyn Kay, Tom Franks and Laurence Smith. Published in 1997 by E & FN Spon. ISBN 0 419 21840 8.

important irrigated crop. For the years 1987, 1990, 1992 covered by the survey, it accounted, on the average, for about 42% of the irrigated area and 51% of water application. [1].

The Anglian region has almost half of the irrigated area in England and Wales [2]. The peak period irrigation water use in the region is considered to exceed all other water uses combined. Studies have shown that not all the supplementary irrigation water demand are reliably met in a dry summer as the water authorities through abstraction restrictions attempt to balance irrigation demand with other competing water uses and environmental considerations [3].

There is increasing awareness of the need for efficient use of available water resources as a means of enhancing the sustainability and reliability of meeting current and future water needs in the UK. Some demand management practices are therefore being adopted by irrigators. One of such is the provision of on-farm storage reservoirs for water abstraction during winter for later use in the drier summer months. Efficient use of the stored water is however necessary for the realisation of the full benefits of the investment. There is also the practice of deficit irrigation the aim of which is to maintain some spare soil reservoir space for any subsequent rainfall. The objective of deficit irrigation may however not be achieved due to possible runoff if significant rainfall occurs soon after irrigation due to the delay in water redistribution from the wetter upper portion of the soil profile to the drier lower regions. The approach may also be problematic for potato growers who may want to maintain as small deficit as practicable at all times for enhanced crop quality.

Another possible approach to supplementary irrigation management would be the use of short-term weather forecast for real-time irrigation decisions, which should ensure more efficient use of rainfall during the growing season. This paper therefore presents a model which adopts a simulation approach to predicting short-term irrigation demand utilising short-term weather forecast.

2   Model description

The main features of the model are illustrated in Fig. 1. The model attempts to keep the entire growing season in perspective during every irrigation decision. To do this, it delineates the season into three time segments namely the past period, the forecast period and the future period. The period from the beginning of the season to any day of irrigation decision is considered as the past period. The short-term horizon for the irrigation decision constitutes the forecast period which is seven days in the model. Actual daily weather data for the season are utilised in the past period to update the state of the crop-soil system while short-term weather forecast data are primarily employed in the forecast period. The future period refers to the time from the first day after the forecast period to the end of the season. In a real-time situation no weather information would be available for this period therefore several years of historical daily weather data are utilised in this period.

The model is presently intended for application to potato irrigation under UK conditions. To describe the crop-soil-water interactions a dynamic physiologically-based crop model capable of simulating crop growth from planting through emergence

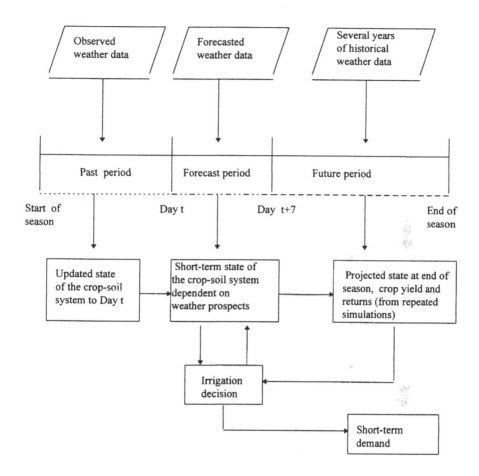

Fig 1 Time periods addressed by the model, eather data inputs to the periods and the flow of the short term demand modelling procedure

to harvest was developed. This involved the modification of an existing potato crop water use model [4] to enable the simulation of tuber yield. The model requires daily data on potential grass evapotranspiration, soil and air temperatures, rainfall and sunshine hours or solar radiation.

The model assumes that irrigation is completed within seven days and also that forecasts for rainfall amount and rainfall probability are available during the forecast period.

The irrigation decision criterion formulation is a modification of an earlier approach [5] which only considered a one-day forecast period. Irrigation decision is made as necessary on any day t regarding the desirability of initiating irrigation on day t+1 and continuing the irrigation to day t+7 or until all the fields of the farm are irrigated.   To arrive at a decision, the expected profit of irrigation $E[\varphi(I,t)]$ is calculated and compared with that for no irrigation $E[\varphi(I^0,t)]$ and irrigation is scheduled only if the expected profit of irrigation is greater.   The algorithm can be stated as follows

$$I_{t+i} = d, \quad i = 1, 2, ., ., ., \eta \tag{1(a)}$$

if

$$E[\varphi(I,t)] > E[\varphi(I^0,t)] \tag{1(b)}$$

else

$$I_{t+i} = 0, \quad I = 1, 2, ., ., ., \eta \tag{1(c)}$$

where η is the number of days to complete the irrigation of all the fields in the farm, and d is the depth of application for each irrigation. Expected profit of irrigation is computed from the expression

$$E[\varphi(I,t)] = \sum_{k=1}^{m} P_{Tk} \frac{1}{n} \sum_{j=1}^{n} \varphi(R_{Tk}, W_{j}, I) \tag{2}$$

where P_{Tk} is the probability of experiencing a rainfall event sequence in the first three days of the forecast period, R_{Tk} is the resulting sequence of rainfall amounts and m is the number of all possible rainfall event sequences. W_j is one of the years the historical daily data of which is used to describe the future period and n is the number of the years.

Similarly, $E[\varphi(I^0,t)]$ is computed by making I_{t+i}, ($i = 1, 2, ., ., ., \eta$) identically zero in Equation (2). The possible rainfall event sequences are eight in number ranging from a sequence of three dry days to a sequence of three wet days.   There are 8n number of simulations required, using in each case a weather year W_j, from the first day after the forecast period to the end of the season for the computation of an expected profit.

The model effectively revises the decision on daily basis once an irrigation cycle has been initiated. Thus a cycle may be terminated on any day after t+1 without necessarily completing it.

The model can be applied to a farm made up of several fields. The irrigator's equipment and labour constraints are considered by specifying a minimum irrigation interval for the fields.

3   Model application

Irrigation in a single farm made up of seven fields was simulated. Each field can be irrigated in one day.   The soil is sandy loam.   The fields are planted to potato on different dates between April 1 to 7 and harvested between August 31 to September 6. The adopted location is March, Cambridgeshire, UK.

The relevant weather forecast data for Cambridgeshire were obtained from Norwich Weather Centre of the Meteorological Office. The centre through its WeatherFax service provides short-term weather forecasts to farmers. The forecasts are produced daily for several regions of the UK for the current day and five days ahead. The several meteorological information provided include sunshine hours, rainfall amount, rainfall probability and maximum and minimum air temperatures. The data on sunshine hours, rainfall amount and rainfall probability required further processing for inclusion in the model. Recorded actual daily weather data for several years representative of Cambridgeshire were also obtained from the Meteorological Office including rainfall data for March.

In the simulations, water application was 25 mm per irrigation with an efficiency of 80%. The maximum desirable soil water deficit was 25 mm subject to a minimum irrigation interval of seven days. Twenty years (1970 - 1989) of historical weather data were used to describe the future period. The actual irrigation seasons simulated were 1992, 1994 and 1995 representing a wet year, an average year and a dry year respectively. Profit of irrigation was calculated as returns less the variable cost of irrigation. The crop price used for each short-term decision was the weighted average of the crop price reported [6] for the five preceding years. It was assumed that the irrigator's price expectation would be influenced more by his most recent past experience. The weights were therefore in decreasing order with the first preceding year assigned the largest weight and the fifth the smallest. Irrigation variable cost included the cost of water, repairs, fuel and labour. Simulations for each of the three years was performed for three irrigation costs of £5.8 /ha mm, £3.4 /ha mm and £1.7 /ha mm (1 ha mm = 0.01Ml).

For the purpose of comparison, simulations were carried out without weather forecast. In this case, only the projection of water deficit in the fields was necessary in order to decide whether to irrigate. No consideration was given to any probable rainfall during the projection. In addition, completely rainfed crop without irrigation was also simulated.

4   Results and discussion

A detailed output of the model (not included) indicated that it generally exhibited good anticipation of rainfall in scheduling irrigations in all the years simulated. The total number or irrigations for the entire farm varied with the years and irrigation costs. The number of applications differed among the seven fields for the respective year-irrigation cost scenarios (Table 1). The number of irrgations was sensitive to irrigation cost in average (1994) and dry (1995) years but not in the wet (1992) year which had the lowest number of irrigations. At irrigation cost of £5.8 /ha mm, there were 52 and 61 irrigations respectively in the average and dry years with each field receiving a minimum of 7 and 8 respectively. The number of irrigations at irrigation cost of £3.4 /ha mm increased to 59 and 73 with the minimum number per field at 8 and 10 respectively but there was no further increase in the number of irrigations at irrigation cost of £1.7 /ha mm.

Irrigation without weather forecast showed no sensitivity to irrigation cost and applied significantly more water than irrigation with weather forecast. At £5.8 /ha mm, water applied in the wet year was more by 69% while in the average and dry years it was more by 52%, and 44% respectively. At £3.4 /ha mm and £1.7/ ha mm it was more by 69%, 34% and 21% for the wet, average and dry years respectively. Soil water deficit also was generally lower for this case than irrigating with weather forecast and was generally highest for the no irrigation case which also had the lowest yields. Irrigation without weather forecast on the other hand produced the highest yields with little variation between the years. In terms of profit expressed as average profits from the seven fields based on the actual crop price, it also produced the highest profits for the average and dry years though this was dependent on the accuracy of the weather forecast. It should be noted that results from irrigation without weather forecast assume that water is always available but this is not likely to be true in drier years.

Table 1. Simulated number of irrigations, average yields and average profits for irrigation with and without weather forecasts, and results for no irrigation for three years and three irrigation costs.

Year	+Total no. of irrigations		Average yield (t/ha)			Average profit (£/ha)		
	*IRWF	IRNF	IRWF	IRNF	NOIR	IRWF	IRNF	NOIR
(a) Irrigation cost of £5.8 /ha mm								
1992	29 (3)	49 (6)	56.56	57.57	53.56	2702	2347	3127
1994	52 (7)	79 (10)	37.87	57.10	26.69	5966	8984	4964
1995	61 (8)	88 (12)	38.85	57.14	27.19	4563	6748	4078
(b) Irrigation cost of £3.4 /ha mm								
1992	29 (3)	49 (6)	56.56	57.57	53.56	2950	2767	3127
1994	59 (8)	79 (10)	42.62	57.10	26.69	7210	9661	4964
1995	73 (10)	88 (12)	40.30	57.14	27.19	5158	7502	4078
(c) Irrigation cost of £1.7 /ha mm								
1992	29 (3)	49 (6)	56.56	57.57	53.56	3127	3064	3127
1994	59 (8)	79 (10)	42.62	57.10	26.69	7569	10140	4964
1995	73 (10)	88 (12)	40.30	57.14	27.19	6027	8036	4078

+ Total number of irrigations for the seven fields (minimum number per field shown in brackets)
* IRWF = irrigation with weather forecast   IRNF = irrigation without weather forecast
   NOIR = no irrigation

Irrigation with weather forecast produced more profit than without weather forecast in the wet year at all the irrigation cost levels despite using less water. The use of weather forecast in such wet years should therefore reduce costs to the farmer from unnecessary irrigations. The profits for the case of no irrigation were however higher for that year compared to irrigation with forecast except at the irrigation cost of £1.7/ ha mm. This suggests that without quality premium considerations, irrigation in a wet year may not pay if irrigation cost is high.

The efficiencies of irrigation water use in terms of yield and average profit per ha mm of irrigation for the cases of irrigation with and without weather forecast (Table 2) clearly show the values to be higher for irrigation with weather forecasts in the wet year at the three levels of irrigation cost thus confirming the advantage of the method for the wet year. At the three irrigation cost levels in the average year, and at the highest cost level in the dry year, both irrigation with and without weather forecast had comparable efficiencies even though there were less water applications with forecast. This is an indication of the usefulness of irrigation with weather forecasts in situations of high irrigation cost and scarce water supply. For the dry year however irrigation without weather forecast gave somewhat higher values at the two lower irrigation costs. The reason is that irrigation without forecasts commenced irrigation early in the season almost immediately after crop emergence in that dry year. The crop benefited greatly from the early irrigation. In contrast, irrigation with forecast started irrigation up to 12 days later on account of poor rainfall forecast early in the season grossly over predicting actual rainfall.

Table 2. Efficiency of irrigation water use for irrigation with and without weather forecasts for three years and three irrigation costs

Year	Yield per unit of applied water (t / ha mm)		Profit per unit of water applied (£ /ha mm)	
	* RWF	IRNF	IRWF	IRNF
(a) Irrigation cost of £5.8 /ha mm				
1992	0.546	0.329	26.09	13.41
1994	0.204	0.202	32.13	31.84
1995	0.178	0.181	20.95	21.47
(b) Irrigation cost of £3.4 /ha mm				
1992	0.546	0.329	28.49	15.81
1994	0.202	0.202	34.22	34.24
1995	0.155	0.182	18.24	23.87
(c) Irrigation cost of £1.7 /ha mm				
1992	0.546	0.329	30.19	17.51
1994	0.202	0.202	35.92	35.94
1995	0.155	0.182	23.12	25.57

* IRWF = irrigation with weather forecast , IRNF = irrigation without weather forecast

5   Conclusion

It can be concluded from the simulated scenarios that the use of weather forecast with the irrigation decision criterion has definite advantages in terms of practical on-farm irrigation management. In the wet year, it was more efficient than irrigation without weather forecast. It should therefore be beneficial in such wet years due to cost savings from unnecessary irrigations. In the average year and under high irrigation cost in the dry year the efficiencies in terms of average profit and yield per ha mm of irrigation were comparable to those of irrigation without weather forecast, therefore irrigation with forecast becomes attractive in situations of high irrigation cost and scarce water supply. Although irrigation without forecast produced more yield and more profit in those years, it required significantly more water applications which may not be met in drier years.

The model, though presently applied to potato, could be equally applicable to other crops. Furthermore, apart from on-farm irrigation decisions, the model should be adaptable for predicting fluctuating supplementary irrigation demand from a resource manager's view point. The performance of the irrigation decision model depends on the reliability of the available short-term weather forecast information.

6   Acknowledgements

This work is part of the research being conducted by the second author for his doctoral thesis at the University of Newcastle upon Tyne under the sponsorship of the Commonwealth Scholarship Commission. The historical meteorological data used were supplied by The Meteorological Office at Bracknell. We are grateful to the Norwich Weather Centre for the provision of the short-term weather forecast data.

7   References

1.   Ministry of Agriculture, Fisheries and Food. (1996) *Survey of irrigation of outdoor crops in 1995 England,* The Government Statistical Service, York.
2.   Weatherhead, E. K., Price, A. J., Morris, J. and Burton, M. (1994) *Demand for Irrigation water*, National Rivers Authority, Bristol, R&D Report 14.
3.   University of Newcastle upon Tyne (1990) *Water resources and demands in the Middle Level*, Final report, Annexe 2 - Agrohydrology, Newcastle upon Tyne.
4.   Hamer, P. J. C., Carr, M. K. V. and Wright, E. (1994). Crop production and water use III: The development and validation of a water use model for potatoes, *Journal of Agricultural Science*, Cambridge Vol. 123, pp 299- 311.
5.   Swaney, D. P., Mishoe, J. W. Jones, J. W., and Boggess, W. G. (1983) Using crop models for management: Impact of weather characteristics on irrigation decisions on soybeans, *Transactions of the American Society of Agricultural Engineers*, Vol. 26, pp 1808-1813.
6.   Murphy, M. C. (1996) *Report on farming in the eastern counties of England 1994/95*, Agricultural economics unit, Department of Land Economy, University of Cambridge, Cambridge.

THE ECONOMICS OF LOW FLOW RIVERS

M.G. POSTLE
Risk & Policy Analysts Limited, Loddon, Norfolk, UK
N. BERRY
Environment Agency, Southern Region, Worthing, West Sussex, UK
R. WESTCOTT
Environment Agency, South West Region, Exeter, UK

Abstract
Over the past few years, an increasing amount of attention has been given to the problem of low flows in some UK rivers. The environmental damages resulting from low flows have resulted in numerous low flow alleviation proposals being sponsored by the regulatory authorities and others. As the costs of alleviating low flows can involve significant expenditure, there is a need to ensure that a balance is achieved between the financial costs and environmental benefits associated with any alleviation proposals. One way of assessing such costs and benefits is through the application of cost-benefit analysis techniques. This paper summarises how such techniques can be applied, gives examples of past applications and indicates some areas for future consideration.
Keywords: Low river flows, cost-benefit analysis, environmental economics, benefit transfer, environmental valuation, willingness to pay.

1 Introduction

Attention is increasingly being given to the problem of unacceptably low flows in some UK rivers. These problems result from a number of different causes, including natural variations in weather patterns, the cessation of historic river management practices and over-licensing of abstractions for drinking water supply purposes. There are a number of methods which can be employed to add more water to a stretch of low flow river including revoking abstraction licences, river bed lining, relocating abstractions and introducing new sources of water to augment flows. In many instances a combination of solutions may be required. As such works can involve

Water: Economics, Management and Demand. Edited by Melvyn Kay, Tom Franks and Laurence Smith. Published in 1997 by E & FN Spon. ISBN 0 419 21840 8.

significant expenditure by the Agency, water supply companies and others, it is important that any proposals are accompanied by a thorough and robust assessment of costs and benefits.

The need to ensure that a balance is struck between the environmental costs and benefits and other economic costs and benefits was recognised by the former National Rivers Authority (NRA). As a result, the NRA initiated research aimed at assisting in the development of systematic approaches for examining the advantages and disadvantages of low flow alleviation projects. The importance of this research has become even greater as a result of a legal duty placed on the Environment Agency which requires that 'takes into account costs and benefits' in all cases where the Agency has discretionary decision making powers (Section 39 of the Environment Act 1995).

The approach which has received the most attention to date is that of social cost-benefit analysis (CBA), which is based on the principles of welfare economics. In a CBA, all the impacts resulting from a decision (or action) are valued in a common unit, that of money, and the decision is considered justifiable if the benefits outweigh the costs. Given that many of the environmental impacts associated with water resource decisions are not normally valued in monetary terms (for example, improvements in the quality of a fishery), CBA requires the application of economic valuation techniques to determine how much individuals would be willing to pay for such gains or to avoid such losses.

2   Valuation of the costs and benefits of low flow alleviation

Economists have defined two categories of value which people hold toward the environment. The first are termed 'use' values which relate to direct and indirect consumption, or use, of the environment, while the second are termed 'non-use' values and relate to an individual's desire to conserve and preserve the environment for the future [2]. Management of low river flows can impact upon a range of different use values and on non-use values, including:

- abstractions for irrigation or other agricultural purposes (e.g. livestock), potable supply and industrial usage (as an input to production or for cooling purposes);
- commercial fisheries and non-commercial fisheries (where these include recreational fisheries);
- in-stream recreation (such as canoeing, sailing, bathing) and out-of-stream recreation (such as walking, picnicking, bird watching, etc.);
- changes in the amenity attached to residential and commercial property values and hence in property prices; and
- non-use values and associated conservation related effects including impacts on aquatic ecosystems and associated flora and fauna.

Obviously out of this general list of potential impact categories, only a limited number may be significant for any given low flow alleviation project. Indeed, the importance of different uses is likely to vary over rivers and regions. For example, in-stream recreation is limited on many rivers due to their physical characteristics; and out-of-stream or bankside recreation will be determined by the degree of access there is to the riverside.

Some of these impacts can be valued through the use of actual expenditure and market price data. This is particularly true where a scheme aimed at alleviating low flows affects industrial and/or potable water abstractions. In some cases, such as impacts on commercial fisheries and agricultural production, market prices may need to be used within a dose-response based valuation approach. The limiting factor to valuation in these cases is likely to be details of the appropriate dose-response relationship, and not information on the economic value of a unit of production.

For those categories where the benefits relate to non-market effects (recreation, amenity and non-use/conservation benefits), techniques such as the travel cost method, hedonic pricing and contingent valuation method are used to develop estimates of economic gain (or loss). The principles underlying these techniques and the analytical procedures involved in their application can briefly be summarised as follows [2, 3]:

- **Travel Cost Method**: The approach is based on the concept that people spend time and money travelling to a recreational site and that these expenditures, or costs, can be treated as revealing the demand for the site. Surveys of site visitors are then undertaken to determine the demand for a site, where visit rates are a function of travel expenditure, income, any entry fees, environmental characteristics and the availability of substitute sites.

- **Hedonic Pricing Method:** The hedonic pricing method is based on the concept that the price paid for a complementary good (e.g. a residential property) reflects the buyer's willingness to pay for a particular environmental good (e.g. adjoining a river). Application requires the use of regression analysis to determine the relationship between the market price of the property and its attributes, of which one (set) relates to associated environmental characteristics. From this, the implicit price for the associated environmental characteristics is derived.

- **Contingent Valuation and Contingent Ranking Methods:** These methods rely on the creation of a hypothetical or experimental market for an environmental good. Individuals are then surveyed to determine their willingness to pay for a specified change in the quality or quantity of the environmental good. The mean willingness to pay value across all bids (including valid zero bids) is then used to provide an indication of the economic value of the specified change.

Table 1 summarises the range of past applications using these techniques which have addressed low flow related impacts.

3   Low flow valuation studies

3.1   Overview
Within the UK, there have been three major studies aimed at deriving economic values for the environmental benefits of low flow alleviation. All three of these

Table 1: Applicability of Valuation Techniques

Benefit Category	Market Price	Travel Cost	Hedonic Pricing	Contingent Valuation
Drinking Water Supply	✓			
Agriculture	✓			✓
Fisheries	✓	✓	✓	✓
In-Stream Recreation	✓	✓		✓
Out-of-Stream Recreation		✓		✓
Amenity/Aesthetics			✓	✓
Conservation/Ecology				✓

studies have relied mainly upon the use of the contingent valuation method to develop estimates of people's willingness to pay for alleviation works. Interestingly, the results of these studies are comparable in terms of the magnitude of estimated benefits, although they have adopted different units of measure and different survey approaches.

3.2   The River Darent Study
The River Darent is fed by groundwater springs and a chalk aquifer and, as a result of groundwater abstraction, has been affected by seasonal low flows for many years. In 1993, the Southern Region of the NRA commissioned a cost-benefit analysis of a proposed low flow alleviation scheme for the river [4]. The aim of this CBA was to provide justification for an Action Plan which had been jointly agreed in principle between the NRA and the water supply company. This Plan comprised revocation of abstraction licences at two boreholes, reduction in abstractions from the chalk aquifer and intermittent augmentation of flows at another location. The analysis was further complicated by the fact that the water company (although considered to be over-abstracting from an environmental perspective) was not taking the fully permitted volumes under its licences. As a result, the CBA examined the benefits of maintaining the current situation as opposed to the company taking fully permitted volumes and the benefits of the Action Plan itself (which would result in 50% of natural drought flow being maintained).

With regard to the CBA, the most important aspect was the contingent valuation survey which was also the first concerning low flow alleviation to be undertaken in the UK. This survey was constructed to elicit willingness to pay values of residents, visitors and the general public for maintaining the current situation and to improving river flows. It was also designed so as to be consistent with US recommendations on the application of the contingent valuation method.

The questionnaire contained questions relating to respondents' use of rivers and the Darent in particular, willingness to pay for low flow alleviation, and questions on the socio-economic characteristics of the respondent households for validation of the survey results. Two willingness to pay questions were used in the survey. The first was aimed at assessing the willingness to pay for low flow alleviation in the top 40 affected rivers in England and Wales. This question helped set a maximum 'budget' for respondents' answers concerning low flow alleviation. The second set of questions was then focused more specifically at the River Darent. In both cases, the questions were asked with regard to respondents' willingness to pay increased water rates as this method of payment was the most realistic.

Table 2 presents the results from the survey. Overall, a large share of the population, whether residents, visitors or the general public, was willing to pay for low flow alleviation [3]. Indeed, when aggregated across the relevant populations, the benefits to residents and visitors associated with the Action Plan exceeded some £2 million per annum, while the non-use benefits to the general public were calculated as exceeding £20 million per annum.

Table 2. Willingness to Pay for Low Flow Alleviation in the River Darent (mean WTP £ per household)

	Residents (within 2 km)	Visitors to the Darent	General Public (within 60 km)
Maintaining Flows in all 40 Rivers	£18.45	£14.31	£17.18
Increasing Flows in all 40 Rivers	£12.32	£ 9.38	£12.92
Maintaining Flows in the Darent	£10.19	£ 4.64	£ 3.85
Increasing Flows in the Darent	£ 6.25	£ 2.75	£ 3.00

3.3 The Misbourne, Wey and Ver Study

The second study aimed at deriving economic values for the environmental benefits of low flow alleviation was that undertaken by Middlesex University [5]. In contrast to the River Darent work, this study focused on developing estimates for the 'use' related benefits associated with angling and informal recreation (it did consider non-use values, but with less statistical reliability). The approach adopted, however, was similar to that undertaken in the Darent study in that it relied on use of the contingent valuation method; although the manner in which the willingness to pay values were elicited was different.

In this case two different populations were surveyed: anglers and informal recreation users. Anglers were stopped outside bait and fishing tackle shops (with 370 interviewed) and the survey found mean willingness to pay values for a day's fishing of around £5.00 to £8.80 per visit for the two rivers respectively. The informal recreation survey involved a larger number of respondents with over 1,000 surveys completed on the Rivers Misbourne and Wey. In this study, respondents were asked a series of questions to establish the level of enjoyment they gained from a visit to the river, the degree to which their enjoyment would change given alleviation of low flow and the value of that enjoyment. This series of questions was used to derive a set of values per visit to the river under different flow conditions. These questions were then followed by a shorter set which asked whether respondents would be willing to pay a sum equivalent to their value of enjoyment to see flows restored. As a result of this latter set of questions, willingness to pay values of £14.40 and £17.30 per household per annum were calculated.

3.4 Low flow alleviation in the South West

The third and most recent study was prepared for the South West Region of the Agency and focused on seven rivers experiencing low flow conditions [6]. The causes of low flow in the rivers examined varies, although most experience low flows as a result of groundwater abstractions for public supply purposes.

The study represents a major piece of work into the valuation of environmental benefits. Overall four separate surveys into the benefits of alleviating low flows were conducted. These examined the benefits to anglers, property owners, informal recreation users and the general, non-user public. The surveys were constructed in such a way as to provide a general set of benefit estimates that may be transferred between low flow sites and they covered a range of environmental issues and Agency initiatives so as to put low flow problems within the proper context.

The survey methods used relied on a number of different environmental economics techniques, including the contingent valuation method, the stated preferences technique and the hedonic pricing method. As a result, a number of different benefit estimates were developed, with these including, the value to anglers of extending the periods when fishing would be possible, per household values for visitors, and values per kilometre of river affected for general river users (i.e. they visit rivers but not those listed in the survey) and non-users. The mean willingness to pay values found in this study are presented in Table 3 for each of these groups. It is of note that the values found by this study are comparable to those found in the two studies quoted above.

Table 3. Willingness to Pay for Low Flow Alleviation in the South West

	Mean (£)	Units
Club Anglers	£26.00	per person per annum
Syndicate Anglers	£71.00	per person per annum
Informal Recreation Users	£2.27 to £3.58	per household per annum
General River Users	£ 0.048	per km river
Non-Users	£ 0.027	per km river

4   The development of standardised approaches

4.1   Benefit transfer

In deriving estimates of the costs and benefits associated with low flow alleviation, there are two approaches which can be adopted. The first is to commission site and problem specific assessments such as the three studies reviewed above. Such studies take a long time to complete and can be costly, requiring the use of specialist consultants. Indeed, derivation of an economic value for just one area of impact may take several months and cost many thousand pounds. It would be infeasible, therefore, to commission detailed valuation studies for each of the low flow problems currently under investigation.

The second approach is to take the results of the above low flow studies and combine them, together with the results of any other relevant valuation studies, into a set of standard values as part of a "benefit transfer" based methodology. Benefit transfer is the process of taking a value or benefit estimate developed for one project decision and transferring it to a proposed project decision. Thus, with respect to low flow alleviation, estimates of the willingness to pay associated with recreation benefits at a particular site are assumed to provide a reasonable approximation of the benefits which would be realised at a different location [7].

The advantage of benefit transfer is that it can be used within desk-top studies to derive order of magnitude estimates of likely costs and benefits. It can also form the basis for the development of a standardised methodology which can be applied consistently across sites. A review of the literature indicates that there are three different approaches which might be adopted in using these types of estimates: the transference of mean unit values;the transference of adjusted unit values; and the transference of a demand function.

The use of mean unit values is obviously the simplest approach that can be adopted. A range of factors may affect, however, the validity or reliability of such an approach. These include whether or not the environmental change measured in the original study is similar to that under consideration; the factors considered in developing the survey and whether or not there are substitute sites or other opportunities which could affect individuals' valuations.

The adoption of an adjusted unit approach involves the analyst adjusting past estimates to correct for biases incorporated in the original study or to take into account differences in socio-economic characteristics, project differences, site differences and the availability of substitute goods. Work in the US has been undertaken to determine what types of adjustments are required (e.g. Loomis, 1992), drawing upon a large pool of previous work in order to inform the adjustment process.

The third approach involves taking the demand function from a previous study, inputting new data relevant to the project in question and re-running the analysis. The advantages of this type of approach are that calculated benefits are based on information on use and unit values which are derived from the same data set. In addition, the approach can take into account factors such as distribution of population around a site, availability and quality of substitute sites, etc. within the demand equation.

4.2   Methodological considerations

Unfortunately, the paucity and nature of existing studies prohibits doing anything more than transferring mean unit values. Care is therefore required in making sure that users are aware of any caveats which must be attached to the end results and that these provide only relative order of magnitude indicators of costs and benefits. This is highlighted, for example, by the study undertaken for South West Region [6]. The consultants concluded that any transfer of the willingness to pay values to other rivers should be done carefully as the values produced were context specific. However, these estimates are not site specific and reflect respondents' general views on low flow alleviation, water quality improvements, etc. in general (although a reluctance to fund additional privatised water company initiatives may have influenced the results).

There are other issues which also need to be considered in developing and adopting a benefit transfer approach. In order to calculate total benefits, it is necessary to combine standard values with estimates of participation rates and, hence, angler numbers affected. Benefit transfer based methodologies, therefore, need to provide guidance on how to develop these estimates and provide other useful data such as participation rates per head of population, average distance travelled to a site, etc. Work carried out to date has often found that the uncertainty surrounding these aspects of benefit assessment are as great as those surrounding any of the transfer values and may have as great, if not more, of an effect on the reliability and robustness of any analysis. Another key issue is the availability of substitute sites and hence of other opportunities for recreational users of a river.

The implications of the above are that a pragmatic approach including thorough sensitivity analysis is essential. This may include considering a lower bound and upper bound estimate for the transfer willingness to pay value and developing ranges for assumptions concerning visitor numbers and participation rates. Through this process some sensitivity analysis can be undertaken and "switching values" can be determined for the various assumptions.

5   Development and testing of a benefit transfer approach

Work has commenced on the development of a benefit transfer based approach for use by the UK Environment Agency, with this including the following activities:

- examination of the appropriateness of transferring data from non-low flow specific studies to a low flow context;
- cross study comparisons to determine how robust the valuations being derived through the various studies are; and
- testing of preliminary desk-top approaches on sites with widely varying characteristics with the aim of determining the feasibility of undertaking such assessments and highlighting any difficulties in applying the methods.

It is hoped that this first step in the development and application of a benefit transfer methodology will help identify those areas where further research is required. It will also provide a good test of the acceptability of such an approach to other organisations in the UK should the results provide the basis for regulatory decision making.

6   Acknowledgements

This paper draws on research and development work which has been undertaken on behalf of, and together with, the Environment Agency by Risk & Policy Analysts Ltd. The authors would like to thank all those in the Agency and, in particular Jerry Sherriff, for continuing support and encouragement over the past few years.

7   References

1.   National Rivers Authority. (1993) *Low Flows and Water Resources Report*, NRA, Bristol.
2.   Department of the Environment. (1991) *Policy Appraisal and the Environment*, HMSO, London.
3.   Postle, M. (1993) *Economic Appraisal Manual*, NRA, Bristol.
4.   ERM. (1994) *River Darent Low Flow Alleviation*, Report prepared for the Southern Region, National Rivers Authority, ERM, London.

5. Middlesex University. (1994) *The Evaluation of the Recreational Benefits and Other Use Values from Alleviating Low Flows*, NRA R&D Note 258, NRA, Bristol.

6. ERM Economics. (1997) *Economic Appraisal of the Environmental Costs and Benefits of Potential Solutions to Alleviate Low Flows in Rivers: Phase 2 Study,* Report to the Environment Agency, South West Region.

7. Postle, M. (1994): *Assessing Water Quality Improvements: is Benefit Transfer Possible?,* Paper prepared for the Joint Treasury and Department of the Environment Group on Environmental Costs and Benefits, London.

FINANCIAL-ECONOMIC ANALYSIS OF DUTCH WATER MANAGEMENT POLICY
An analysis of three policy variants

M.J. VAN DRIEL and P.M. LICHT
Institute for Inland Water Management and Waste Water Treatment, Department of national affairs, Lelystad, the Netherlands

Abstract
Annual costs of current Dutch watermanagement policy will increase from 8.3 billion guilders in 1995 to 10.4 billion guilders in 2015. Most of the costs are spent on measures to reduce emissions; 47% in 1995 and 60% in 2015. The increase in costs of 1% per year results in an increase in the burden of costs for households with 3% per year, to 260 guilders per inhabitant in 2015 (f 1.0 = 0.52 $).

If watermanagement policy goals are to be reached annual costs will rise to 34 billion guilders in 2015. Annual costs for households will rise to 1000 guilders per inhabitant. In this case for almost all sectors of industry the annual burden of costs would increase at a rate outstripping their overall expected economic growth, having a negative impact on the Dutch overall economic development.
Keywords: Costs, economic analysis, water management policy, watersystems.

1   Introduction

The Dutch Aquatic Outlook (DAO) contains a technical and scientific survey of Dutch water systems directed both at the evaluation of current water management policies and at the analysis of development in the policy field [1]. The results of the DAO serve as a contribution to the debate on future policy which will lead ultimately to the publication of the fourth National Policy Document on Water Management in 1998.

The project approaches water management from the watersystems point of view. The condition of these water systems is defined in quantitative and objective terms by means of a limited set of chemical, biological and physical and functional target variables. A specific way of summing up all distance-to-target ratios is used as a

Water: Economics, Management and Demand. Edited by Melvyn Kay, Tom Franks and Laurence Smith. Published in 1997 by E & FN Spon. ISBN 0 419 21840 8.

measure for the water system quality. Project data are accessible by way of the project informationsystem generally known as the 'Water Dialogue'.

This paper presents the part of the DAO were the financial-economic consequences of water policy in the period 1985-2015 are examined.

The paper starts with a description of the items that belong to water management policy and the target variables that are used to present the financial-economic consequences. Next the annual costs are presented per policy theme, and per sector. Than the annual costs are related to the economic development per sector. The paper finishes with a short presentation of benefits and effects on employment and value added due to water management policy.

2   The different variants used in water management policy

Watermanagement policy is for this purpose limited to the next policy items:

- Emissions to surface water
- Hydraulic (re)design and habitat restoration of watersystems
- cleaning up and maintenance of Aquatic soils
- hydrology (regular quantitative management of surface water and groundwater)
- combatting groundwater depletion
- management and maintenance of fairways (including see-shipping and maritime affairs)
- management and maintenance of flood protection
- water-based recreation
- Organisation and research of water management

The financial-economic consequences of water management policy were examined for the CURRENT POLICY, USE POLICY and SYSTEM POLICY variants.

The CURRENT POLICY variant is restricted to measures of which the decisionmaking process has been completed, and that are actually being implemented. This does not necessarily mean that with those measures the goals of the current water management policy will be realised.

The USE POLICY variant concentrates on a hydrology and watermanagement as much as possible focused on the needs of user-functions. Measures to improve the water system quality are only included in the variant if the financial abilities of a specific sector allows this. Therefore it is assumed that the development of the costs may not exceed the economic development rate of the European Renaissance scenario of the Dutch central planning bureau [2].

In the SYSTEM POLICY variant the hydrology and watermanagement is designed to meet the demands necessary to restore and develop ecological values according to the goals of current watermanagement policy.

The CHANGE POLICY variant is aimed at an integral optimisation of user-functions and watersystems. This demands structural societal changes. No overall estimates were drawn up for the CHANGE POLICY variant, since the long-term economic changes (up to 2045) assumed in the variant make it impossible to make any predictions in this respect.

3   Economic target variables and calculation method

The financial-economic analysis is examined for a restricted number of target variables. There are four economic functional target variables: annual costs, investments, employment and value added.

The variable 'annual costs' is a good indication of the financial consequences for sectors of industry and households. For more than 200 measures annual costs are calculated by adding capital costs (interest and depreciation costs), operational costs (personnel, energy, etc.) and levies paid (sewage-levy, water-treatment levy) minus subsidies received for a specific measure. The annual costs for household consist primarily of levies paid.

The method used to calculate annual costs in order to make future predictions of the development of costs is the "methodiek milieukosten" [3]. This method is widely accepted in the Dutch government environment:

- depreciation method: lineair;
- depreciation period: electro-mechanical investments 10 years and 25 years for construction investments;
- interest level: real interest rate on capital markets for government investments and a 0.75 per cent point extra for non-government investments.
- measures are reinvested
- Constant prices: price level 1995

Savings, for example water and energy, are subtracted from the annual costs. If the savings should exceed the costs of a measure, than this measure is not considered a measure of watermanagement policy. This is because these 'profitable' measures would also be taken without watermanagement policy.

Besides costs the watermanagement policy generates benefits for user-functions of water as well. These benefits are not included in the estimates of annual costs, but, if known, presented separately. Finally the economic spin-off of watermanagement policy on sectors of industry and households is estimated by calculating the influence of investments and costs on the variables 'employment' and 'value added'.

4   Annual costs of individual themes

Table 1. Total annual costs for the different variants (billion guilders, prices 1995)

	1995	2015
Current policy	8.3	10.4
Use policy		10.6
System policy		34.4

General

Table 1 gives a summary of the total annual costs per variant. Table 1 shows that the CURRENT POLICY and USE POLICY do not differ much. This is the reason in this paper not to focus on the USE POLICY. The annual costs of the CURRENT POLICY will increase 1% per year in the period 1985-2015. In the SYSTEM POLICY the annual costs increase to 34 billion guilders in 2015, which represents a 7% increase annually.

Emissions

In 1995, action to deal with emissions accounted for some 47 per cent of the total costs of water management policy. By 2015 this proportion is expected to increase under the CURRENT POLICY variant to almost 60 per cent. Under the SYSTEM POLICY variant, the proportion would increase to 84 per cent. Two-thirds of these extra costs as compared with CURRENT POLICY would be caused by the extremely thorough treatment methods at all sewage treatment plants, which would cost around 17 billion guilders a year.

In addition, the SYSTEM POLICY variant includes expensive large-scale measures relating to sewerage (maintenance and replacement, disconnection of metalled surfaces, in existing urban areas, and connection of decentralised discharges), stringent policies on fertilisers and treatment of industrial effluent at source.

Hydraulic (re)design and habitat restoration

Under the CURRENT POLICY variant, the estimated costs of the hydraulic (re)design and habitat restoration theme are low as compared with those of the emissions theme, at around 115 million guilders a year. Under the SYSTEM POLICY variant, these costs would be doubled.

Aquatic soils

The costs of the aquatic soils theme are relatively high in 1995. The reason for this is the construction of two large-scale storage sites (in the 1995-2000 period). Through to the year 2000 the budget of the Ministry of Transport and Public Works will include an annual sum of around 50 million guilders for cleaning up aquatic soils in waters under State management [4]. The estimate under the CURRENT POLICY variant assumes that this amount will continue in the estimates even after 2000.

Under the SYSTEM POLICY variant, the costs of aquatic soils policy in 2015 would be around 1 billion guilders a year. This estimate takes account of the additional costs of removing contaminated spoil from maintenance dredging.

Groundwater depletion

The estimated costs of the groundwater depletion theme will increase gradually under the CURRENT POLICY variant to almost 330 million guilders in 2015. Under the SYSTEM POLICY variant the costs of combatting groundwater depletion would amount to 590 million in 2015. The main causes are water management measures, reduction of groundwater abstraction for drinking water supply companies and industry, and extra agricultural losses as a result of increased damage due to water logging in the short term.

In the long term (2015), due to the autonomous decrease in agricultural areas, combatting the groundwater depletion will not have a negative impact on agriculture. This is because it is assumed that the so called Ecological Main Structure (EMS) will be realised. Areas that would suffer most from an upward shift of the groundwater level are located in this EMS.

Fairways

Expenditure relating to the fairways theme show a rise to 1.5 billion guilders in 1995. The main reason for this is extra spending relating to the Multi-Year Plan for Infrastructure and Transport [4]. From 2000 onward, the cost under all the variants would stabilise at a level of just over 1.4 billion guilders.

Flood protection

Costs relating to the flood protection theme are declining because the huge Delta Works are now nearing completion. The share of this theme in the total costs of watermanagement policy has declined from over 30 per cent in 1985 to around 10 per cent in 1995. The estimates take no account of the future sea level rises.

Under the SYSTEM POLICY variant, the costs would' be higher in 2015 than in 1995 because of the large-scale introduction of foreshore protection in the Lake IJssel area.

Water-based recreation

The costs of the water-based recreation theme are low in comparison with those of other themes. It must, however, be remembered that they do not include costs relating to nautical dredging problems in yachting harbours. These are accommodated under the aquatic soils theme.

Organisation and research

Spending on organisation and research have remained reasonably stable since 1990, at a level of just over 1 billion guilders a year. The costs would likewise remain constant under the USE POLICY and SYSTEM POLICY variants, since it is assumed that no extra measures would be taken in this area.

5   Annual burden of costs per sector

This paragraph describes and evaluates the annual costs for sectors of industry, households and authorities for the three variants. Total costs for sectors equal total costs for themes and are therefore not repeated here. Table 2 gives the distribution of annual costs between households, sectors of industry and authorities.

Table 2. Distribution of costs between sectors (in percentages)

	1985	1995	2015 CURRENT POLICY	2015 USE POLICY	2015 SYSTEM POLICY
Authorities	58	37	22	25	18
Households	19	28	38	35	46
Sectors of industry	23	35	40	40	36

The burden of costs to the authorities, households and sectors of industry take account of levies paid and subsidies received by them.

Authorities

Since 1985, the government share in the burden of costs has declined. Reasons for this include the declining costs of State flood protection measures, the declining contribution of the municipalities, due to the trend towards passing on the rising costs of managing the sewerage system to households in the form of sewerage charges, and the rising costs of the sectors of industry and households

Households

Under the CURRENT POLICY variant, the annual burden of costs of water management policy are expected to have risen by 2015 by over 100 guilders a head as compared with 1995, to a total of approximately 260 guilders. This means an increase of 2.8 per cent a year. This increase in burden of costs is less than the annual growth rate over the preceding period (1985-1995) which totalled 4.8 per cent. The development of annual costs in the CURRENT POLICY variant equals the expected economical growth rate of almost 3% annually in the European Renaissance scenario [2].

Under the USE POLICY variant, the increase in burden of costs would be somewhat less, amounting to 245 guilders a year.

Under the SYSTEM POLICY variant, the proportion of total burden of costs met by households would rise to 46 per cent. The main reason for this is the introduction of four-step treatment at sewage treatment plants and extensive sewerage measures, the costs of which would be passed on in their entirety to households and companies. Annual charges to households are expected to rise over 1000 guilders per head by 2015. This substantial increase in charges is equivalent to an annual growth rate of over 10 per cent.

In the SYSTEM POLICY variant the increase of annual costs is far above the economic growth rate. As a consequence households would be paying a larger share of their household spending on their water account.

Sectors

The extent to which the burden of costs have an impact on the sector is estimated by examining the share of annual costs of water management policy in the value added of sectors. A higher share will have a larger negative impact on the economic functioning of a sector. It should be mentioned however that this is only an indication, and that the actual strength depends on many other factors, such as market position (national or foreign markets), policy in other countries (same or less restrictive environmental policy) and the capacity of a company to withstand cost-price rises internally. A good analysis requires thorough investigation per company, which has not been conducted within the DAO.

The analysis of annual costs development compared to the economic state and development of sectors has taken place on a high level of aggregation. Due to lack of data it was not possible to analyse into further detail. The analysis therefore gives an indication of the sectors that encounter a relatively high or low cost development.

Agriculture

Despite the rising burden of costs to agriculture under CURRENT POLICY, the trend in burden of costs over the period 2000-2015 will lag behind the overall economic trend in the industry. Not all sectors of agriculture will face these burden of costs equally. The stockbreeders and glasshouse growers are expected to bear the major part of the burden.

Under the SYSTEM POLICY variant, the costs to agriculture would rise to almost 2.9 billion guilders. This would mean that the annual burden of costs would increase at a rate outstripping the overall economic trend in the agricultural sector.

Sectors of industry

Within industry, the burden of costs are at present borne mainly by food, drink and tobacco industries, the oil industry, the chemicals industry and the metal industry.

Under CURRENT POLICY, the burden of costs to industry will rise by around 1.9 per cent a year over the 1995-2015 period. This increase in burden of costs will occur in virtually every sector. Except in the case of the oil industry, the trend in burden of costs will lag behind the general economic trend in the sector concerned.

Under the SYSTEM POLICY variant, the annual burden of costs to industry would rise by 2015 by almost 3 billion guilders as compared with CURRENT POLICY. This burden of cost increase would affect virtually all sectors of industry and would be in part the result of increases in the water treatment levy and sewerage charges generated by stringent measures in these areas. As a result, the trend in burden of costs in the food, drink and tobacco industries, the textile, paper and paper products industries and the oil industry would outstrip the economic trend. '

Other sectors

Under the SYSTEM POLICY variant, burden of costs would also increase greatly for the following sectors of industry: the drinking-water supply companies, the building industry, trade and services and the road industry.

6   Benefits to human use functions

The water management policy will benefit water users as well as the environment. Agriculture, for example, will benefit from watermanagement directed in part at providing it with the right water supply and drainage conditions. The estimated charges made to agriculture as presented above include a levy directed in part at supplying funds for the waterboards to manage water for the benefit of agriculture. This watermanagement will therefore have a positive impact on agricultural losses due to drought damage and damage due to water logging.

Navigation also stands to gain from present watermanagement policies. For example, the analyses of our shipping model show that implementation of planned projects to clean up aquatic soils will reduce the annual operational and delay costs of the inland shipping fleet by 60 million guilders. Watermanagement policies represent therefore, not only burden of costs but also benefits of the shipping industry.

Water-based recreation also stands to benefit from energetic watermanagement policies which will increase the attractiveness of the Dutch watersystems for recreational activities. The CURRENT POLICY and USE POLICY variants will slightly increase recreational spending and job opportunities. Given the extensive measures included in the SYSTEM POLICY variant, this would increase to 25 million guilders extra spending and 90 extra jobs in 2015.

The drinking-water supply companies will feel the positive effects of improved surface water quality, although these benefits are not quantified further with the DAO.

7   Indirect economic effects

The costs of watermanagement policies in one sector will have an impact on other sectors (suppliers and customers). If a producer is able to pass on the cost-increase to his customer than the financial consequences will finally pour down to the end-user, most likely the consumers. In that particular case the consumer has to spent less on other goods. If for a specific sector the market does not allow passing on the cost-increases this could lead to a decline in production in this specific sector, which in turn causes negative effect to the suppliers to this specific sector. This problem is analysed using an input-output matrix of the Dutch economy [5].

The analysis shows that macro-economic effect of the CURRENT POLICY and USE POLICY are slight. This is different in the case of the SYSTEM POLICY variant. The costs here would be considerably higher. Under this variant both employment and Gross

National Product would be depressed by about 1 per cent. This effect would occur primarily in the period after 2015.

At a sector level the effect differ greatly. For the period 1985-2015 the sectors that supply the investment goods, like metal industry and building industry, encounter a positive effect on employment (0,3-1%) and value added (3-5%). Other sectors on the other hand show a negative effect, especially trade and services(-0.5- -2.9%) for employment resp. value added. And to a lesser extend food, drink, and tobacco industry (-0.5- -1.5) and the chemical industry (-0.2- -0.7%)

8   References

1.   Luiten, H., et al. (1996) *Achtergrondnota: Toekomst voor water*, Rijksinstituut voor integraal zoetwaterbeheer en afvalwaterbeheersing, Lelystad, WSV, nota nr. 96.058.

2.   CPB (1992) *Nederland in drievoud*, Centraal Plan Bureau, Den Haag.

3.   Min. VROM (1994) *Methodiek Milieukosten*, Ministerie van Volksgezondheid, Ruimtelijke Ordening en Milieu, Den Haag.

4.   Min V&W (1995) *Rijksbegroting 1996, Hoofdstuk XII*, Ministerie van Verkeer en Waterstaat, Den Haag.

5.   Beumer, L., et al. (1997) *Beleidsanalyse WSV, Economische aspecten deel II, werkgelegenheid en toegevoegde waarde*, Nederlands Economisch Instituut in opdracht van het Rijksinstituut voor kust en Zee en het Rijksinstituut voor integraal zoetwaterbeheer en afvalwaterbeheersing, Rotterdam/Den Haag, RIKZ-97.015, RIZA-97.006.

ECONOMIC ASPECTS OF IRRIGATION SYSTEMS MANAGEMENT IN DAGESTAN

AIDAMIROV DJABRAIL

Land Reclamation and Water Management of the Republic of Dagestan, Makhachcala, Russia

Abstract

The cost of a water management organisation for all types of operation and maintenance works which are to be executed to ensure a specified level of economic fertility of lands and their soil and hydrogeological conditions are calculated based on a standard method. These costs include repair and operation and maintenance measures, upkeep of observation network, staff of watchmen and repairmen, data processing, arrangement of water delivery and removal.

The share of intra-farm operation and maintenance costs paid by water users makes up from 2 to 6% of productivity increase from the irrigated lands. The state pays 30% of the costs.Eventually it is planned to include the full cost of water relative to basin water management cost. (?)

Keywords: irrigation systems, water management, agricultural crops, maintenance service, cost of water.

Dagestan relies heavily on irrigated agriculture which accounts for 70% of all agricultural production. The efficient operation of irrigation systems is therefore one of the country's main targets.

The territory of the republic with its diversity of soil and climatic conditions is distinctly different from other regions of Russia from the point of view of intense agricultural production. It is characterised by vertical zoning, severe topography and the influence of the Caspian Sea and adjacent desert plains.

Under conditions of an arid climate crops such as rice, vegetables, fodder and cereal crops, grape, etc. are grown under irrigation. Land reclamation and irrigation are interrelated, and their successful management is the main factor in solving food problems.

Water: Economics, Management and Demand. Edited by Melvyn Kay, Tom Franks and Laurence Smith. Published in 1997 by E & FN Spon. ISBN 0 419 21840 8.

The operational efficiency of irrigated agriculture is mainly determined by the economic relations between the operation and maintenance services and the water users.

The radical changes of the formal economic system taking place in the country has resulted in large-scale structural changes and a more practical reorientation of the economy. Such economic principles are being applied to the use of water to ensure savings.

The present interrelations between the operation and maintenance organisations and water users, the order of planning, are inconsistent with the immediate task of irrigated agriculture to meet the demands of the society in improving productivity with minimum operating costs. In essence, the operation and maintenance service is working for itself, and the higher its expenses, the better are appraised the results of its operation.

At present, there is a firmly established opinion among the major part of the agrarian economists of Russia that the new principles of management can be realised in the best possible manner with the introduction of a water charges. While not rejecting this notion in principle, it should also be noted there are difficulties in implementing this. The introduction of water charges will require all irrigation systems, including small distributors, to be equipped with water metering devices and this will involve a great number of technicians to control the irrigation water discharges. On the other hand, the productivity of irrigated lands depends on:

1. Prompt water supply to agricultural crops in strict accordance with irrigation regime and prevailing agrometeorological conditions;
2. Establishment of land reserves needed to be reclaimed at the expense of optimizing the water supply.
3. Implementation of reclamation measures to improve the physical, mechanical and chemical properties of soils;
4. Timely technical maintenance of the whole irrigation systems.

Water supply alone, without these measures will not secure conditions for the introduction of systems of irrigated farming required to obtain good crop yields. Moreover, the delivery of planned volumes of water with inadequate drainage will result in the deterioration of soil and hydrogeological conditions.

However, the water required for the irrigation of agricultural lands, is dependent not only on the type of irrigated crops, but also on the technical support and the system itself. In one climatic zone, for example the water demand varies from 6 to 17 thous.m^3/ha net. Under these conditions it is obvious that water users will not be working under similar economic conditions.

Enormous difficulties are expected when trying to determine the volumes of excess water used.

Some prefer a system based on a standard per hectare system of cost. The cost would be borne by the water users and operation and maintenance organisations. The basis of such a system would be a fixed share of standard costs for obtaining specified productivity of reclaimed lands in a certain zone.

Presently the cost of the water management organisations for all types of operation and maintenance work to be executed for the provision of a specified level of economic fertility and soil and hydrogeological conditions are calculated on the basis of a standard method. These costs include repair and maintenance measures, maintaining observation network, service staff, data processing, organisation of water delivery and removal.

Exclusive of 30%, paid into budget by the state, the share of operating costs paid by the water users comes from 2 to 6% of productivity increase of irrigated lands. Thus, for each hectare of serviced irrigated area the water managers receive 2-6% of the value of production increase from the water users and one-third from the state.

The share of the costs is calculated by a standard method. It assumes the area is supplied with water in strict correspondence with the plan and quality of lands. If the farm was deficient in production through its own fault, it must still pay a share to the service. If the irrigation managers are at fault, the share is deducted from them. In this way the interests of the irrigation managers and agricultural enterprises are taken into account.

Eventually it is proposed to include a rent assessment into the cost of water in relation to the basin water management costs. Costs in this case may increase significantly but it is hoped that their full compensation will be possible because of increases in the purchasing prices for agricultural production. It is hoped that the workers of the operation and maintenance service of irrigation systems will become more interested not in the volumes of implemented works, but in conscientious maintenance of reclaimed lands and ensuring their maximum productivity, since part of this income will be dependent on farm productivity.

Because of the lack of financial assets of water users and the inadequate level of equipment for irrigation, during the first period of transition to paid water use, it will be necessary to compensate for the costs of maintenance and repair of irrigation systems belonging to operation and maintenance services. The water charge in this case will include cost price of water delivery, taxes and penalties for the violation of water use.

In 1997 the government of the Republic of Dagestan adopted a decision, as an experiment, to introduce payment for the delivery of water to non-agricultural water users. The calculations, substantiating the size of water charges were prepared by the Ministry of Economy. According to their calculations the cost per 1 m^3 of water received from the canal is $3,7. As this takes place in 1997 by this experiment the operation and maintenance service of irrigation systems should receive the income of $208,4 million dollars due to realization of water from the canal to non-agricultural users.

FINANCIAL AND ECONOMIC CONSIDERATIONS IN INTERNATIONAL WATER MANAGEMENT
International Water Management

K. LEENDERTSE
River Basin Management, Delft Hydraulics, ,Delft, The Netherlands

Abstract
This paper addresses financial agreements and non-agreements between States regarding the use of shared water resources. A distinction is being made between water use and water system use. The first considers physical extraction of water from the source for economic or social benefit whereas the latter concerns economic activities that take place in the vicinity of water systems. This distinction allows adequate management approaches and structures and subsequent financial measures and/or compensations between States based on the actual use. The paper provides an introduction to the economic valuation of transboundary use of water for consumptive or productive use and its financial consequences, whereas water system use is addressed from the point of view of transboundary management of shared water bodies, and the financial consequences for individual States. Examples include: a proposal for the valuation of water in dispute in the Middle East; consequences of the joint Hungarian-Slovak attempt and failure of the Gabcíkovo-Nagymaros dam, and the apparent success of the Canadian-US International Joint Commission; non-address of financial matters in the Meuse parlementarian conference; charging for water use in the draft EU directive establishing a framework for European Community water policy; financial consequences of the implementation of the Lake Victoria environmental management project for the riparian States; financial incentives for resolving water disputes.
Keywords: access, agreements, finance, incentives, international, management, shared waters, valuation

1   Water as an economic asset

Growing awareness of scarcity of water as a resource leads to increased attention to water as an economic asset. Economic development and environmentally sustainable

Water: Economics, Management and Demand. Edited by Melvyn Kay, Tom Franks and Laurence Smith. Published in 1997 by E & FN Spon. ISBN 0 419 21840 8.

use of the resources depends to a large extent on considering water as a scarce resource and using economic principles for its management [1]. However, it has also been clear that markets for water either typically do not exist, or are highly inefficient, or necessitate specific considerations. Common access to water resources causes severe constraints in applying market-like instruments to manage the resource. The concept of the "tragedy of the commons"[1] does apply to water resources use, although the consequences for management principles need to be treated with care. This is mainly due to the availability of information to the users, as well as common property perceptions by users' groups, which may lead to the application of traditional or non-institutional management regimes. These management practices reflect a perception of resource ownership by the users and of responsibilities for a sustainable use.

In this respect, identification of the type of water resource and its use will assist in proper water management planning. Here, a distinction should be made between water supply and the use of water systems for water related activities. The first considers physical extraction of water from the source for economic or social benefit, for example for irrigation or drinking water, whereas the latter concerns activities that take place in the vicinity of water bodies, such as rivers, lakes or in coastal areas. It is important to make this distinction because the supply of water as tradable economic good for consumption, irrigation, industrial use, and the like, may require different management approaches and measures than for access to and use of surface water systems for economic (such as fishery, tourism, industry, navigation, etc.), social or cultural activities. It may also lead to compensations between States based on the actual use. A key-issue in this respect is ownership of the resource.

Definition of ownership of the resource, be it an individual, an institution, a group of people, or a State, can make the exploitation of the resource economically feasible. This water ownership is equivalent to a right to the monetary value that water represents [2]. In other words, control over the source and the extraction of water, as well as for example exclusive right to make use of a coastal area, can be quantified in monetary terms.

Water is a scarce resource, increasingly under pressure of growing population densities and multiple uses of the same resource. In theory, the stronger the competition between the users over the resource, the higher the value of water. In practice, however, the value of water is also highly dependant on access to the source, which may also be a determining factor for the value of land in the vicinity of water resources. These values of water are not free market values. Many factors influence the value of water, such as historical users' rights and traditional practices. Water also has a social value and environmental functions. Also, water is necessary for human life, which makes demand for water to a certain extent insensitive to price changes and price elasticity is therefore low. Price fluctuations will be supply driven, and depending on the availability of water and distribution system.

In surface water systems, however, it is not the water scarcity that determines the monetary value of the resource, but the rate of accessibility to the water system. Where access is controlled, it is also more scarce and therefore of higher monetary value. This demand for access to the resource is more price elastic. The availability and price of access is a determining factor in the assessment of the economic feasibility of an activity.

In both cases, water supply and water system use, control over the resource and limitations to the free use of or free access to the resource determine the extent to which water as a scarce good can be quantified. The unit of measure in the first instance will be the price per unit of water whereas in the latter occasion the price will be set by the demand for limited access and the unit will be the price per unit of access.

For example in Zimbabwe, there have been experiments with tradable fishing licences on some small water bodies. The value of the licence, and therefore of the access to the water body, depends on the expected returns from fishing and the competition among fishermen. However, more fishermen on the lake means lower average catch per unit of effort because of the lake's maximum sustainable yield of fish quantities and by species. Hence the introduction of the licensing system and the value of the licence depending on the number of displaced fishermen (as a consequence of the regulation), their purchasing power, and the internal rate of return on fishing activities. Thus, with a fixed number of licences, price fluctuations are demand driven.

2   Financial considerations in shared waters management

Water resources do not take into account national borders. Increasing demand and consequent growing scarcity can cause conflicts between States sharing a river basin, a watershed area, a coastal area or lakes. On the other hand, these same factors can provide strong incentives for regional economic and joint management of water resources [6]. We will examine some cases in which bilateral or multilateral management of shared water resources has been, successfully or unsuccessfully, implemented and to which extent financial aspects of such arrangements have been considered.

We saw that growing scarcity leads to increasing monetary value of unit of water or unit of access. Strangely, the price of access to, or unit of, water is hardly considered when resources management arrangements are being made between States. Yet, taking into consideration the monetary value of water may prove to be an objective parameter, and eventually a solution, when scarce water is being disputed among States.

3   Valuation of water in dispute

This principle has been elaborated by Franklin Fisher, an economist at the Massachusetts Institute of Technology, in an application to the Middle East [7]. Reaching an agreement over water rights and water allocations in the Middle East is an important condition for the success of the peace process in the region. The waters of the Jordan and Yarmouk rivers and of the Mountain Aquifer shared between Israel, Jordanians and Palestinians are subject to various claims. By valuing these waters in dispute, parties can evaluate the possibility of trading off water for non-water benefits.

Questions of ownership and scarcity rent of water have been dealt with in estimating a monetary value per unit of water. Apart from obvious advantages of such an analysis for the evaluation of the feasibility of water management projects, such as construction works, it can also assist in resolving disputes over ownership rights by producing estimates of the monetary value. Important in this context is Fisher's argument that it will be easier to negotiate responsibility for an amount of money than a scarce natural resource. This particular case had three objectives: in the short run, the results of the study were to be used to generate compensation for water usage; in the medium term, the estimate of the true economic value of water in dispute should facilitate negotiations over water rights; and eventually it was the intention that optimal allocation of the region's scarce water resources could be reached. The valuation was based on the fact that countries that own water and use their water themselves do not get the water at no cost, because they give up the money they could make by selling the water to others. The study also dealt with the two main elements, property rights and who uses the water, as analytically separate questions. In the case of Gaza, no matter who owns the water, the high population density and little fresh water sources will make the provision of sufficient water expensive. The upper limit to the value of water is the cost of desalination in Gaza itself. As long as extraction and transportation costs from another source do not exceed the cost of desalination, water would be imported.

In considering the users and locations of water, Fisher concluded that it is the scarcity of water that gives it its value: "where water is not scarce, it is not valuable". Two of the major conclusions of the study were: the value of water in dispute is not high compared to the economies involved; and, there will be no crisis of water for consumption or productive use in the region, but better conveyance facilities are required. It is argued that valuation of water in monetary terms, rather than quantities, allows for a water settlement among the parties.

However, some recent studies have questioned the equity and justice associated with market allocations [8]. The conclusions from these studies suggests that economic considerations alone may not provide an acceptable solution to water allocation problems, and especially concerning disputes between States. Market structures and arrangements may be different from regular markets, because there is a relative small number of parties with different objectives and perspectives. Here too, historical rights may be a determining factor in assessing the value per unit of water. However, it is generally agreed that there is a need for joint efficient water use and management arrangements to ensure sufficient water supply from shared sources between States.

4   Collaboration between States

To improve collaboration between States for more efficient use of resources, however, is not an easy task. Back in 1977, the Hungarian and Czechoslovakian governments signed an agreement to construct and operate jointly the Gabcikovo-Nagymaros hydroelectric system. It was agreed that the construction and operational costs would be shared equally between the two countries, with main structures considered as mutual property, and that produced electric energy would also be divided equally among them. The two partner States were to organise construction and operation in such a way that they would not influence adversely the water quality, environmental interests, or the fishery activities in the area. In 1989, Hungary suspended and later annulled the agreement under pressure of environmental protests. Czechoslovakia studied the new situation and formulated seven possible solutions, one of which could be realised without collaborating with Hungary. This option would imply an increase in costs of around 50 percent. An argument between Hungary and Slovakia has started in which Hungary blames foreseen environmental damage on the unilateral decision by Slovakia to divert the Danube, with detrimental effects on the Hungarian wetlands, while Slovakia blames Hungary for breaking the contract after twelve years of mutual implementation, up to a stage at which effects on the environment are irreversible and attempts to return to the pre-project situation may even have more damaging effects.

At this moment, the argument is being handled by the International Court of Justice. Whatever the outcome of the case, the financial consequences of this dispute are considerable. It has been estimated that the costs for restoration would equal the construction costs and in that case there would be no compensatory revenues [9]. On the Slovakian side alone, the theoretical annual loss for not delivering electricity was estimated at US$ 1 billion for each year the project was not operational. Cost recovery from the Gabcikovo plant would cover around 40 percent of this financial loss. On the Hungarian side, huge claims from Slovakia and commitments to Austria are still pending while no immediate revenues may be expected.

A more successful attempt to prevent and resolve disputes between neighbouring countries on the use of water resources, has been the establishment of the International Joint Commission (IJC) between Canada and the US. The purpose of the Commission is to prevent and resolve disputes over use of shared waters and to provide advice on matters affecting the shared environment. These advises are given upon request by the two federal governments [10]. The Commission was established as an independent international organisation under the Boundary Waters Treaty of 1909. It approves projects, such as dams or water diversions, that affect water levels and flows across the boundary. Under the 1978 Great Lakes Water Quality Agreement, it assesses progress to restore and maintain the chemical, physical and biological integrity of the Great Lakes. The Commission also investigates and reports on transboundary air and water pollution, persistent toxic substances, exotic species and other matters of common concern along the international boundary. The functioning of the Commission is financially supported by the governments of Canada and the US, but otherwise all water management related activities are covered on an ad-hoc basis, depending on the project, between the two States and mainly through public-private financing.

5   Multilateral water arrangements

Also in Western Europe, inter-Statal efforts have been made for improved management of shared resources, and in particular transboundary management of river systems. One example is the establishment of the Meuse parliamentarian conference, which held its second meeting in November 1996 [11]. Within the framework of the Interparliamentarian Consultative Council of the Benelux, the conference discusses issues related to management and control of the Meuse, and the interactions with the natural, physical and economic environment. It also looks into transboundary impacts of implementation of specific measures. The need for international co-operation and co-ordination of has been stressed on various occasions. However, up to this moment, the financial implications of implementation of management and control measures for the riparian States and how these relate to the anticipated benefits for each of the States, have not been considered. Bilateral deals are struck between States on a reciprocal basis, regardless of the effects for other countries in the river basin.

This will change drastically if and when a European Council directive establishing a framework for European Community water policy will come into effect. To this extent, a proposal has been prepared and is currently being reviewed. The proposal does not only consider co-financing by the European Community from three different sources, but also foresees as a tool the actual charging for water use to end-users. The philosophy behind this proposal is based on the fact that the efficiency of water use and the effectiveness of environmental provisions can be improved by ensuring that the cost of water is a genuine reflection of the full economic costs involved. This will have far reaching consequences. Under the new directive, the administrative management unit will be the river basin rather than regional or national administrative units as has been the case in many occasions so far. The Member States, however, will still be held responsible for imposing the actual charges, ensuring that there will be no hidden subsidies. The economic analysis of the proposed directive itself runs short in quantifiable information. The analysis therefore is limited to an identification of the nature of implementation costs of the directive and advantages associated with the proposal. The costs include those to the Member States, to private households as well as to industry and agriculture. It is assumed that the advantages in terms of environmental and sustainable policies outweigh the costs incurred. It is further foreseen in the proposal that responsible parties for management of the river basin shall undertake economic analyses of water use to provide basic information needed to calculate the full economic costs necessary for charging these costs to the end-user.

6   Financial commitments and incentives

It is a generally observed problem that financial consequences of implementation of water management measures and setting up of institutional structures for participating States are not reckoned in an appropriate manner. This induces considerable risks when a management structure is put into place. For example, the implementation of the Lake Victoria Environmental Management Project (LVEMP) involves considerable costs to the riparian States: Kenya, Tanzania and Uganda [12]. The LVEMP is a comprehensive programme aimed at rehabilitation of the lake ecosystem for the benefit of the people who live in the catchment area, the economies of which they are a part, and the global community. It should maximise sustainable benefits to riparian communities, and conserve biodiversity and genetic resources. Although Kenya,

Tanzania and Uganda control respectively 6, 49 and 45 percent of the lake surface, their shares in the project are 35, 29 and 36 percent. The budget is composed of GEF grants, IDA loans and government contributions. The shares of the States in each of these components do not differ significantly from the overall shares. Around 45 percent of the 77 million dollar budget consists of low-interest loans and 10 percent Government contributions. This is not only leading to substantial commitments by the participating States for the implementation of the project but may also raise doubts about continuation in the post-project period of those project components that would ensure implementation of lake-wide management, such as the institutional framework. One of the main risks identified in the project document is that the strength of commitments by the three Governments will fail to sustain the regional environmental management programme which will be developed in the course of the project. Provisions have been made within the structure of the project to reduce these risks during the project implementation but sustainability of the institutional framework, through the set-up of three co-ordinating secretariats, may be questioned.

A suggestion which may contribute to sustainable arrangements between States in water management has been the introduction of financial incentives to circumvent impasses in negotiations [13]. Financial incentives to stimulate co-operation between States may come from outside the region, such as from financial or bilateral aid agencies, or from within the region, from a better-off country in need of water supply or access to water systems to a neighbouring country with abundance of water. One increasingly common linkage being made is that between water and energy resources, as for example the assistance in funding hydroelectric projects in return for electricity or water. It has been noted that in some cases such reciprocal arrangements have survived dramatic political changes within a region. However, the political will to co-operate remains crucial in the establishment of international water management arrangements and agreements between States.

6 Conclusions

In retrospect, the following aspects of international water management may be highlighted:

- first, in transboundary water management arrangements there is a general lack of attention for financial matters, affecting not only the co-operation between States but also the availability of water to end-users and their access to water resources;
- secondly, valuation of water supply and access to water resources, by determining the price per unit of water or unit of access, may be instrumental in order to come to effective agreements between States, especially when scarce water is being disputed among States. However, cautiousness is advised when applying traditional market-like instruments and mechanisms in allocating water resources, and even more so when considering common access to water;
- thirdly, to reach effective use of shared water systems, collaboration between States in managing these resources is imperative. Arrangements between States may have to be institutionalised in agreements for economic effects to become clear. Economic assessments may be done on the basis of the total water body rather than being limited to national borders. An integrated approach will take into consideration the different effects on, for example, the economy, the ecosystem,

etc, of different uses of the resource and allow for prioritisation of specific activities;

- finally, financial commitments of States for the implementation of joint management projects and programmes should be in line with their respective capacities. Financial incentives may be considered as a tool to stimulate co-operation between States.

7   References

1.   Briscoe, J. (1996) Water as an Economic Good; the idea and what it means in practice. In: *The future of irrigation under increased demand from competitive uses of water and greater needs for food supply.* 16th International congress on Irrigation and Drainage; Cairo, Egypt: 177-202. International Print-O-Pac Ltd., New Delhi, India.

2.   Fisher, F.M. (forthcoming). *The Economics of Water Dispute Resolution, Project Evaluation and Management: an application to the Middle East.* Massachusetts Institute of Technology, Cambridge, USA

3.   Hardin, G. (1968) The Tragedy of the Commons. *Science* 162: 1243-1248

4.   Wade, R. (1986) *The Management of Common Property resources: collective action as an alternative to privatization or state regulation.* Discussion Paper no. 54. Research Unit, Agriculture and Rural Development Department, The World Bank, Washington D.C., USA.

5.   Willmann, R. (1992) *Community-Based Resources Management: experiences with forestry, water and land resources.* FAO/Japan Expert Consultation on the Development of Community-Based Coastal Fishery Management Systems for Asia and the Pacific: Kobe, Japan, 8-12 June 1992. FAO, Rome, Italy

6.   ODA. (1996) *Water for Life; water and British aid in developing countries.* Overseas Development Association, London, UK.

7.   see ref. 2.

8.   Bingham, G., Wolf, A., Wohlgenant, T. (1994) *Resolving Water Disputes; conflict and co-operation in the United States, the Near East, and Asia.* Irrigation Support project for Asia and the Near East (ISPAN), Arlington, USA.

9.   Liska, M.B. (1996) Controversies in Developing the Slovak - Hungarian Section of the Danube. In: *Proceedings of the International Conference on Aspects of Conflicts in Reservoir Development and Management*: 775-784. City University Print Unit, London, UK.

10.   Craven, M. (1996) Preventing and Resolving Disputes - the Challenge of the IJC. In: *Proceedings of the International Conference on Aspects of Conflicts in Reservoir Development and Management*: 767-774. City University Print Unit, London, UK.

11.   Waterloopkundig Laboratorium (1996) *Van zorgen om de Maas naar zorgen voor de Maas*; 2^e Parlementaire Maasconferentie. Delft Hydraulics, Delft, The Netherlands

12.   World Bank (1996) *Lake Victoria Environmental Management Project*: Staff Appraisal Report. The World Bank, Washington D.C., USA.

13.   see ref. 8.

1.   Hardin [3] introduces the concept of "Tragedy of the Commons" in relation to overgrazing as a consequence of open access to communal grazing land. Each herdsman faces the trade off between the benefit or income he can receive from selling his animals, and the cost incurred from overgrazing the common pasture as the number of animals increases. The key argument is that for the rational herdsman the cost of an additional animal in terms of added overgrazing is shared by all herdsmen while the income from selling an additional animal is purely his. Each herdsman, thus, has an incentive to continue adding animals until the final destination - ruin - is reached.

The theory thus explained may apply to the exploitation of any resource whose access is free, such as tropical forestry, capture fisheries, and also the use of water resources. However, Hardin's parable is criticised on primarily two accounts. First, it is argued that Hardin assumes that the individual herdsmen have no information about the status of the pasture and its imminent collapse. Otherwise, it will be against the self-interest of the last herdsman to add another animal leading to the final collapse of the resource. Second, it is argued that Hardin fails to distinguish between common property and open access. His parable may not apply to the former because, as the resource is held and shared by a specific group of people, it is unreasonable to assume that the group would not react to increasing levels of overgrazing by imposing certain restrictions on its use [4]. In this concept, the group is the "owner" of the resource and applies restrictive measures to its use [5].

SECTION F

POLICY FRAMEWORK
LEGAL AND INSTITUTIONAL ISSUES

THE ECONOMICS OF REGULATION IN WATER SUPPLY, IRRIGATION AND DRAINAGE

R. LASLETT
London Economics[1]

Abstract

Water supply, irrigation and drainage are all industries which can demonstrate natural monopoly characteristics. While private provision can realise considerable productive and investment efficiency advantages, it is likely to lead to higher water prices than under public provision. Options for economic regulation are examined. Despite varied country experience an apparent trend towards long contracts and the embodiment of the regulatory agency within the contract framework, may be leading to a convergence of apparently contrasting approaches.
Keywords: water supply, irrigation, drainage, regulation, water policy, monopoly

1   Introduction

Water supply, irrigation and drainage are such important basic needs that they have often been regarded as too important to leave to private firms to provide. And when private firms are involved, it is widely felt that their activities should be strictly regulated to make sure that they work for the public good rather than their own profits. Meanwhile economists tend to stress the advantages of free markets and the problems of regulating effectively. They point out that regulation itself can easily fail to serve the public interest, especially when viewed in a dynamic context where lack of proper incentives can lead to inefficient performance.

This paper looks first at what regulation is, and where the balance lies between these two competing points of view. Most of the argument is presented in relation to water supply, where the debate has been mainly engaged, but I also show how it applies to irrigation and drainage. There can be competition for water resources between supply for domestic and industrial use on the one hand and for irrigation on the other. Equally, the quality of drainage and sewerage outfall water can be very important for the cost of processing water to acceptable standards for municipal

Water: Economics, Management and Demand. Edited by Melvyn Kay, Tom Franks and Laurence Smith. Published in 1997 by E & FN Spon. ISBN 0 419 21840 8.

supply. For these reasons it makes sense to bring them under a single regulatory framework, certainly in conceptual terms and sometimes also in a single institution.

2   Economic regulation and water supply

Water supply has traditionally been regarded as a part of the economy where government intervention, and indeed government ownership, are well justified, and where involving the private sector is very difficult. At bottom, this is because water supply is often seen as a "natural monopoly". That is to say it would be excessively costly to arrange physical access for competing supplies in most places[2]. As a result of the sunk costs it has invested in the pipe network, an established supplier has such a large cost advantage that it cannot realistically be dislodged by a competitor.

In most markets, it is not advisable to leave the provision of natural monopoly services to the private sector without some form of state control over their prices and quality of service - monopolists are liable to abuse their privileged position at the expense of the customer. The traditional response in much of Western Europe (as well as in the developing world) has been for the state to take an ownership role in water supply[3]. The state as owner can pursue financial objectives other than maximum profits, and can in principle ensure that the monopoly utility behaves in some key respects as if it were a competitive firm. State ownership has other attractions too. Particularly important in the case of water is the objective of social protection. In many countries, prices and investment have been set to favour priority regions or economic groups. But state ownership has drawbacks as well as advantages, in both the developed and the developing world. The most serious are that:

- without private shareholders, all investment has to be financed from the government budget or borrowings, and government may hold back necessary investment as it tries to balance the budget;
- it leads to political interference in the management issues; and
- without the discipline of a profit objective, firms tend not to make much effort to reduce costs over time, and their performance deteriorates relative to world standards.

These problems led to a realisation in the 1970s and early 1980s that there were advantages to not having water supply under government ownership. This was partly based on the experience of the USA where important public utilities[4] had long been in private ownership but under public regulation.

Public regulation can mean a wide variety of things, depending on the context. Governments routinely regulate the behaviour of firms in respect of environmental protection, safety standards, minimum wages, accounting standards and basic adherence to the rule of law. Economic regulation governs pricing and quality of service and specifically aims to alter the behaviour of monopoly suppliers to make it more like that of competitive firms.

Economic regulation is a response to the *principal-agent problem*. The principal (government) licenses the agent (firm) to supply water on its behalf for the social good. However, the firm has different aims than the government - it wants to expend as little effort as possible in carrying out the task, and to make as much profit as possible. Regulation aims not simply to keep prices down, but to do so by ensuring that costs are kept to a minimum, by making the firm run as efficiently as possible. The principal-agent problem exists for firms of all types, not just regulated ones. There is always a difference of interests between owners (the principals), who are interested in profits and managers (the agents), who are interested in salaries.

Private firms have developed a number of instruments to address the resulting problems, including incentive mechanisms such as stock options for managers, monitoring mechanisms such as the annual reporting requirement for quoted firms, and market pressures such as the threat of hostile takeover. While such mechanisms exist for privately-owned companies, they are missing for state-owned ones. This means that privatisation can make the task of regulation easier. Most discussion concerns the ways in which regulation is still required, but private ownership will usually serve to improve the effectiveness of monitoring and control over management efficiency.

In the case of water supply, water is of paramount importance and Government involvement in regulating water quality is widespread. There are often hard choices to be made between the objectives of price regulation (which broadly requires that costs be kept to a necessary minimum) and those of quality regulation (which may require that additional costs be incurred to maintain or improve quality). This means that there are advantages to having separate policy responsibility for price and quality issues. However it does not necessarily follow that the institutions of price and quality regulation need be separated, as they are in the UK where price regulation is carried out by the Office of Water Services and quality regulation by the Environment Agency. Such separation can lead to conflicts between regulatory agencies that are not easy to resolve. There are several possible ways in which economic regulation can be carried out in practice. For instance:

1. The French monopoly electricity utility EdF is in government ownership and is regulated under a "Contract Plan which governs the investment, pricing and operational targets of the firm, but leaves it to management to ensure the targets are met.
2. In the water industry in Argentina, Scotland and the Philippines for example, regulation takes the form of long-term concession agreements. Under these concession agreements, the assets used in water supply remain in public ownership but investment and operations are undertaken by a private contractor.
3. Regulation can also take the form of a contract between a private firm and a government, for instance in Indonesia's telecommunications industry, in which all the investment and operation are undertaken by a private company, but the assets eventually revert to public ownership.
4. In the electricity and telecommunications industries in the USA and elsewhere, regulation is carried out by an independent agency while the utilities are privately owned. This model has been adopted with modifications in the UK, Norway, Sweden, Australia, New Zealand and several developing countries in Latin America, notably Chile and Argentina. It is under contemplation in South Africa, Portugal, Spain and in several of the transition economies in Eastern Europe.

Within the long-term contract type models (2 and 3 above), there are important differences depending on the length of contract. In Scotland with 7 year concessions, prices are controlled by the terms of the contract. In Buenos Aires with its 95 year concession contract, this is not sufficient and the contract provides for the establishment of a regulatory body.

Whatever method is chosen, one of the keys to a successful economic regulatory system is that the regulatory agency be independent both from the utility - which has a clear business motivation to influence regulators' decisions about prices - and from the government - which may have political aims to push. Without independence, the regulatory agency may be "captured" by the firm it regulates, and become incapable of articulating a separate point of view. Equally, many countries lack the concept of a separation of powers to back up the creation of regulatory agencies independent of the government. In countries with a unitary state, responsible for all aspects of public life it may be very difficult, if not impossible, to create a regulatory agency that is genuinely independent of political interference.

In France itself, the response has been to combine the functions of the state as owner and regulator. In countries with a unitary state but lacking a strong tradition of public service for the public good, it may be very difficult indeed to establish an independent regulatory agency.

To sum up, price regulation is only needed in a market if the product or service being supplied is not being produced or distributed under competitive conditions. Otherwise, the competitive process itself will generally produce better economic results, in three areas:
- efficiency in production, in the sense of making sure that current inputs are used most productively;
- efficiency in consumption, in the sense that prices to consumers reflect the marginal costs of production; and
- efficiency in investment, in the sense that projects are chosen and carried out in the best possible way.

In most markets, non competitive supply leads to inefficient consumption: a monopoly supplier will set unit prices higher than marginal costs and as a result of the price-sensitivity of demand sell less than a competitive supply industry would. An interesting feature of the market for water supply is that this is not necessarily the case, because water use is typically little reduced, even by quite high prices. Furthermore, the monopoly water supplier often has an even better option than raising the unit price for extracting profits from consumers, namely two-part pricing. Under two-part pricing, the price of units of water supplied can be set at the marginal cost (leading to efficient consumption decisions), while suppliers profit through high access charges.

Thus the characteristic problem of unregulated monopoly water supply is not that of inefficient consumption, but of an inequitable sharing of the benefits between supplier and consumer. For equity reasons, water supply has often been placed in the public sector, allowing direct policy intervention in pricing. And where it is conducted by the private sector, it is usually under some form of regulation. Other aspects of efficiency do remain of concern in the water supply industry. Whether in public or private ownership, a monopoly supplier will be less responsive to consumer needs and less likely to attend to efficiency in investment and production than competing suppliers would.

3 The economic regulation of irrigation and drainage

Having described the function of economic regulation and the way it applies to water supply, one can move on to ask whether irrigation and drainage are like water supply in respect of their economic structure and need for regulation? The main question is whether a natural monopoly exists, which in turn has to do with the cost structure of supply.

3.1 Regulation and irrigation

Irrigation can display one of several possible cost structures, depending on whether access to the resource is mainly a question of collecting water or whether transporting and distributing it are also important. And there are important interactions with other uses of water.

At one end of the spectrum, where the task of irrigation is mainly one of collecting water - as when it is taken from farm dams, boreholes or by means of individual abstraction from surface water - sunk costs are low and there is unlikely to be a natural monopoly problem. The technology of pumping on this scale is unlikely to have significant economies of scale beyond the level of individual farms. There are therefore many potential competing suppliers, and often each farmer can supply himself at a cost not greatly above that of purchasing supply on the market.

At the other extreme, where the task of irrigation involves a substantial element of transporting and distributing the water, the situation is very different. There are large scale economies and sunk costs in long distance transportation, and as a result the supplier is likely to have considerable market power; it is a natural monopoly like a reticulated water supply company. Natural monopoly problems can also arise where there are substantial water-processing costs, for instance desalination, which require equipment with large fixed costs. As a result, irrigation through large scale water collection, transportation and distribution is economically similar to water supply for domestic and industrial use. There is the same need for a regulatory framework, which could take any of the forms noted above.

One significant difference is that water quality is of less concern in irrigation than in household uses, so the cost of water processing is lower and the need for water quality regulation is much less pressing (though not necessarily absent altogether). As a result, quality regulation (which as we saw in the case of drinking water is best regulated separately) can be integrated jointly with prices in the case of irrigation.

Individual abstractions often pose an economic problem: not that of regulating the monopoly power of a supplier, but that of managing access to a limited resource efficiently. There is an incentive for over-use because each farmer tapping into an underground aquifer for example has an individual interest in pumping out as much water as he needs for his farm. But it may be that overall abstractions are depleting the aquifer in an unsustainable manner. In this situation, the farmers involved would do better to collectively reduce their abstractions to a sustainable level. But individually none of them has any incentive to do so. Since the problem is that the private cost of the water is less than the social cost, the issue can sometimes be resolved through a tax on abstractions to reflect the social costs of depleting the aquifer. In practice, co-operative solutions where farmers agree and police their level of abstractions, and or regulatory solutions where a central authority plans and enforces abstraction limits are more frequent.

In principle it would seem that a supplier transporting the water over a distance would not have the free-rider problem. A firm with sole access to a water resource is able to plan its exploitation on a sustainable basis, and its profit-based business decisions will correspond closely to the interest of society (though it may have a different - usually higher - discount rate than the government, and be inclined to deplete non renewable resources faster than would be socially optimal).

But as was pointed out at the start of this paper, it is rare in practice for a supplier to be the sole user of a water resource. There are usually competing users: local farmers taking water directly, hydroelectric stations, municipal or industrial abstraction. The need remains for a tax, regulatory or co-operative solution. The regulatory solution is probably the most workable in this case, because of the scale of the problem and the size disparity between the actors involved.

3.2   Regulation and drainage

In the case of drainage, the situation again depends on the cost structure, but interactions with other water uses are also important. Where land can be drained by individual farmers (eg through fitting porous pipes), there is no monopoly power, and correspondingly no need for regulatory intervention. Where drainage requires the construction of transportation systems to remove the surplus water and prevent flooding of adjacent land, or waste water requires extensive treatment to allow its safe discharge into other watercourses, the cost structure is similar to that of a monopoly. But it may be difficult to control individual farmers' access to transportation systems, particularly if they are open channels at or below ground level. Because of this, it may be impossible for a privately-owned firm to build and operate such drainage facilities profitably.

Here again there could be either a tax-based, a co-operative or a regulatory solution. In the tax based model, the public authorities would levy a compulsory contribution on all farmers benefiting from the scheme, to fund the cost of drainage infrastructure. In the co-operative model, farmers would band together to fund the costs voluntarily, but individually they would not want to pay, even though collectively they would benefit. In the regulatory model, a distinct entity would levy contributions to fund infrastructure.

In this case the tax-based and regulatory solutions are the most workable. Large scale drainage by private firms will probably only be feasible where access to the transport network can be controlled, eg when the water is evacuated in pipes or pumped up to raised channels. In these cases, the drainage firm has market power over the farmers and the need for regulation to control monopoly power may emerge again.

3.3   Conclusion

The cost structure of irrigation water supply is similar to that of municipal or industrial water supply when water has to be transported and distributed over a distance, and possibly when it has to be processed extensively before it can be used. The cost structure of drainage is similar when waste water has to be transported over a distance and/or treated, and where access by non-contributors can be controlled.

In these cases a regulatory system is required that closely resembles those needed in the case of municipal or industrial water supply, in order to prevent the abuse of monopoly power and the inequitable distribution of econ omic rents (a much larger share of the rents going to the irrigation/drainage company than to farmers). The rest of the discussion is relevant only to these cases.

4   How is the price of water regulated around the world?

A variety of regulatory formulae have been extensively discussed. The main ones are rate of return regulation (broadly used in the USA), and price cap regulation (extensively used in UK and being adopted in developing countries). Each has advantages and drawbacks, particularly in respect of incentives for efficiency, and each in practice is modified to try to moderate its drawbacks.

The principle of rate of return regulation is to set prices periodically (usually every year) so as to give the regulated firm a fair rate of return on capital investment in the provision of the service. Its main advantages are:

i)   That the rule is clear and commonsense in principle, and
ii) That it leads to stable profits over time and is therefore politically acceptable and attractive to investors.

The main drawbacks are:

i)   That if the period between price settings is too short, there is little incentive for the firm to minimise its operating costs, and
ii) That there is no incentive for the firm to develop efficient capital investment programmes.

The latter has led regulators to intervene in detail to review what capital expenditure is necessary, and this can lead to protracted and costly exchanges of information and even to legal disputes between the regulator and the regulated firm.

The principle of price cap regulation, by contrast, is to set prices (or a formula for the escalation of prices) at longer intervals, perhaps every 5 years. The price cap is based on the best-practice costs of providing that service. The main advantages are:

That it gives strong incentives for the firm to operate and invest efficiently, because the firm keeps the additional profits it derives from reducing costs, and
i)   That it can be less intrusive, because regulatory review is less frequent.

The main drawbacks are:

i)   That it leads to large variations in profits both between companies and over time, that may be hard to accept politically and for investors, and

ii) That it gives strong incentives for firms to conceal possible efficiency gains in the run up to a periodic price review.

There is relatively little information on the regulation of water supply in Eastern and Central Europe. London Economics has recently completed a study of private provision of water services in Ostrava and Northern Bohemia (Czech Republic) and Pecs and Kaspovar (Hungary) [1]. The model for private involvement in all these cases is based on the French concept of delegated management in which the municipality owns the assets while a private company is responsible for management and operations. A 15-30 year operations contract with the municipality defines how the operating company is remunerated and how tariffs are set, generally by annual negotiation on the principle that the operating company should cover its costs and earn a reasonable profit. The tariff is based on financial forecasts for several years ahead, and the calculations are reviewed by and negotiated with the municipality. In one of the cases (Pecs) the tariff is based on a formula with clauses to trigger renegotiation if circumstances change (eg major investment requirements arise).

The main challenge in such a regulatory framework is to align the incentives of the operating company with the public good. The relatively common and simple rule of allowing profit to be a certain percentage of costs does not do this for example - it encourages firms to inflate costs. A fixed profit provides no incentive either way. A profit based on the difference between actual and best practice costs provides the correct incentive but is hard to achieve.

5   Conclusions

Water supply, irrigation and drainage are all industries which can demonstrate natural monopoly characteristics. While private provision can realise considerable productive and investment efficiency advantages, it is likely to lead to higher water prices (particularly connection charges) than under public provision. Economic regulation is an option worth examining in this context. It can be implemented in various ways, though the political and institutional context of developing and transition countries would seem to favour regulation through contract rather than regulation by an independent agency, because it is hard to guarantee the independence of a regulatory agency.

There are various forms of regulatory body in Western Europe and the USA. While the French model of regulation by contract is politically easier (because it does not involve privatisation), the trend towards long contracts and the embodiment of the regulatory agency within the contract framework (as in Buenos Aires), may be leading to a convergence of apparently contrasting approaches.

6   References

1. London Economics, *Private Provision of Water Services in Central and Eastern Europe - Ensuring Value for Money and Affordability*, 1996, Report to the European Bank for Reconstruction and Development.

---

[1] I am indebted to my colleagues David Ehrhardt and Ian Alexander for useful discussions. Responsibility for the content remains my own.

[2] The main exception being at the frontier between different suppliers' areas.

[3] Of course, in the centrally planned economies state ownership over large and important firms was universal and the water sector was no exception.

[4] Though in the USA water supply itself has often remained in municipal ownership.

DEMAND MANAGEMENT FOR EFFICIENT AND EQUITABLE USE

J.T. WINPENNY
Overseas Development Institute, London, UK

Abstract
Demand management (DM) is becoming the dominant note of policies towards the water sector. It offers clear economic, financial and environmental benefits, compared to supply augmentation, and, carefully implemented, meets equity concerns too. The paper outlines the main elements in DM policies, summarised as enabling policies, incentives and projects/programmes, with examples relevant to European agriculture.
Keywords: demand management, efficiency, equity, prices, incentives, environment.

1   Introduction

In the world as a whole, agriculture accounts for around two-thirds of water use, and a higher proportion in arid and semi-arid countries.[1]

The fact that agriculture is such a dominant and, by many accounts a profligate, user of water has led many people to believe that relatively small savings in its water use would be easy to achieve. This would release enough water to satisfy other sectors. It would also permit an expansion of irrigation based on existing water supplies. Looking ahead 20-30 years, in many regions, including the Far East and South Asia, most of the required increase in food production is expected to come from irrigated agriculture [2].

When this is taken together with the constraints - hydrological, economic, financial and environmental - facing the development of new water sources - there is a strong prima facie argument for expecting DM to be the dominant policy paradigm in irrigated farming. This is likely to be just as true for European agriculture as for tropical and semi-arid cultivators.

Water: Economics, Management and Demand. Edited by Melvyn Kay, Tom Franks and Laurence Smith. Published in 1997 by E & FN Spon. ISBN 0 419 21840 8.

2   The case for demand management

2.1   What is demand management?

Demand management (DM) is a policy for the water sector that stresses making better use of existing supplies, rather than developing new ones. One way of visualising DM is to compare it with its alternative, supply augmentation, which was the prevailing policy in many countries until recently. To caricature this approach, faced with evidence of future shortages of supply to meet growing demand, the typical response was to:

commission a comprehensive study of water resources;
project the demand for water on an "unconstrained" scenario;
consider the various supply augmentation options:
recommend that which met projected demand at the least cost;
implement the scheme through public agencies, and at subsidized prices.

The keynotes of the traditional approach to water problems were thus: centralized planning and prescription, public agency, supply augmentation, subsidy, and reliance on adminstrative and legal instruments for allocation of supplies and pollution control.

In contrast, DM stresses: waste reduction, economy in use, the development of water-efficient methods and appliances, creation of incentives for more careful use for both suppliers and users, improved cost recovery, reallocation from low to high value uses, a role for the private sector, greater devolution of water management to consumers and user groups, and the greater use of economic instruments (prices, markets) alongside other methods of matching supply and demand.

DM typically includes measures to relate the value of water to its cost of provision, and motivate consumers to adjust their usage in the light of those costs. DM entails treating water more like an economic resource, as opposed to an automatic public service.

2.2   Benefits of demand management

The benefits of DM can be summarised as follows:

in countries or regions facing implacable hydrological limits, DM recognises water scarcity as a fact of life and creates the conditions in which users can appreciate its real value;
by making better use of existing supplies, DM obviates the need for costly new investments (in many countries the cost of new schemes is increasing exponentially);
DM avoids the environmental disturbance that is inherent in many new supply schemes;
DM is normally part of a process that facilitates improved cost recovery, which helps to fund the operation and improvement of water systems, and reduces its dependence on subsidies.

The benefit of DM for the efficient use of water is obvious, and is discussed in section 2.3 below. Its impact on equity, however, is less obvious - indeed, fears about the effect on poor or deserving groups lie behind much of the opposition to water conservation and its instruments, such as metering and volumetric charging. These concerns can, however, be addressed, and are discussed in section 2.4.

2.3 Efficiency

DM promotes the more "efficient" use of water, by discouraging waste and seeking to increase value-added from existing supplies. However, sceptical voices have been raised about the real scope for improving water efficiency, especially in agriculture [3]. In any case, "efficiency" means different things to different people. This section reviews some of the ways in which the term "efficiency" is used, focussing on some of its technical, social and economic meanings.

To practitioners of irrigation *engineering and hydrology* "efficiency" is primarily a matter of ensuring that as much as possible of the available water is actually applied to beneficial use. There are three possible levels at which this criterion can be applied, namely farm, scheme and basin.

At farm level the clearest measure of efficiency is the amount of water taken in (e.g. at the sluice gate) and applied to the fields to produce a given amount of crop. An irrigator concerned to improve efficiency has various options: levelling land to ensure more even application; reducing leaks from conveyances or pipes on the farm itself; minimising evaporation from standing water by switching from furrow to sprinkler or pipe; calibrating water intake to crop requirements more carefully by means of drip systems, etc. Also, for a given water intake, crop yields could also be improved by investment in drainage, to reduce waterlogging.

From the perspective of irrigation schemes or projects the relevant efficiency concept is the relationship between water delivered to farmers and water received from the source at which project jurisdiction starts - the intake, reservoir, canal or well. The main scope for improving efficiency is in minimising transmission leaks and evaporation from storage and canals [4].

At basin level efficiency normally refers to the % of catchment yield actually applied to "productive" uses rather than lost to evaporation, unrecoverable groundwater, pollution, or the sea. On this criterion, the policy aims would be to increase re-use, increase storage of water from wet season/years to dry, minimise losses, etc.

In a basin with many individual water users (usually farmers) water is likely to be re-used to a high degree: one user's return flow may be another's intake. Even water draining into a groundwater aquifer may be re-usable. Hence, water at a basin level may be fully used, even though there is wasteful use at a farm or scheme level. The Nile Basin is often cited as a case in point. In these circumstances, water savings made by an individual farmer may not represent savings to the whole basin; if that farmer used the savings to grow more crops there would be less return flow available to others. More generally, water savings that are available for productive use have been termed "wet" savings, those that are not recoverable are "dry" [5].

The **economic** approach to efficiency is to examine the relationship between resources used up in providing a good or service and the value of the latter. If the *economic benefits of an activity exceed its economic costs, the activity is deemed* efficient and should proceed. For instance, installing drip-feed irrigation systems would be justified if its benefits (increased yields, savings in water charges, etc.)

exceeded its costs (capital plus recurrent). This rule applies irrespective of the actual volume of water savings.

In determining the level of output/activity, the marginal rule comes into force: output should expand until the benefit of the marginal unit of output equals the (marginal) cost of providing that unit. For example, where water is charged for volumetrically, the farmer should increase its application until the cost of the extra cubic meter (or other unit of measurement) equals the estimated increase in crop yield from that unit of water.

Where benefits cannot be measured in monetary terms, or where several options are available to meet a particular standard of service, the cost-effectiveness criterion applies. Under this criterion, the preferred option is that which satisfies the objectives at least cost (where cost includes recurrent and capital costs, discounted to the present). For example, the "cost of conservation" criterion compares the different options for saving water with each other, and with the cost of a new supply source, and the preferred option is that with the lowest cost per cubic meter of water saved, provided this is less than that of developing a new source.

Where water is plentiful, and there are no obvious users of any savings, increasing irrigation efficiency may not be justified. A more difficult decision concerns sunk costs: in an uneconomic irrigation scheme, there may still be a case for investing in efficiency if the water has more valuable alternative uses (though in this case there would be an even stronger argument for closing it down!).

Secondly, economists are concerned with the uses to which the water is put. Ideally, if water is becoming scarce, its use at a national level (or that of a river basin, whatever the relevant standpoint) should be optimised in the sense that no greater economic value could be obtained from its reallocation. To apply this criterion, the value of water in all its different uses should be known, and reallocation between uses should be done up to the point where its marginal benefits are equalised. One obvious pre-condition is the existence of the means of conveyance of water between different kinds of users, in different locations.

Although the full satisfaction of the principle is a remote ideal in most cases, it can and does operate in partial form; "high value" users pre-empt or bid water away from "lower value" users, who tend to be farmers. The important practical point of all this is to recognise the many possible values of water in other sectors, including its environmental uses, and to place more store on how water is used in its passage down a river basin, than on using every drop before it gets to the sea.

2.4 Equity
In any discussion of subsidies to irrigation, it is relevant to note that the average level of income and wealth in the farm population is higher than the national average in practically all European countries [6].

Within agriculture, DM would enable available water to be distributed more evenly and equitably. In many systems available resources - the water itself, administrative and professional time, and financial subsidies - are concentrated on the more established and privileged landowners and cultivators. Reducing subsidies, raising water tariffs, and improving the rate of revenue collection could release funds for the expansion, maintenance or improvement of the supply network. The reduction of waste and physical losses in the distribution system, or improvement of on-farm efficiency, would enable a given volume of water to literally go further. These factors would improve the position of marginal or "tail-ender" farmers, and all those who receive unreliable supplies. The only qualification to make to this point is that improving the technical efficiency of irrigation may burden poorer farmers with more on-farm costs.

Making more active use of water pricing is often imagined to disadvantage poorer operators. But there are ways of taking the sting out of irrigation charging while preserving farmers' incentives to make more efficient use of water. For instance, if cost recovery was part of a package including improvements to the reliability of supply, this would permit farmers to move into higher value crops, where reliability is crucial. There are also cases (e.g. California, Israel) whereby farmers are allowed 80% or 90% of their previous, or normal, water supply at a cheap rate, with a sharply rising unit rate for anything in excess of that [7].

3   Policies for demand management

3.1   The policy hierarchy
A comprehensive policy towards DM should comprise action at three, mutually reinforcing, scales: setting enabling conditions by central government policies; the creation of specific incentives for water users; and the implementation of spending projects and programmes.

Enabling conditions would include: institutional and legal changes, utility reforms, privatisation, re-examining macroeconomic and sectoral policy as it affects water consumption, etc.

Incentives comprise both market-based and non-market measures. Market-based incentives include: active use of water tariffs, pollution charges, groundwater markets, surface water markets, auctions, water banking, etc. Non-market solutions include: restrictions, quotas, norms, licences, exhortations, public information campaigns, etc. **Projects and programmes** may include: canal lining, leak detection and repair, the promotion of water-efficient appliances, land levelling and sculpturing, simple treatment and recycling of wastewater, etc.

3.2   Enabling conditions
The government needs to create a policy climate encouraging everyone concerned with water to start treating it as a scarce and valuable commodity - an economic good. This may entail legal and institutional reforms, and a reexamination of economic policies affecting key water-using sectors, such as agriculture.

Legal reforms may be necessary to remove ambiguities over the ownership of water, the conditions under which it can be transferred, and the powers of governments to acquire or allocate water in the public interest. Institutional reforms should allow for the operation of markets and the involvement of private operators in different parts of the sector. In irrigation, public authorities can be placed onto a more commercial basis, more powers delegated to local tiers or users, or sold (or even handed over) to private farmers. Consultation with users and other stakeholders should be routine and automatic.

Policies towards agriculture will be critical, in view of the heavy water consumption of this sector. Decisions to open up new irrigated areas, and to allocate existing water for irrigation projects, need to take a realistic view of future water resources and the growing demands on them from other sectors, otherwise the resulting schemes will be uneconomic and unsustainable. The understandable pursuit of national food security has to be judged in this light.

The best intentioned and designed reforms in the water sector will be frustrated if key economic signals prove to be countervailing. The impact of more rational pricing of irrigation water will be negated by artificially high farm support prices: equally, raising charges will impose a double penalty on farmers if crop prices are depressed in the interests of urban consumers. In short, reforms planned for the water sector should be consistent with other key economic signals.

3.3   Incentives

The permissive effects of enabling conditions may be sharpened by the creation of incentives for the more rational use of water. These may be positive or negative, market or non-market. They will be categorized below as: tariffs; water markets; and non-market inducements.

Although water tariffs are common for household supplies in a number of countries at all stages of development, they are less common in an irrigation context and in any case are usually seen as a means of cost recovery rather than a way of actively managing demand. The principles of economic tariff setting are well established and accepted, and are similar to those in use in the power sector. They can be summarized as setting prices according to Long Run Marginal Costs. This usually entails adjusting the structure of tariffs to include a fixed and variable element, with the latter rising for successive increments ("progressivity").

The use of economic tariffs is facilitated by metering, which is not always feasible or sensible for agricultural supply, where other proxies for the volume of water may have to be used. There are cases where the volume of water supplied to farmers is closely monitored, and where prices increase sharply for marginal supplies. Volumetric charging is more difficult to apply to groundwater users - though there may be good environmental reasons for levying some charge on groundwater extraction, e.g. through the price of power used for pumping..

Where farmers have well-established customary or legal rights to the use of water, reallocation can only occur if those rights are bought out (with compensation) or where markets develop between buyers and sellers of water rights. Groundwater markets are long-established and widespread in certain parts of the Asian sub-Continent. Surface water markets exist in some Western states of the USA and Eastern Australia, principally to transfer water from low-value irrigated farming to urban consumers. Sometimes the transfers are semi-permanent arrangements, e.g. the efforts of the Los Angeles MWA to acquire long term water rights from its agricultural neighbours. In the UK there are proposals to enable farmers holding abstraction licences to sell them to other farmers in their district [8].

Water auctions, although unusual, are well-established in parts of Spain, and have been tried in Australia. Water banking has also been tried: as a response to the recent drought, the state of California bought up water rights to farmers to hold in reserve for urban and industrial use (and most of the stock was drawn down for these purposes).

Demand management can, and often does, rely on non-market devices, either on their own, or working in tandem with economic measures. Education and publicity campaigns can help to convert the public to the need for water conservation, though the message will be convincingly brought home by the use of tariffs.

Some societies have successfully combined prescriptive norms, approximating "best practice" or reasonable usage in each case, and penal charges for users exceeding these norms [7].

3.4   Projects and programmes

The third level of the tier consists of direct intervention by the government, water utility or irrigation authority to bring about the necessary conservation. This category of measure usually entails public spending and absorbs administrative and technical resources. Many of these interventions can be thought of as projects. Examples include canal lining, programmes to reduce leakage and the promotion of water-efficient irrigation methods. In practice, a large number of technological options for increased water efficiency are available for consideration [4].

4   The balance of policies

This paper has concentrated on DM as the thrust of water policies; but this is not to deny the place of supply-augmentation schemes in an integrated sector programme.

Each country has unique water options: its choice between demand- and supply-oriented measures will depend on its level of development, hydrological situation, political and social institutions, management skills, financial resources, popular attitudes to water, and other factors. It is therefore unsafe to make general recommendations.

On the other hand, the general response of countries to their water problems is likely to be determined by certain common parameters. One view, using the twofold criteria of water need and its availability per head, places countries in one of five categories [9]:

i)   high level of demand and high rates of utilisation The main future options are demand management and the avoidance of water pollution. Centralised management and allocation is likely to be called for;

ii) high level of present demand, but a low rate of utilisation of available supplies. Options are further development and modernisation of resources, and evening out regional imbalances.

iii) low current demand levels, but high utilisation rates, with serious challenges as population grows. This calls for policies on all fronts - supply augmentation, storage, demand management, reduction of pollution, recycling, etc.

iv) low demand levels and low rates of utilisation. This situation confers the greatest degree of freedom in managing future growth.

v) overconsumption of renewable resources, usually by extracting fossil groundwater. Sustainability needs to be restored, usually at considerable expense.

Although the national balance of policy may well be determined by such general considerations, specific options can be compared and chosen using more precise criteria, such as cost-benefit analysis, cost-effectiveness, cost per unit of water supplied or saved, etc.[7]. This is true whatever option is being considered: conservation, leak reduction, metering, recycling, reallocation, or the development of new supply sources [10].

5   References

1.   Postel, S. (1993) Water and agriculture. Chapter 5 *of Water in Crisis: a guide to the world's fresh water resources*, ed. Peter H. Gleich. Oxford Univ. Press.
2.   Dyson, T. (1996) *Population and food: global trends and future prospects.* Routledge, London.
3.   Frederiksen, H.D. (1996) Water crisis in developing world: misconceptions about solutions. *Journal of Water Resources Planning and Management,* March/April.
4.   Mei, X.; Kuffner, U. and Le Moigne, G. (1993) *Using water efficiently: technological options.* World Bank Technical Paper No 205. Washington, DC.
5.   Keller, A.; Keller, J.; and Seckler, D. (1996) *Integrated water resource systems: theory and policy implications.* IIMI, Colombo.
6.   Hill, B. (1996) *Farm incomes, wealth and agricultural policy.* Avebury Press.
7.   Bhatia, R., Cestti, R. and Winpenny, J. (1995) *Water conservation and reallocation: best practice cases in improving economic efficiency and environmental quality.* World Bank/ODI.
8.   Rees, J., Williams, S., Atkins, J.P., Hammond, C.J., and Trotter, S.D. (1993) *Economics of water resource management.*   National Rivers Authority/Foundation for Water Research, UK.
9.   Falkenmark, M. and Lindh, G. (1993) Water and economic development, in Peter H. Gleick (ed.) *Water in crisis: a guide to the world's fresh water resources.* Oxford University Press
10.   Winpenny, J. (1994) *Managing water as an economic resource.* ODI/Routledge.

DEMAND MANAGEMENT OF IRRIGATION SYSTEMS THROUGH USERS' PARTICIPATION

D. GROENFELD and P SUN
Environment and Natural Resources Division, Economic Development Institute of the World Bank, Washington, DC

Abstract

In most developing countries, irrigation related projects are constructed and managed by government authorities such as irrigation departments. Conventional wisdom has assumed that only the state was capable of handling large modern projects requiring heavy capital investment, complicated technical inputs, and the legal mandate to distribute water, and collect fees. Recent experience challenges this wisdom. The record of government management is poor; irrigation systems often poorly maintained with steadily deteriorating infrastructure. Yet some of these same systems show dramatic improvement when their management is transferred to water user associations (WUAs) who then select their own managers.

This paper examines the potential advantages of this appoach to demand management.

1 The Mexico experience

One of the most impressive examples of demand management reform has been in Mexico where the financial crisis of the early 1980s, and the ensuing structural adjustment, stimulated a variety of reforms in the agricultural sector. Among the most significant reforms was a policy to transfer management responsibility of large-scale irrigation districts from the government to associations of water users, newly created for this purpose. During the financial crisis leading up to these reforms, investments in the irrigation sector had fallen dramatically, resulting in deterioration of the schemes, and poorly maintained irrigation and drainage canals, roads and infrastructure.

The program to transfer management of the irrigation districts to water users was adopted out of necessity. The government simply lacked the funds to carry out basic management functions. In 1989 government administration of the water sector was restructured with irrigation functions consolidated under a newly established National

Water: Economics, Management and Demand. Edited by Melvyn Kay, Tom Franks and Laurence Smith. Published in 1997 by E & FN Spon. ISBN 0 419 21840 8.

Water Commission. This agency had a mandate to move quickly to financial autonomy from the central treasury. Transferring the costly burden of operations and maintenance to farmer associations was seen as the quickest and most effective way of attaining financial health.

In 1990, the first irrigation district was transferred to the users. By 1995, more than 2/3 of the country's 3.2 million ha network -- divided into 80 irrigation districts, had been transferred to 316 irrigation associations. The work involved countless meetings at various levels, from discussions with leaders of producer and marketing associations to one-on-one discussions with users. The transfer program was initially focused on the most productive irrigation districts, with the most commercially oriented farmers. The most important criterion for selecting districts was the potential of the user organisation to become financially self-sufficient, with users paying the fees to cover the costs of operations, maintenance, and administration.

What could the government offer the farmers as an incentive to accept higher costs for their irrigation? In fact, there was a carrot as well as a stick. The carrot was management autonomy and the transfer of mechanized equipment from the agency to the farmer association. The farmers would become the owners of this equipment, and would be free to set their own rules for when to clean the canals, how to distribute the water, and which technical staff to employ for this purpose. The canal would be theirs on a 20-year concession, which is in practice a transfer of ownership. But there was also a "stick". If farmers refused to take over management, the government could offer no assurance that the canal network could be kept in repair. The government in effect threatened to default on its conventional understanding with farmers regarding levels of subsidy in the irrigation sector. Many farmers, and particularly the commercially oriented ones, could not accept the risk that the irrigation infrastructure might collapse. They preferred to take over the management, and with a few exceptions, they haven't looked back. They are paying much more for their water without the government subsidy, but the reliability and responsiveness of their new management structure is well worth the price. For them it has been a "win" situation, and for the government as well.

The basic arrangement in Mexico is that large irrigation districts of even 100,000 ha, which comprised a single management and hydraulic system, are sub-divided into units called "modules" of about 5,000 ha to 20,000 ha, depending on the hydraulics and the farm sizes. Typically there will be between 1 to 4 thousand farmers in each module. The farmers elect a management board, and that board hires professional staff to operate the irrigation system. Usually the staff includes a high level engineer who serves as the operational manager, plus several engineering field assistants, and office technicians (notably for accounting). The management board is accountable to the farmers who elect them every 2 or 3 years. The basic organisational structure is spelled out in the 1992 national water law.

Each irrigation association at the module level has a legal contract with the water district -- whose management remains in government hands, for a specified share of the total water supply. This is carefully measured at the point of management transfer, normally at the head of a primary or secondary canal. The contract between the association and the government irrigation district provides the users both rights and responsibilities over the 20-year period of concession.   The contract also specifies the proportion of irrigation fees that will go to the association and the proportion that will go to the government.   In some cases, module-level associations are federating to form a district-level association which reduces the government role to merely providing water to the headworks of the irrigation system, but with no management role in the system itself.   In such cases, the government's share of irrigation fees is further reduced.

The so-called "Mexican Model" has received international acclaim, particularly through World Bank seminars and study tours.   The rapidity and thoroughness of the transfer process offers a dramatic example of water policy reforms being implemented on the ground.

2   The case of Turkey

Following Mexico's lead, Turkey, adopted a similar policy of irrigation management transfer in 1993.   More than half the systems administered by the government in Turkey were transferred to local user associations within the first 3 years of the program.   In Turkey, the irrigation policy has allowed transfer of irrigation systems to farmers since 1954, but few requests were made to the government to do so.   The government was providing adequate service, and heavily subsidized capital improvements to the systems; why should farmers take over?

In 1993, partly in response to the success of Mexico's transfer program, and motivated also by a mandate to streamline the Department of State Hydraulic Works (DSI), the Accelerated Transfer Program was launched.   Under this program, the DSI actively encouraged farmers to form a legal entity, an association, to take over management of the system from the government.   The newly formed associations receive their water supply from the DSI just below the headworks, or from branch canals in large schemes, and from this point the farmers manage their own affairs.   In return, they pay no water fee to the government, and receive some technical and financial assistance in maintaining the secondary canals within the system.   This is a case of genuine management transfer, but with continuing subsidies and management guidance from the state.

3   Participation, transfer, and demand management

Participatory irrigation management (PIM) refers to the involvement of irrigation users in all aspects of irrigation management, and at all levels.   All aspects include planning, design, construction, supervision, financing, operation, maintenance, monitoring and evaluation, and the process of establishing policy for the irrigation sector. All levels include field ditch, quarternary, tertiary, secondary, main canal system, headworks, project, and the irrigation sector as a whole. Potentially, farmers can manage very large-scale, technologically intensive systems if they are provided with adequate support services and the right incentives.

The concept of PIM is closely related to, but distinct from, the concept of "irrigation management transfer" (IMT).   The term, IMT, refers to a shift from public sector management to user management, which is typically the central feature of PIM. However, the transfer of management to users is only part of the PIM paradigm; the users should also *participate* in a meaningful way in the management of their system. The act of transfer by itself does not ensure that rank-and-file farmers will participate, but it does establish conditions that render participation easier. Similarly, government-managed irrigation systems can adopt a participatory mode, encouraging inputs from farmers and adopting a demand-oriented management style. The Philippines offers an example of a strong participatory orientation with little real management transfer.

The concept of "demand management" as used here implies a formal transfer of management responsibility and authority and the management participation of all segments of the farming community. The new managers of the irrigation systems are not a third-party, but the users themselves. While technically a form of privatization, this particular feature of user-management is more accurately referred to as *userism*: management by a private entity comprised exclusively of users. The nature of the user organisation varies from country to country, and even within countries and may be called a water user association, farmers' council, irrigation union, irrigation district, etc. For consistency, we use the term "water user association" (WUA) to refer to any such organisation. The legal framework of a WUA typically comprises (1) an enabling law at the national or state level, (2) bylaws defining its rules and functions, and (3) a transfer agreement defining and authorizing the WUA to undertake functions previously in the domain of the state irrigation agency (Salman 1997).

4   The rationale for demand management

Countries that have adopted the approach of demand management have done so for both negative ("push") and positive ("pull") reasons. On the negative side, many countries are experiencing increasingly high shortfalls in revenue collections, while maintenance budgets are correspondingly cut and the costs of bringing systems up to operating standard rises at an increasing rate. Transferring this cost burden to the users makes sense from the perspective of the government agency, and in extreme cases, such as Mexico during the debt crisis, farmers may agree that transfer is also in their interest. But once the transfer is made, farmers need no more convincing. Of the thousands of systems transferred in Turkey and Mexico, only a handful have reverted to government management. Farmers see the benefits of organizing themselves to undertake user-controlled, demand-oriented management.

When management is transferred from the supply-oriented irrigation agency to demand-oriented farmers, a solution to irrigation management problems -- whether poor maintenance or poor water distribution -- becomes suddenly more possible. It is the irrigation users who have the greatest incentive to resolve problems that affect them directly. Placing farmers in management control brings the incentives and the authority into harmony. And environmental issues such as drainage, which may be neglected by an irrigation-focused agency, receive priority attention when farmers are in charge. The mandate of farmers -- to operate economically productive agriculture on a sustainable basis -- is far broader than the mandate of any irrigation agency. If there is a problem that may be marginal to irrigation, but central to production, the farmer irrigation managers will not argue that it's none of their business. This logic is certainly not unique to irrigation. The literature of business management is full of examples of the benefits of employee ownership, of management participation by workers, and the concept of total quality management that encourages managers to be concerned with the overall production process. PIM is not a revolutionary concept, but rather a way of bringing irrigation management into line with proven management theory.

5   Impacts of participatory irrigation management

After management transfer is implemented, what are the likely results? The impacts may be viewed from three perspectives: (1) farmers, (2) the government (and more generally, society as a whole), and (3) the irrigation agency which formerly controlled system management. These impacts are presented in Table 1 and summarized below:

From the **farmers' perspective**, the major positive impacts are often intangible, but nonetheless real: a sense of ownership, increased transparency of processes, improved irrigation service and reduced conflicts among users. The main negative impact would be increased water fees.

From the national **government perspective**, the key positive impact would be the reduction of subsidies to irrigation, particularly with regard to O&M costs. This impact is often the primary incentive for a government to initiate PIM policies. Another potential benefit to the government, but one which is not clearly documented, is increased irrigated area and/or reductions in overall water use.

From the **Irrigation agency perspective** , a significant negative impact is often a sharp reduction in regard to field personnel and O&M staff. In Mexico, 5000 out of 7000 government personnel were released from the irrigation department, while the Philippines has reduced NIA staff from 19,353 to 10,368 because of the turnover program. This might be a painful process. But once the turnover thoroughly takes place in the system, it is important for the agency to shift its role from direct irrigation management towards guidance, monitoring and technical support to WUAs. This new role of the government agency is exactly in line with the demand management approach.

There are some benefits that apply to all three interest groups. Reduced conflicts among water users, for example, is appreciated by farmers, as well as by the irrigation agency and by the larger government. But clearly the three parties do not have an equal interest in PIM. The clear winners appear to be the national governments, or society as a whole, who benefit from reduced subsidies to the irrigation sector. Farmers also have much to gain in terms of transparency, more reliable information about future irrigation timing, better service, etc. But what about the irrigation agencies? They appear to be the losers, since the successful introduction of PIM transfers authority from the agency to the farmers. Moreover, the agency becomes accountable to the farmers for delivering water on schedule, making repairs to the main system, etc. While there are also some benefits that agencies can derive from

PIM -- such as less political interference -- some of those benefits -- such as reduced opportunities for rent seeking -- are actually costs to those officials who had previously benefited from those reduced practices.

Given these varying, and in some cases conflicting, sets of interests, where is the stimulus for PIM most likely to be initiated?   Based on five case studies, comissioned by EDI and the International Irrigation Management Institute (IIMI), on Mexico, Argentina, Colombia, Turkey, Philippines, the following conditions were identified as important in stimulating PIM policies:

(a) *national budgetary crisis*. Without a crisis, there is no urgency to initiate changes: This was applicable to the cases in Mexico, Turkey, Colombia, and Philippines. In Argentina, however, the PIM process was part of a privatization of the economy as a whole in early 1990s.

(b) *top level political will and commitment*.   Without strong political commitment at the highest levels, the initiative to transfer management from the public sector agency will not go far. This condition was applicable in four of the five case study countries. The exception was the Philippines, where a bottom-up approach was adopted.   The Government and the National Irrigation Administration (NIA), collaborating with NGOs have promoted farmers participation in irrigation management in a gradual manner since the mid-1970s.

(c) *good physical condition of irrigation infrastructure*.   Unless the system works, farmers will not be willing to take over management. This was generally applicable in all five cases. However, there is considerable room for negotiation in the kinds of repairs that are required prior to transfer, and even on the timing of the repairs.   In Mexico, the government promised farmers to rehabilitate certain portions of their systems after management turnover. In some systems in Colombia, the government transferred poor infrastructure to WUAs. As a result, it made those WUAs financially unmanageable, and created serious second-generation problems.

(d) *a workable legal framework*. The legal framework needs to be "workable" though not necessarily enacted specifically for supporting the transfer program. The legal aspects should include establishment of WUAs, water rights, the new role of irrigation department, and supportive measures to WUAs. In Turkey, existing laws were adequate, although not ideal, to support the transfer of irrigation management to various entities such as WUAs, Cooperatives, and community enterprises.   In most cases, a new set of laws will be required, although the transfer process might begin while the new law is taking final form. In Mexico, the first systems were transferred in 1990, although the new water law did not take effect until 1992.

Mode of Implementation.. Implementation modes range from the very bottom-up or gradualism approach that has been used in the Philippines to the highly top-down or big-bang approach used in Colombia and Mexico. In terms of promotion of PIM, Colombia has invested much less time and effort than the other countries. As a result the process of transfer was very quick but the degree of farmers' participation was very limited. At the other extreme, the Philippines has spent more than 15 years to promote PIM with intensive use of institutional development officers and farmer organizers to serve as catalysts.

In practice, both top-down and bottom-up approaches are needed in the process of adopting PIM. Top-down is particularly needed in the beginning stage including awareness campaigns, formulating a legal framework, reforming irrigation agency structure, and rehabilitating infrastructure. However, a bottom-up approach is essential when WUAs are to be organized, during which time thorough understanding and participation of farmers are required.

Size of a WUA. With the exception of the Philippines, sizes range from 2,000 to 40,000 ha and cover at least a secondary canal system or even an entire irrigation project. The number of farmers ranges from several hundred to several thousands. These IAs are run as businesses with technical staff hired by the IA management board and supervised by a hired manager. Salaries are paid from irrigation service fee collections, and in many cases the IAs purchase their own transport and maintenance equipment.

The exception is the Philippines, where IAs are always below 1,000 ha and usually 100 to 300 ha. Most of the labor is voluntary, provided by the users, and very few, if any, of the irrigation staff are hired professionals. There are diseconomies of scale for these small sized WUAs. They can not afford to own specialized maintenance equipment. To some extent, the small size is due to smaller land holdings (generally less than 1 ha per farm) but more importantly the small size is linked to the limited scope and objectives of the transfer program itself.

6   Is PIM sustainable?

Will these large WUAs prove to be a permanent feature in the irrigation sector, or is their creation a management fad that will soon require a government rescue? Based on the five cases cited above, the evidence suggests that if a few basic conditions are met, WUAs can enjoy long term sustainability. These are: (1) *financial autonomy* and (2) *management transparency.*

Financial autonomy implies that the WUAs are not directly subsidized by the government, and thus are protected from shifts in government policy and funding priorities. The principal source of revenue for most WUAs is the irrigation service fee. Setting an appropriate fee structure and establishing an effective collection system are critical to the financial health of the association.

Management transparency is necessary both to keep the association honest, and to inspire confidence in its members that their irrigation service fees are being well spent. A number of steps can be taken to increase transparency in WUA management, including regular financial audits, wide dissemination of budgets and financial plans, and broad representation among users on the association's board of directors.

7   Conclusions

Participatory irrigation management offers an approach that ensures response to users' demands. Under PIM, the technical operators of the system are directly accountable to the system managers, who themselves are directly accountable to the irrigation users who elect them. Furthermore, the principles of demand management extend naturally into the government agency as well. An organized association of irrigation users becomes an equal management partner with the government agency. The growing body of experience with user-oriented demand management around the world suggests that irrigation users can also make the best irrigation managers.

Table 1. Potential impacts of IMT[1]

FARMER PERSPECTIVE

Positive Impacts	Negative Impacts
Sense of ownership	Higher costs
Increased transparency of processes	More time and effort required to manage
Greater accessibility to system personnel	Less disaster assistance
Improved maintenance	No assured rehabilitation assistance
Improved irrigation service	Less secure water right
Reduced conflicts among users	Decreased agricultural productivity
Increased agricultural productivity	

GOVERNMENT PERSPECTIVE

Positive Impacts	Negative Impacts
Reduced costs to government	Less direct control over cropping patterns
Greater farmer satisfaction	Need to reduce staff levels, sometimes over union opposition
Reduced civil service staffing levels	Reduced ability to implement new agricultural policies through the irrigation agency
Reduced costs to the economy (greater economic efficiency)	

IRRIGATION AGENCY PERSPECTIVE

Positive Impacts	Negative Impacts
Fewer conflicts to deal with	Reduced bureaucratic and political influence
Reduced operational involvement	Uncertainty over agency role
New responsibilities	Reduced opportunity for rent seeking
Reduced opportunity for rent seeking	Reduced control over water resources
Reduced political interference	
Reduced O&M staff levels	

8   References

1   Economic Development Institute (1996), *Handbook on Participatory Irrigation Management*, Washington, DC: EDI/World Bank.

2   Salman M.A. Salman (1997), *The Legal Framework for Water User Associations: A Comparative Study.* World Bank Technical Paper No. 360.

3   Svendsen, Mark; J. Trava and S. Johnson (forthcoming), *Lessons from the International Workshop on Participatory Irrigation Management: Benefits and Second Generation Problems.*   EDI Report, Washington, DC: EDI/World Bank.

---

[1]   Adapted from Svendsen, Trava, and Johnson (forthcoming)

WATER AS AN ECONOMIC GOOD: INCENTIVES, INSTITUTIONS, AND INFRASTRUCTURE

R. MEINZEN-DICK and M. W. ROSEGRANT
Environment and Production Technology Division International Food Policy Research Institute
Washington, D.C., USA

Abstract

Increasing water scarcity around the world, together with evidence of its inefficient use in many contexts, has given impetus to the call to treat water as an economic good. But the objectives of policies to treat water as a commodity are often not clear, and their implementation is far from simple. This paper examines several of the major "economic" approaches to water: raising user fees, efficiency pricing, and water markets, in terms of the incentives they can create for demand management, and the infrastructure required to make them effective. The latter includes not only technical apparatus, but also the institutions to measure water use, collect charges, and adjudicate disputes or third party effects.

Keywords: irrigation sector policy, water markets, efficiency pricing, irrigation fees, demand management

1 Introduction

Currently, 28 countries with a total population of 338 million are considered water stressed (with freshwater resources in the range of 1,000 - 1,600 cubic meters per capita per year), and 20 of these countries are water scarce (with less than 1,000 cubic meters per person annually). Water shortages will increase dramatically in the next 30 years: by 2025, it is projected that 46 to 52 countries with an aggregate population of around 3 billion will be water stressed [1]. Countrywide, regional, and seasonal water scarcities in developing countries pose severe challenges for national governments. These challenges are exacerbated by degradation of soils in irrigated areas, depletion of groundwater, water pollution and its impact on human health.

Managing scarcity by augmenting supplies is increasingly difficult because of limitations in renewable water supplies, increasing costs of developing new water sources and transfer schemes, and degradation of water quality due to pollution or

Water: Economics, Management and Demand. Edited by Melvyn Kay, Tom Franks and Laurence Smith. Published in 1997 by E & FN Spon. ISBN 0 419 21840 8.

salinity. Mechanisms for demand management are therefore gaining prominence. This has drawn attention to wasteful use of already developed water supplies, and the massive subsidies and distorted incentives that govern water use. Technical measures to improve the efficiency of water use--such as lined canals, improved pipe systems, drip irrigation, and computerized timing of water deliveries--are available, but are unlikely to be adopted unless there are sufficient incentives for governments or private users to save water.

One of the simplest and yet powerful ways of creating incentives for efficient use of a resource is through economic means: putting a price on it or letting market forces govern its allocation. As part of the widespread trend toward privatization and market development in many sectors around the world, there have been a number of calls to treat water as an economic good [2,3].

But water is not a conventional commodity. Many characteristics inherent in the resource (notably widespread externalities, natural monopolies, and the vital role it plays in life itself) cause market failures. Technical and social considerations often mitigate against simple application of economic principles for water allocation. This paper reviews the potential advantages, feasibility, and limitations of two major economic tools for demand management: water pricing and water markets.

2   Water pricing

2.1   Raising user fees

With a superficial analysis, it is easy to equate the low irrigation charges and inefficient use of water resources as cause and effect, leading to the conclusion that if charges were raised, farmers would use water more efficiently, leading to better system performance. This view is exemplified in an article in *The Economist* [4]: "One reason for waste is that irrigation water is everywhere hugely subsidized. . . If water is cheap, it will be wasted. Price it properly, and people will treat it as the precious commodity it is."

A closer examination shows that, given the present structure of irrigation fees, this is not so simple. In most cases of surface irrigation in developing countries, individual farmers pay not based on amount of water consumed, but based on area (nominally) irrigated. Thus raising irrigation fees does not provide an incentive to conserve water (unless it is raised enough to dissuade farmers from irrigating, or induces a significant shift in cropping pattern, as discussed below).

This is not to say that pricing reforms are not important. It does mean that we must be clear what the objectives of that reform are, and the implications for structural change and implementation. Raising the level of irrigation charges is important in many countries (e.g. India, Pakistan, or Egypt) to reduce the fiscal deficits of governments that are finding it difficult to maintain current levels of subsidies over large areas.

However, for irrigation fees to provide incentives for water conservation, some form of volumetric water charges is required. As long as farmers pay a flat fee per area irrigated, raising irrigation charges will not give them an incentive to conserve water use. To the contrary, an increase in water charges may give the impression that they are entitled to *more* water, because they are paying more.

Moreover, raising irrigation charges is often politically unpopular. For example, the water rates prevailing in 1996 in several states in India have not been raised since the early 1980s. This coincides with the rise of farmer political lobbies that have demanded low irrigation fees--demands that populist governments have not refuted. The states that have recently attempted to change water rates in real terms have faced considerable opposition, and several have reverted back to low charges, or have been unable to collect the new higher fees.

2.2 Volumetric pricing

Charging each farmer a volumetric price for the water consumed requires a combination of technology for measuring and administrative apparatus to record and collect the fees. The technological requirements for this are available, but not trivial. Most flow meters require delivery through a pipe, and there is some head loss that reduces water availability at the farm level. Both of these conditions are problematic in the flat terrain and open tertiary channels that deliver water to individual farmers in areas such as India or Pakistan. Even where technology is not constraining, the cost of installing meters and--more importantly--monitoring them and collecting fees from many smallholders is prohibitive. This implies that volumetric user charges are most likely to be cost effective where farm sizes are larger, so that there are fewer delivery points to monitor.

Effective demand management requires that users can get water on demand (and turn it off when it is not needed). Much over-use is due to uncertainty about when farmers will next receive water, due both to natural variability and to poor management of irrigation systems. The low tolerance of many crops to under-watering creates indivisibilities in demand [5]. This means that farmers will take extra water when it is available, rather than risk losing the crop by not having water when they next need it. The marginal cost of water would have to be very high for a farmer to risk losing what has already been invested in a crop by cutting water use, especially at critical points in the season (when water is usually most scarce). Conversely, many systems do not physically allow farmers to refuse water deliveries when they do not want it. This is most problematic in field-to-field irrigation systems, but applies also to many systems with scheduled water deliveries. In short, it is difficult to create demand management by simply changing water prices in irrigation systems that are designed with a supply orientation: "it is unrealistic to think of resource pricing to shape user resource demands if the supplier can neither deliver according to a schedule, not protect the user against excess supplies" [6].

2.3 Volumetric wholesaling

"Wholesaling" of water from main canals to user groups provides a compromise between flat rate and volumetric charges for all users. These systems, which are found especially in western India, have measured water deliveries at the turnout point from the main system to the area served by a group of users. The user group pays for water volumetrically, and then redistributes it among individual users and collects fees from them. However, in most cases of volumetric "wholesaling," the users' association is billed based on water consumed, but collects from individuals based on area irrigated [7]. This creates an incentive for the group to collectively conserve water, but individual farmers do not have an economic incentive to use water efficiently. Unless the group is able to monitor consumption closely or exert normative pressure on its members (coercive or normative incentives), this form of volumetric pricing will not induce conservation.

It is instructive to look at the experience of Mohini Water Distribution Cooperative in Gujarat, India--one of the best-known examples of this arrangement. There, the wholesaling mechanism has not led to water savings. On the contrary, Mohini farmers planted 85 percent of the area under water-consumptive sugarcane, rather than the authorized 18 percent [8]. In recent years the association has not even been able to collect fees from its members due to a leadership struggle, and has defaulted on its payments to the government [9]. This shows the critical importance of strong and effective user groups--which often do not exist.

2.4   Other pricing options

Where individual volumetric measurement may not be feasible, some form of proxy for water consumption may work. The timed rotation periods, such as those of *warabandi* systems in North India and Pakistan, can provide such a proxy if farmers are charged based on hours of water delivery, rather than on area irrigated. Pumping charges also offer a proxy measure for water deliveries, and can be modified to include a charge for the water, along with energy and pump maintenance. Charging based on the number of irrigations actually received on an area would require more record-keeping on the part of the billing authority, and does not deal with how generously farmers irrigate each time, but it may be seen as equitable by farmers, especially in areas where the number of irrigations is variable.

Even area and crop-based charges can provide incentives for water conservation if the charges are proportional to the water consumption (or average number of irrigations) of the crops grown. While this also does not take into account how efficiently farmers irrigate, it provides incentives for farmers to shift cropping patterns away from water-consumptive crops. Perry [10] found that in Egypt crop-based charges were almost as effective as volumetric charges in inducing shifts in cropping pattern. This can provide significant reductions in water use because the differences in water use are usually greater between crops (e.g. sugarcane, paddy, cotton, and millets) than between farmers growing the same crop.

3   Water markets

3.1   Types of water markets

Although the policy attention to water markets is relatively recent, the design of mechanisms and institutions for market allocation in water is an old subject. In many parts of the world, water markets have evolved or are evolving either informally or formally, as instituted by law. There are actively functioning markets in New Mexico and California in the United States, and in Chile. In developing countries, informal water trading and water markets are expanding in Tamil Nadu, Pakistan, Indonesia, and Jordan. The development of water markets in these countries has generally been most extensive for groundwater, either within agriculture or from agriculture to municipal use. Most cases of spontaneously-developed water markets involve the sale of given quantities of water, delivered at a specific point in time. They are usually localized, and can be considered "spot markets." Markets for tradable water rights have a broader potential effect, but require considerably more physical and institutional infrastructure.

3.2   Potential advantages

The concept of market allocation hinges on providing the necessary economic incentives for water users, in their final demand for irrigation, industrial uses, or residential uses, by considering all the opportunity costs of water not only in its current use but also in its other numerous and often competing uses. In this it provides much greater flexibility than even volumetric pricing, because the opportunity cost of water, and hence its allocation, is not administratively fixed, but can vary over space and time to reflect changes in supply and demand. In comparison to the rigid non-transferability of water allocation in centralized management, market-like transactions permit the selling and purchasing of water across sectors, across districts, and across time.

Markets in tradable water rights may also have lower information costs than marginal cost pricing, because the market, composed of irrigators with expert knowledge of the value of water as an input in the production process, would bear the costs and generate the necessary information on the value and opportunity costs of water. Moreover, in existing irrigation systems, the value of prevailing water use rights (formal or informal) has already been capitalized into the value of irrigated land. Imposition of higher water prices is seen as expropriation of existing rights that creates capital losses in established irrigation farms, and, as described above, is often met with strong opposition from established irrigators. The establishment of transferable property rights would formalize existing rights to water, and therefore may be politically more feasible [11].

3.3   Conditions for water markets

For markets to yield an efficient outcome, economic theory says that there must not only be well-defined property rights that allow for the internalization of all externalities, but also perfect and competitive markets and zero enforcement costs [12]. However, water rights have been difficult to define and enforce, particularly for the individual user. The nature of irrigation water resources creates numerous sources of market failures [3].

In the real world, transaction costs are rarely equal to zero. Costs typically associated with market exchange in the form of information gathering, physical conveyance losses, monitoring and enforcement costs could substantially exceed the benefits to be earned from trade. Rosegrant and Gazmuri [13] argue, however, that where the difference in value of output per unit water is high, markets in tradable water rights could still lead to considerable efficiency gains and benefits great enough to offset the transactions costs.

The pervasiveness of externalities such as pollution, overdraft and lower water tables, waterlogging, and other adverse, often irreversible environmental effects, is one of the strongest challenges to be met if effective water markets are to be developed. Even in the presence of externalities, markets could still be made workable if these effects are identified, agreements as to their effects are reached, and compensations are paid to the aggrieved parties. In fact, water markets may offer effective solutions to some of the externality problems, such as groundwater overdrafting, because appropriately defined tradable rights may cause farmers to internalize and thereby eliminate these externalities. Although water markets can assist in solving some externality problems, the presence of other forms of externalities and third-party effects means that the establishment of markets in tradable property rights does not imply free markets in water. Rather, the system would be one of managed trade, with institutions in place to protect against third-party effects and possible negative environmental effects that are not eliminated by the change in incentives.

Water markets require infrastructure that can transfer water from one user to another, and measure quantities consumed by each. Although this is not necessarily sophisticated technology, it should allow flexible deliveries to meet changes in demand over time and space. For example, simple proportional division devices are used to divide water into assigned shares as a proportion to total canal flow in traditional systems in Nepal or Bali, as well as in modern systems in Chile. Nor are large-scale transfers necessary: from country experiences, flexibility of transfer could exist even if only a small volume of water is subjected to reallocation [13].

Property rights that are completely specified, exclusive, transferable, and enforceable are a prerequisite for efficient market operation [12]. In the case of water, this would involve clarifying the rights of individuals and groups in many cases. A clear and equitable initial distribution of water rights is needed if markets are not to incur opposition for appropriating certain groups' rights. It is therefore essential that the process of establishing water rights takes into account not only formal legal rights, but also local perceptions of water rights, including rights to return flows. However, one of the advantages of markets is that they provide for compensation to those who give up water, whereas public allocation may divert water from one use (or user) to another without compensation.

Markets also require supporting institutions that can broker agreements, enforce contracts, and arbitrate disputes. This means that water markets are likely to function better where the judicial system is seen as fair and accessible, and where markets are well established for other goods and services. In New Mexico in the United States, prospective transactions are openly advertised as required by law. There are legal provisions to compensate those who are adversely affected for damages that would arise from the transfer or alteration in the water rights on appeal. However, legal processes can be long and difficult, and claiming compensation for third party impacts through the legal process may be nearly impossible [14].

4   Discussion and conclusions

4.1   Feasibility

Both water pricing and water markets require a combination of physical and institutional infrastructure to be feasible and effective.   While the physical infrastructure may seem difficult or costly to install, in most cases it is the institutional infrastructure that is most constraining.

In the case of volumetric pricing, there are trade-offs between measuring technology and monitoring requirements.   For example, continuously-recording meters with closed pipes may only have to be read once a month or once a season, but simple marked measuring sticks in open canals need to be read and recorded very frequently.   The question of whether a country or agency has the capacity to measure water use and collect the fees, and whether the costs of collection will outweigh the revenue collected, is a significant issue.   Wholesaling requires fewer measurement and collection points, but needs local institutional capacity to collect and create incentives for conservation.   Proxy measures for volume of water deliveries are appropriate where the cost or feasibility of technology and institutions for volumetric charges are problematic.

For water markets there are the additional requirements of defining water rights and transferring water from one user to another.   For water rights, simple technology such as proportioning weirs may be adequate, provided there are social or legal sanctions to uphold those rights.   For water transfers, the technology to convey water from one place to another is often costly, and the institutions to broker trades, regulate the market, and adjudicate and compensate for third party effects require serious attention.

4.2   Efficiency effects

Water markets induce water users to consider the full opportunity cost of water when making their water allocation decisions.   Consideration of the full opportunity cost will provide incentives to increase the economic efficiency of water use through use of less water on a given crop, investment in water-saving irrigation technology, shifting of water applications to more water-efficient crops, changes in crop mix to higher valued crops, and transfers of water to higher valued non-agricultural uses. The actual magnitude of the efficiency gains from water markets will be a function of the degree of water scarcity and the effectiveness of institutions and infrastructure.

4.3   Equity effects

There are strong equity concerns about the outcome of water markets.   Because water is vital to life, as well as to livelihoods, there are strong social norms that argue against it being treated as a simple marketable commodity.   Pursuing efficiency through market allocation may not be politically or socially acceptable if equity considerations are not met.

This means that markets are likely to be more appropriate where there are opportunities for those who give up water to either invest in more efficient water uses, or earn income from other sources.   Where this exists, it is relatively easy to show gains from trade for both buyer and seller. In other cases, if selling their water rights jeopardizes the livelihoods of certain groups, water markets are more likely to encounter opposition. However, the possible negative equity effects of water markets can be mitigated through appropriate policies. Market-based allocation, combined with reduction in the massive capital and operating subsidies on irrigation and water supply, which usually favor better-off producers and urban consumers, would free-up budgetary resources to target water subsidies to the poorest sectors of the population in both rural and urban areas.

4.4   Final considerations
Economic instruments such as pricing or markets are most likely to be appropriate where there is some degree of water scarcity, so that water has a relatively high value. This approach also has higher payoffs where there are significant differences in the value of output per unit water that provide potential for gains in efficiency of water use. This is most likely to occur when there is competition between agriculture and industrial or municipal demand, but can also be found when there is a heterogeneous cropping pattern, with some high-value crops.

Note, however, that under conditions of extreme scarcity, rationing may be necessary, to make sure that everyone has access to minimum quantities for domestic consumption. Certain uses (e.g. ecosystem preservation) and users (e.g. the poor) may need to be protected through special allocations or targeted subsidies on water use.

Because water is essential for life, there are stronger norms and values regarding water than for most resources or commodities. These need to be taken into account in developing any form of water allocation policies, if the policies are to be adopted, rather than being circumvented or encountering serious opposition.

Economic tools provide only one (albeit powerful) set of incentives for water conservation. There are other possibilities, notably coercive and normative incentives. Effective demand management in a context of growing water scarcity is likely to require combinations of all of these approaches to create appropriate incentives for water users.

5   References

1.   Engelman, R., and LeRoy, P. (1993) *Sustaining Water: Population and the Future of Renewable Water Supplies,* Population Action International, Washington, D.C.
2.   ICWE (International Conference on Water and the Environment) (1992) Development Issues for the 21st Century. The Dublin Statement and Report of the Conference, World Meteorological Association, Geneva, Switzerland.
3.   World Bank (1993) *Water Resources Management.* World Bank Policy Paper, World Bank, World Bank.
4.   *The Economist* (1992) The first commodity. March 28-April 3, pp: 11-12.
5.   Meinzen-Dick, R. et al. (1997) Sustainable water user associations: Lessons from a literature review, in *User Organisations for Sustainable Water Services,* (eds. A. Subramanian, N.V. Jagannathan, and R. Meinzen-Dick), World Bank

Technical Paper Number 354, World Bank, Washington, DC.

6.   Moore, M. (1991) Rent-seeking and market surrogates: The case of irrigation policy, in *States or Markets? Neo-liberalism and the Development Policy Debate,* (eds. C. Colclough and J. Manor), IDS Development Studies Series, Clarendon Press, Oxford, U.K., pp. 297-305.

7.   Meinzen-Dick, R. and Mendoza, M.S. (1996) Alternative water allocation mechanisms: Indian and international experiences. *Economic and Political Weekly,* Vol. 31, No. 13. pp. 25-30.

8.   Patil, R.K. (1987) Economics of farmer participation in irrigation management. *ODI/IIMI Irrigation Management Network* (NP 87/2d). pp 3-17.

9.   Kalro, A.H., and Naik G. (1995) Outcomes of irrigation management transfer and financial performance of water users' associations in India: Some experiences. Paper presented at the Workshop on Irrigation Management Transfer in India, December 11-13, Ahmedabad.

10.   Perry, C.J. (1996) *Alternative Approaches to Cost Sharing for Water Service to Agriculture in Egypt.* IIMI Research Report, No. 2, International Irrigation Management Institute, Colombo.

11.   Rosegrant, M.W., and Binswanger, H. (1994) Markets in tradable water rights: Potential for efficiency gains in developing country water resource allocation, *World Development,* Vol. 22, No. 11, pp. 1613-1625.

12.   Coase, R. (1960) The problem of social cost. *Journal of Law and Economics,* Vol. 3, No. 1, pp. 1-44.

13.   Rosegrant, M.W., and Gazmuri Schleyer, R. (1994) *Tradable Water Rights: Experiences in Reforming Water Allocation Policy,* Irrigation Support Project for Asia and the Near East, Arlington, Virginia.

14.   Campbell, L. (1996) Stream adjudications, acequias, traditional water users and water rights in Northern New Mexico. Paper presented at International Association for the Study of Common Property meetings, June 4-8, 1996.

ECONOMIC INSTRUMENTS IN WATER MANAGEMENT
A more efficient approach

P. KESSLER
Ministry of Environmental Protection, Wiesbaden, Germany

Abstract
Economic instruments in water management are defined as all price-related and/or regulatory instruments which harness the commercial self-interest of actors. The main economic instruments in water management and their functions in environmental protection are reviewed. Finally, a positive outlook is given.
Keywords: Water price, fees, charges, tradeable permits, water markets.

1   Introduction

The call for economic instruments is based on the recognition that free access to a resource tends to result in excessive use. Excessive use can be counteracted by defining and allocating user rights and by putting a price on resource utilization. External costs, such as damages inflicted on nature, are internalized and distributed over the price. The ultimate aim of the application of economic instruments is to achieve more efficient utilization of the resources. These thoughts can be demonstrated both on forestry and on fishery, and thus also on water resources [1].

Regulatory instruments on the other hand, require a considerable degree of monitoring, are mostly built up on end-of-pipe technologies and tend to inhibit innovation.

The number, types and varieties of the economic instruments available make a survey difficult [2]. To restrict this variety of instruments, to which new solutions are continuously added, at least with regard to their function, this paper will be based on a definition developed by a working group of the German Agency for Technical Cooperation (GTZ) and the German Foundation for International Development (DSE):

"All price-related and/or regulatory instruments which harness the commercial self-interest of actors (i.e. industry, farmers, transport users or the population at large) for environmental goals" [3].

Water: Economics, Management and Demand. Edited by Melvyn Kay, Tom Franks and Laurence Smith. Published in 1997 by E & FN Spon. ISBN 0 419 21840 8.

The terminology used in the relevant literature is confusing. In the following, I will use the terms "price" and "levy" as general terms. The word "fee" will be used for a compensation for special services, such as water supply. A charge is characterised as being earmarked while a tax goes in to the general budget without any restrictions in terms of appropriation.

In what follows, the economic instruments that play a role in water management will be examined more closely.

2   Water utilization rights

Whether water rights should be classed with the economic instruments or whether they are in fact a prerequisite to their use may be open to argument. A starting point for pricing is however needed. That environmental utilization rights as such will appear in most listings of economic instruments is due to the fact that in many countries resources are still subject to public use so that there is no basis for a market-oriented approach. Only the assignment of environmental utilization rights to juristic persons or village communities will provide the basis for mobilizing their commercial self-interest to conserve a resource. Only when water rights are firmly established and the parties entitled to their use pay an appropriate price for the use of water, will they take good care of this resource. To that extent, water and land rights are rightly counted among the economic instruments.

3   Drinking water price

While in most central and northern European countries the water price is structured to be largely cost-covering, this is not the case in most other countries. Drinking water is either a free commodity or the prices imposed for its consumption are so low that they cover the costs only partly. With regard to the different prices of water, the interrelationship between the cost of water and water consumption becomes readily apparent. In the USA and Canada, where the price of water is low, consumption is the highest worldwide; the water crisis in Spain must be traced at least in part to the fact that in that country the price of water only covers a fraction of the costs, so that no economic incentive for an economic use of water exists.

4   Wastewater fees

Wastewater fees are commonly based on the quantity of freshwater consumed. So, with regard to the consumption behavior, wastewater fees are governed by the same rules as freshwater prices or fees.

Local authorities in Germany are only now beginning to charge separately for freshwater and wastewater. In this way, runoffs from roofs and sealed pavements discharged into stormwater drainage systems are taken into account, as these burden the sewerage. There exists however also an ecological interest in achieving as much water infiltration into the ground as possible right at the source, to recharge the groundwater. Here, a separate metering of freshwater and wastewater may be helpful. In this way, pricing will also influence the consumption pattern.

5   Charges with an incentive function

Charges that are tied to resource utilization functionally fall into the category of incentive charges. They are meant to influence consumption behavior. Apart from that, they also have a financing function: One or the other function will dominate, depending on the structure.

The proceeds gained in this manner are used for promoting measures of environmental protection as specified in a given charges act. They are earmarked and do not flow into the general budget, and are therefore available in the full amount to the environmental department. Parts of the proceeds are used to varying degrees - within the framework of their intended use - to pay for the administrative expenditure involved with imposing the charge. This applies both to staffing and expenditures on materials. In times of dire financial straits and severe cutbacks in expenditure in the environmental sector; this particular advantage of the earmarked charges should not be underestimated. However, incentive charges have one decisive disadvantage - more so than eco taxes: If they bring about the intended changes in consumption and production patterns, the proceeds will possibly decrease in time. Once the intended purpose has been fully achieved, which however is conceivable only in exceptional cases, the water incentive charge will make itself redundant.

5.1   Water utilization charges

Water utilization charges have by now been introduced in most German federal states (Baden-Württemberg, Berlin, Bremen, Hamburg, Hessen, Schleswig-Holstein, and Thuringia). The charges are based either on water withdrawal of any kind or on groundwater withdrawal. The charges imposed range from a few pennies per cubic meter to DM 1.10 per cubic meter for use of cooling water in accordance with the Hessian groundwater charges act. As is the case with other economic instruments, here, too, the structure of the charge imposed depends on the objective to be gained and on the framework conditions. Where drinking water supply is obtained almost exclusively from groundwater (Hessen: 95%) and when the excessive use of groundwater causes economic and ecological harm, it makes sense to tie the levy to groundwater withdrawal. In other cases it may be more expedient to impose levies on all kinds of water withdrawal.

5.2   Wastewater charges

Wastewater charges exist in several industrialized countries like France, Germany and the Netherlands. Every discharger is liable to pay for discharges and the amount depends on the quantity and quality of the effluent.

Noteworthy among the more recent studies conducted on the subject appears to be a report prepared by Andreas Kraemer entitled "Evaluating Effluent Charges in Germany" [4]. While earlier investigations concentrated more closely on the incentive effect of the wastewater charge, the availability of funds and the costs involved with raising and administrating the charge, Kraemer's investigation

emphasizes in particular the strengthening of the administration and regulatory law. Beyond its principal function, i.e. the incentive effect, and the financial function, the levy is viewed as an instrument that
- finances jobs and other expenditures on materials
- advances research
- improves environmental monitoring
- broadens the information basis.

These statements also apply to other charges, e.g. charges imposed on the generation of hazardous waste. With this charge introduced in some German States, not only staff and expenditures on materials are financed. Its implementation has also prompted the administration to considerably improve its information policy on waste flows. Here, the close interlinkage of the economic instruments with regulatory law becomes apparent.

5.3   Other charges in water management
When it comes to the choice and structure of economic instruments there are virtually no limits to the imagination, as has already been indicated. Decisive is that the objectives are met. An economic instrument must be "made to measure," as the objectives and the framework conditions are subject to change.

What applies to economic instruments in general, also applies to charges in particular. Apart from the water utilization fee and the wastewater charge, other less well-known levies have been imposed to protect the water or man - or they are at least being discussed. The Hessian water law provides for the implementation of a flood protection charge to finance replacement retention areas in the event of flood basin loss. So far, this particular basis of authorization has not been invoked as both the computation and the implementation of this charge appear to be difficult. It is also feared that the solution, to impose a charge, would weaken the prohibition to build in flood retention areas so that in the end no substitute retention area would be assigned.

The nitrate levy, a product levy that used to be much discussed, would also benefit water protection. A ground-sealing or land levy, too, would counteract ground sealing and promote groundwater recharging. The levy for the conservation of nature that has been implemented in some German states also serves this goal - in addition to other goals.

6   Administrative fees

Administrative fees imposed by the environmental administration cannot so simply be counted among the economic instruments, as their incentive effect would be very difficult to prove. They can, however, be counted among the economic instruments inasmuch as they not only cover the administrative costs, but are also oriented on resource utilization. In Hessen, for example, the fee charged for a water right depends on the type of permit (ordinary permit, higher-level permit, grant) as well as the amount of water removed per year.For agricultural irrigation, the area to be irrigated plays a decisive role.

These often rather high administrative fees not only have a considerable financial effect.   They also achieve an internalization of ecological cost.   Water supply enterprises or agricultural producers must apportion these administrative fees to the price of their products.   It is however this very internalization of external costs that makes them economic instruments.

7   Compensation and tradeable permits

Permit trading plays an important role in air pollution control.   Here, pollution quotas are traded in the form of certificates or licenses.   While in the USA, trading of this kind has been practiced nationwide and in the federal states for a long time, Europeans have not progressed beyond the experimental stage.

The thought to deal with "pollution rights" can also be applied to water.   In a study for the Fox River in Wisconsin, for example, a system of transferable discharge permits for industrial and communal treatment systems has been proposed.   The study showed the potential for cost savings of 29% to 66% when compared to the system of uniform emission standards and application of the Best Available Technology.

In practice, only the "bubble concept" appears to have come into use.   According to this concept, a transfer of discharge rights is permitted when a change from one emission site to the next is involved within complex industrial plants, provided the total pollution load is not increased.

For certificate trading, a dynamic model, too, is conceivable, for which the water quality standards are raised annually by a certain percentage with the result that the price of the certificates rises each year due to increasing scarcity.

8   Water markets

"Water markets" are usually understood to mean the trade with water utilization rights.   Such trading, of course, presumes first of all that the water rights or the water utilization rights are clearly defined and assigned (see 2. above) [5] [6].

In Spain, water rights are transferable within a river-basin under tight government restrictions [7].   In Australia, the State of Victoria's legislation allows the transfer of irrigation water rights among land owners belonging to the same or to different irrigation districts.

The transfer of water rights has been eagerly practiced in the western part of the United States [8], in Chile and, informally, in India [9].

In the western states of America, the trade with water utilization rights has been common practice for a long time.   As trading is subject to the laws of the federal states, its structure varies greatly from one state to the next.   Common to all of these legal system is that the purpose for which the water is utilized may not be changed without state approval.   This leads for example to the strange situation that farmers, due to their commitment to use the water for a specific purpose and the low water price, grow rice in the Californian desert, while expensive desalination plants must be erected in the cities nearby to supply their citizens with water.   Since the 80s, however, water consumption for irrigation management in the western part of the USA has been steadily on the decline.

The trade with water rights in the USA in general is subject to numerous administrative restrictions. To the advocates of water markets, this is the reason why the advantages of water markets are not fully exploited.

The requirement to obtain approval traditionally serves the protection of other holders of water rights. Of late, environmental protection and in the widest sense public interest in general are being added as further criteria for approval.

There is general agreement that water markets can be dangerous. One particular danger is that of monopoly formation. Water markets can only function under a system of water utilization rights, an administration that ensures the inviolability of the water rights and a good infrastructure system for receiving, storing and distributing the water.

9   Outlook

Water is scarce in many countries, not only in arid and semi-arid zones, but in countries where water resources have been traditionally abundant. Several European countries are beginning to become concerned about water.

Large-scale technical solutions are now proposed to overcome these water supply problems. Viewed under ecological aspects, these proposals are regarded as being long out of date. In the Mediterranean area, for example, it has been proposed to solve the water supply problems in some parts of the country by building pipelines that would transport the water from areas with heavy rainfalls. In other proposals, plans for large dams are again being revived.

Experience has taught us that technical measures like these are expensive and cause enormous ecological and social damages, without solving the problem. In particular socio-economic instruments and especially cost-covering water prices will cause agriculture, industry and the population to handle water with more care. To avoid social and economic hardships, transitional solutions are certainly indicated, in particular in poor countries. However, if the objective to achieve sustainable water management is to be reached, there is no other way than to charge an adequate price for the utilization of this resource.

10   References

1.   Panayotou (1993) *Green Markets,* San Francisco
2.   Gale, R., Barg, St. and Gillies, A. (1995) *Green Budget Reform,* London
3.   Deutsche Gesellschaft für Technische Zusammenarbeit (GTZ) GmbH (1995) Market-Based Instruments in Environmental Policy in Developing Countries, Eschborn
4.   Kraemer, R.A. (1995) Evaluating Effluent Charges in Germany, mimeo, Berlin
5.   Holden, P. and Thobani, M. (1996) *Tradable Water Rights,* mimeo
6.   Rosegrant, M.W. and Binswanger, H.P. (1994) *Markets in Tradable Water*

Rights, World Development, Vol. 22, No. 11 pp. 1613-1625
7.   Burchi, S. (1991) Current Developments and Trends in Law and Administration of Water Resources - A Comparative State-of-the-Art Appraisal,   Journal   of Environmental Law Vol. 3 No 1, Oxford
8.   Mac Donnell, L.J. (1990) The Water Transfer Process as a Management Option for Meeting Changing Water Demands, Boulder
9.   Shah, T. (1993) Groundwater Markets and Irrigation Development, Bombay

THE FEASIBILITY OF TRADEABLE PERMITS FOR WATER ABSTRACTION IN ENGLAND AND WALES

J MORRIS, E K WEATHERHEAD and J A L DUNDERDALE
Cranfield University, School of Agriculture, Food and Environment, Silsoe, UK
C GREEN, S TUNSTALL
Flood Hazard Research Centre, Middlesex University, Enfield, UK

Abstract
There is widespread concern in Britain about the impact of increased abstraction of water for irrigation on the environment, water availability and reliability of supply. The existing licensing system has been criticised as being insufficiently flexible to meet changing user requirements and ensure that water is put to its best, most economically efficient use.

This paper, drawing on a study funded by the Royal Society for the Protection of Birds, examines whether the introduction of tradeable licences or permits would be feasible, would improve the allocative efficiency of water amongst irrigating farmers and whether this would have any impact on nature conservation. These aims were investigated within a study area consisting of two neighbouring sites; an irrigated farmed Internal Drainage Board area and a nationally significant wetland nature reserve.

Hydrological models were used to estimate water requirements for irrigation and nature conservation. The views of farmers, regulators and environmentalists were obtained through focus groups and individual interviews. An analysis of irrigation costs and benefits was undertaken to determine the value of water for spray irrigation.

The exploratory study concluded that a system of tradeable permits is feasible and would emphasise the economic value of water. It could, however, increase the risk of greater abstraction. Continued close regulation would needed to avoid negative impacts on both farming and environmental interests.
Keywords: Tradeable permits, water abstraction, irrigation economics

1   Introduction

The increasing demand for water in the UK has resulted in widespread concern over the impact of increased water abstraction on water resources and the environment [1].

Water: Economics, Management and Demand. Edited by Melvyn Kay, Tom Franks and Laurence Smith. Published in 1997 by E & FN Spon. ISBN 0 419 21840 8.

In many catchments, water resources are already actually or theoretically over-committed and additional licences for groundwater and surface abstraction during the summer are unobtainable. Restrictions and bans have been introduced in some catchments. It is often agricultural users who are most affected.

This paper examines the practicability of introducing tradeable water permits. Water requirements were estimated for agriculture and nature conservation in an area sharing a common water source. The views of various stakeholders were sought regarding the advantage of tradeable permits relative to current licensing arrangements. Estimates were derived of the value of water as a traded commodity. The paper draws on work funded by the Royal Society for the Protection of Birds [2].

2   Water resource objectives and characteristics

The overall objective of water resource management is to ensure that water abstraction does not exceed the level of available supply, that water is put to its most beneficial use, and that the social and environmental impacts of water use are adequately addressed.

Compared to conventional economic commodities, water has a number of complicating features. It is:

- a 'fugitive', re-usable good;
- a common property/public good;
- a stochastically supplied resource;
- subject to economies of scale in provision;
- an essential, life supporting commodity with no substitute; and,
- associated with many non-market, environmental qualities.

For these reasons water resources are usually managed in accordance with perceptions of public interest, rather than left to unconstrained use. There is, however, much interest in treating water as an economic commodity.

3   Water resource management options

There are three broad management options, the advantages and disadvantages of which are summarised in Box 1:

Command and control systems use quota and regulation to define how much, when, by whom and for what purpose water can be abstracted. Under the present system in England and Wales, almost all abstractions from surface or groundwater for spray irrigation require a licence granted by the Environment Agency under the Water Resources Act 1991.

The possession of a licence for irrigation gives the abstractor a legal right to take a specified quantity of water from a specified source for the duration of the licence or until the licence is revoked, except when drought restrictions are in force. Licences, however, do not guarantee water quantity or quality.

Box 2. Water resource management options

COMMAND AND CONTROL SYSTEM	
Water abstraction controlled by regulation authority using non-tradeable abstraction licences	
Advantages	**Disadvantages**
sustainable quotas	not necessarily economically efficient
predictable	inequitable
regulator/administrator orientation	static, inflexible
	slow and expensive to adjust
COMMON RESOURCE MANAGEMENT OPTION	
Water and related infrastructure owned by community based organisations	
Advantages	**Disadvantages**
local control, sustainability	multiple layers of administration
reduced central bureaucracy	external strategic issues neglected
general good	local conflicts of interest
ECONOMIC INSTRUMENTS	
Use of pricing and market mechanisms	
Advantages	**Disadvantages**
pricing reflects benefits of use and costs of supply	market failure: economic, social, environmental consequences
allocation efficiency	income redistribution effects
policy instrument	market domination
revenue yielding to regulation agency	regulation difficult

Two types of spray irrigation licence exist. The 'Licence of Right', issued under The Water Resources Act 1963 gave water rights in perpetuity to water abstractors according to previous usage.   Post 1963, virtually all new irrigation licences have been 'Temporary Licenses', valid for a specified period such as 10 years.   The granting of a new licence is subject to an assessment of reasonable need for a specified purpose, available water, no adverse environmental effects, and no derogation of other user rights. These licences are subject to various conditions including restrictions or bans if there is an intolerable risk of water shortage and related environmental consequences. The Water Resources Act 1991, Section 57, may be used by the Environment Agency to impose restrictions during drought periods.

Licences specify the maximum abstraction permitted per year, per day and in some cases, per hour. These conditions are independent of each other. Licences can also restrict the period during which abstraction is permitted, for example during the winter or summer season only. Once water becomes discrete from the natural hydrological system, such as impounded in a reservoir, it is not subject to restriction on use.

Most abstraction licences for spray irrigation can only be used on specified parcels of land. In order to alter this specification, the licence holder must apply to the Agency. Any request by the abstractor to vary the licence requires the Agency to review the whole licence, and it may impose new conditions before agreeing to the variations.

Applicants for new or renewed licences pay an administration fee. Annual abstraction charges are based on a formula reflecting licensed quantity, water source, the loss of water to the hydrological system (high in the case of spray irrigation), season, and region. These charges are set to recover the total costs of administering the licensing and regulation system. There is no charge for water *per se*. Pricing does not reflect user benefits, nor the value of retaining water in the aquatic environment.

Farmers have been very critical of the existing licensing system, mainly because it is bureaucratically complicated and inflexible, especially regarding locations of abstraction and use.

Common property systems, whereby users themselves decide on the management of the resource, have been a traditional form for many irrigation and drainage systems throughout the world. They are typically small states in miniature with their own rights of taxation, courts and councils. Some of the early Internal Drainage Boards (IDB) and Drainage Commissioners in Britain reflect the concept of local, democratically constituted management. This model is currently fashionable in former centrally planned economies (such as Bulgaria, former Yugoslavia, Romania) as a means of transferring the ownership, operation and maintenance of irrigation and drainage systems to Water User Associations.

The model could have some merit for water resource management within defined hydrological areas. It is consistent with the EU commitment to subsidiarity and participation, although perhaps contrary to the actual centralising trend in the UK. Its introduction would, however, imply more equal access to water within the potential membership. This might be confounded by the existing licensing system, such that some changes in water law would be needed. It also requires users to work together and accept common liabilities. Some elements of the regulatory function could be passed to water user groups or Non-Government Organisations (NGOs), such as IDBs.

Economic instruments involve actions which influence resource allocation and use through price incentive and market mechanisms. They include interventions by the resource management agency in the form of taxes and subsidies, and the setting of prices at levels which, to varying degrees, reflect supply cost or benefits derived by users. They may include the construction of a market for buying and selling water rights, usually subject to some overall regulation and quota limits, often determined by social and environmental criteria.

3   Tradeable permits

The concept of a water market whereby access to water is traded between buyers and sellers has been promoted by resource economists as an effective means of reallocating scarce water to its most productive use. The relevant environment protection agency determines a total water quota and constructs or facilitates a market for the buying and selling of permits. Depending on circumstances, the starting point might be equal distribution of available quota amongst participants, distribution according to prior rights, or open auction. Permits will, it is argued, move to their highest bidder. High prices will encourage participants to adopt permit saving strategies, such as water saving technologies. Permits work best where there are many buyers and sellers within a defined relatively homogenous catchment area.

The argument for water permits for irrigation reflects the desire to ensure the economically optimum allocation of water amongst irrigators, rather than allocation being fixed by historic licences.   There is some anecdotal evidence of informal trading in water amongst farmers, which includes pumping and application to neighbouring farms, and land letting including water for irrigation, but no direct transfer of water permits.   In the USA tradeable permits have facilitated the release of water from agriculture to domestic and industrial use [3,4].   In some cases, municipalities have purchased 'water farms' for this purpose.   Farmers have also purchased additional water, demonstrating the relatively high potential returns to water in the irrigation sector.

The concept of quota is not new to UK agriculture, neither is that of tradeable quota or permits. The introduction of milk quota in 1984 in an attempt to limit milk production, quickly led to a brokerage market in the sale and leasing of milk quota, with prices reflecting the profitability of the dairy sector. A market has recently emerged in beef and sheep quota which were introduced in 1994. Quota are regularly advertised in local and national press. At the point of initial allocation of permits and on subsequent transfers, the regulating authority (MAFF and their agents) have siphoned off a percentage (15%) of the quota to reduce total production quantities over time. This siphoned quota goes into the national reserve which is used to help new starters in the market.

The experience of trading in milk and livestock quota indicates the feasibility of tradeable water permits, and the possible form this might take. That is:

- permits distributed according to prior rights;
- possibly some siphoning in return for improved security of supply;
- supervision by a regulatory authority;
- open market trading under a brokerage/commission agent system; prices of permits finding their own level.

If irrigation licences were replaced by tradeable permits, they could conceivably be sold or leased on a similar basis to milk quota. The extent to which a market for tradeable permits could develop would be controlled by the availability of the water resource and the level of demand. The unit price of each trade would be agreed by the buyer and seller engaged in the transfer.

4   Organisation of a tradeable permit system

For water markets to function effectively, water rights must be clearly specified and legally enforceable. Water supplies would have to be reasonably reliable and deliverable on a measured basis. As long as the total permits issued are within the available hydrological capacity, trading should reallocate supplies to higher value uses without resulting in over-abstraction.

The main argument in favour of a market for tradeable permits *within* the irrigation sector is that this encourages the transfer of water from less efficient to more efficient producers, thus improving resource efficiency. In theory, the price of tradeable permits would be 'bid up' by those with the greatest ability and willingness to pay for extra water.

In theory, willingness to pay will reflect the marginal value product (added value) of water. Thus tradeable permits will go to those who can use water most efficiently and therefore can pay most at the margin. These will be farmers who engage in high value, water-responsive cropping, are constrained by existing supplies, and can achieve greatest added value as a consequence of acquiring additional water. In practice, these are likely to be the larger specialist contract growers achieving economies of scale.

Tradeable permits could, lead to a concentration of the water market in a few hands. Whilst this may be economically efficient, it may result in socially inequitable income distribution within the sector. This could be further accentuated by the ability and willingness of larger producers to pay for extra water at prices which reflect the relatively high average value of water rather than value at the margin. In practice, therefore, water may not necessarily go to the most efficient users, but to those in a position to pay.

Tradeable permits, in that they demonstrate the value of water to users, may help to induce farmers to adopt water conservation technologies, including water storage. They could, of course, do this in the absence of tradeable permits.

The transfer of permits could be permanent or temporary. If permits were leased, the permit could revert back to the original holder under the original terms and conditions at the end of the lease period. Under a permanent trading agreement, new owner holds the water rights. It is likely that, on trading, licences of right would be converted to temporary licences.

Four methods of trading may exist:

- the *buying and selling of land* between farmers in order to acquire an abstraction licence and to change abstraction points. This type of trading is already practised.

- agreed *transfers between abstractors without land ownership change* agreed by the regulatory agency on a case by case basis where water supply conditions are suitable, especially if this also involves a downward variation in quantity.

- *long term trading* whereby the regulatory agency administers a scheme in which licences are bought or sold on a permanent basis. Each trade would require approval from agency;

- *a free market* with licences bought, sold or leased in a water market over a short timescale but with the Agency retaining overall regulatory powers.

Box 2 contains a number of inter-related key questions relevant to the design of a trading system for water permits.

5   Study area

The feasibility of introducing tradeable permits for water was examined in the context of a selected study area which comprised two neighbouring sites, sharing the same river water source. One was an intensively farmed IDB area of about 12,000 ha. The other was a 300 ha wetland Nature Reserve of national designated scientific and habitat importance. In times of low flow, the water requirement of the Nature Reserve is given priority.

Box 2 Design criteria for tradeable permit systems

• What is the total amount of water on offer ?
• What is the entitlement under the permit ?
• Who starts with the initial holding ?
• Who enters the bidding ?
• What is the geographical range over which the permit is tradeable (over how wide an area can the permit be traded) ?
• Are permits limited to one class of abstraction ?
• How is the system to cope with future increases or decreases in available supply (e.g. as a result of climatic change) ?
• Will there also be charges for abstraction in addition to a system of permits ?
• Will there also be charges for discharges ?
• What are the anticipated behaviours of abstractors ?
• How can the total volume of abstraction be limited to a level which protects the environment ?
• Who administers the scheme ?
• Will owners of traded temporary licences be given priority for repurchase?

Irrigation needs were calculated using an Irrigation Water Requirement model [5], taking account of local climate, soils and cropping practices. For the estimated irrigated crop area, the theoretical dry year (5th driest in 20) water requirements exceeded licensed quantities by a factor of three: 80% of these requirements were for potatoes. This confirmed water deficits in dry years. Over the period 1991 to 1994, however, agency records suggested that between 41% and 74% of the licensed quantities remained unused. Theoretically, this 'surplus' in licensed quantity could be available for trading. However, the estimate of unused quantity is thought unreliable because of possible under- recording and declaration, restrictions placed on abstractors in dry years, and licences which in any one year were locationally mismatched with the needs of crops in rotation.

The hydrological requirements of winter wildfowl, breeding waders and wet grassland flora on the Reserve were estimated. Summer water requirements per ha were similar to those of the irrigated areas. By modelling water requirements for ditch and adjacent field flora and aquatic invertebrates were estimated for the April to August period. It was concluded that failure to meet the Reserves needs would have major ecological impacts. Thus during dry summers, the irrigated area and the wetland site were theoretically in competition for water, but the latter was given priority in practice.

5   Views on tradeable permits

The views of stakeholders with interests in tradeable permits were obtained through a combination of focus groups and personal interviews.

Farmers:

> criticised the existing licensing system as inflexible and bureaucratically complex. They felt that there are considerable potential gains in efficiency to be gained from freeing up the existing licence system, for example, by removing the specification of the land to which the water is applied, by reducing the time taken to process licence applications and by examining the relevance of licence conditions.

expressed concern that, given limited water availability, a market in licences would lead to increased abstraction and more frequent and possibly earlier introduction of restrictions. This could reduce the potential value of tradeable permits and simultaneously compromise the reliability of water to those who did not wish to trade. Farmers thought tradeable permits could be made to work but felt that they would not address the fundamental problem of water shortage in a deficit area.

generally accepted the priority given to environmental criteria (in terms of prior access, restrictions and bans), mainly because they had no choice, and that this would prevail under a trading system.

thought willingness to sell would be confined to retiring farmers or those with or intending to construct reservoirs who have water surplus to their requirements. (At present, a water licence adds 20% to 30% to land value). Selling a permit could help to fund a reservoir.

preferred trading within an exclusively agricultural market. They were concerned that they would not be able to compete with non-agricultural users (conservation, water supply companies) in an open market. They thought a brokerage system would work best.

thought tradeable permits would encourage them to think about the value of water and ways of using it more efficiently, including greater precision in scheduling and application.

The Regulators:

considered that tradeable permits could encourage greater water use efficiency, with water being put to best economic advantage;

identified risks of over abstraction and negative environmental impacts, but these could be controlled through an appropriate form of regulation in those areas where trading is likely to be significant;

saw greater risks of abstraction restrictions coming in earlier because licensed quantities previously not taken up might be traded;

were concerned about the practicalities of implementation at a local level unless there is some form of devolution to ensure trading is not costly and bureaucratic;

Local drainage organisations thought that trading would make the management of the irrigation/drainage system much more difficult. Farmers without licences who relied on sub surface irrigation from retained water in ditches could be compromised if ditch levels fell due to increased abstraction.

Environmental organisations, especially RSPB, were favourably disposed to tradeable permits if they led to improvements in water use efficiency, and on condition environmental safeguards were built in. They were concerned about the risk of over-abstraction if currently unused licensed quantities are taken up. They could however, become participants in the water market, purchasing rights to protect or enhance the aquatic environment.

6   Economic aspects of water as a traded commodity

The focus group discussions confirmed that the value of water and hence willingness to pay for licensed quantities lies between the cost of supply and the benefits derived, although farmers found it difficult (and strategically risky) to place values ($£/m^3$) on irrigation water itself.   Willingness to pay will be affected by the perceived risk abstractions may be restricted in dry years.

With respect to *cost based criteria* the lower boundary cost ($£/m^3$) equates to that of the licence application plus summer abstraction charge.   The upper boundary equates to the cost of the licence application, winter abstraction charges and the costs of water lifting, storage and release.

At current abstraction charges, demand exceeds supply.   It would not be rational for farmers (except in the short term) to pay a price ($£/m^3$) for a direct abstraction licence which exceeds the cost of on-farm storage. Higher prices would encourage investment in winter storage thus increasing supply of water for irrigation, assuming permission for winter storage may be obtained.

With respect to *benefit based criteria*, the value of water is evident in crop yield and quality benefits due to an additional unit of irrigation water, net of application costs. Deriving estimates of value-added are very difficult due to their dependability on the crop, its growth stage, the depth of irrigation water applied and a complex of weather, soil, crop husbandry and market factors.

Table 1 shows the boundaries for water pricing based on average cost and benefit criteria.[1,2] In practice, farmers would base their pricing decisions on the marginal cost of obtaining or the marginal benefit of using a specified extra quantity of water.

Table 1. Irrigation water values (assuming average crop responses and 1996/7 prices)

Cost based:

	$£/m^3$
Existing charges (summer)	0.023
Ex-reservoir	0.30-0.50

Benefit based (net of irrigation cost excluding water): Potatoes	$£/m^3$
yield (value of yield assurance net of irrigation costs)	0.20 - 0.50
quality (value of quality assurance)	0.90
total	1.10 - 1.40
Sugar beet	0.03 - 0.30
Potatoes v cereals	0.80

Using cost as the pricing criterion, farmers might be expected to pay between the existing charges for summer abstraction (£ 0.023 /m³) and the cost of water stored in a reservoir (between £ 0.30 and £ 0.50 /m³ according to the need for lining).

Using benefit as the criterion, farmer willingness to pay will depend on the crop, the extent to which water is currently a constraint, and the cost of applying it once it has been purchased. A farmer might be willing to pay up to £ 0.50 /m³ for water applied for yield response on potatoes through an existing system. If additional capital investments are required the price limit would be £ 0.20 /m³. Quality benefits on potatoes could add another £ 0.90 per m3. Where the alternative to irrigated potatoes is a crop of rainfed wheat, the price limits might be about £ 0.80 per m3. In the long term, over the investment life of an irrigation system, it is the lower of the two benefit estimates (i.e. after total costs) which is relevant.

Farmers found it difficult to determine at what price they would buy and sell licensed water quantities, due to lack of information on current costs and benefits. Those who wished to buy more water expressed a willingness to pay which ranged between three and six times the current abstraction charges. Where water was purchased from neighbours, they expected to pay prices which at least covered the cost of water delivered to them, including storage costs if relevant.

7   Conclusions

The exploratory study concluded that scope exists for improving the flexibility of the existing licensing system to better suit the needs of modern irrigation practices. Licences could, however, be traded in open market or auction transactions, managed by a broker or agent. This could increase allocative efficiency and, because water would command a market value, could encourage water saving practices. The Environment Agency would retain overall regulatory powers, but some functions could be handed to local organisations such as IDBs where appropriate..

With respect to environmental impacts, careful monitoring of water regimes will be required to protect the hydrological environment and avoid unacceptable risks. Care is required to define the hydrological unit within which water can be traded in order to avoid deficits and related environmental consequences. Trading could result in greater take up of licensed quantities and reduced flows and levels in the river and drainage system. This could have consequences for the aquatic environment (as well as for non-irrigating farmers) and could be detrimental to the ditch flora and fauna of the wider countryside. Controls on abstraction would be needed to protect environmental standards. Tradeable permits could however, allow environmental organisations to participate in the water market, thus providing greater flexibility in the development of wetland sites.

This exploratory study suggests tradeable permits could encourage the perception of water as an economic commodity which commands a price reflecting cost of supply and benefit in use. Farmers would need to obtain and analyse relevant data on the economics of irrigation in order to guide their participation in a water market.

8   References

1.   Weatherhead, E.K., Place, A.J., Morris, J. and Burton, M. (1994) *Demand for irrigation water*, NRA R&D Report 14, HMSO, London.

2.   Morris J., Weatherhead, E.K., Knox, J.W., Gowing, D.J.G., Dunderdale, J.A.L., Green, C. and Tunstall, S. (1996) *The practical implications of introducing tradeable permits for water abstraction.* Report to the Royal Society for the Protection of Birds, by Cranfield University, School of Agriculture, Food and

Environment and Middlesex University, Flood Hazard Research Centre. Cranfield University, Silsoe.

3.   Howe C.W., Schumeier D.R. and Shaw W.D. (1986). Innovations in water management from the Colorado-Big Thompson Project and Northern Colorado Water Conservancy District. Scarce Water and Institutional Change (ed) K.D. Frederick, *Resources for the Future*, Washington.

4.   Michelson A M (1994) Administrative, institutional and structural characteristics of an active water market. *Water resources Bulletin* Vol 30, No 6

5.   Hess, T.M. (1994) *Irrigation Water Requirement Model*, Silsoe College, Cranfield University, Unpublished

MANAGING WATER AS AN ECONOMIC GOOD
Rule for reformers

J. BRISCOE
The World Bank, Washington DC, USA

Abstract

This paper extends a previously-developed framework for the management of water as an economic resource, by showing how positive and negative externalities can be taken into account. The main focus of the paper, however, is on assessing the lessons of experience which emerge from successful reforms. The following emerge as a tentative set of "rules for would-be reformers":
initiate change only when there is a powerful, articulated need for reform;

- have a clear strategy for involving all interested parties in the discussions of reform, and for addressing fears seriously, with effective, understandable information;
- pay attention to general principles, but be sensitive and innovative in adapting these in different institutional and environmental contexts;
- do not advertise water markets as a silver bullet or a panacea, but ensure that they are part of an effective water resource management system;
- start with the relatively easy problems to get experience and build momentum for reform;
- acknowledge that there are no perfect solutions, and don't let the best become the enemy of the good;
- pay close attention to prescribing institutional arrangements which will address legitimate third-party issues, but which will simultaneously minimize transactions costs.

Keywords economics, managing transitions, politics, strategy, water markets, water resources management.

1   A basic conceptual framework for water as an economic good

A recent paper to the World Congress of the International Commission on Irrigation and Drainage (ICID) [1] outlined the theoretical underpinnings of the idea of "water

Water: Economics, Management and Demand. Edited by Melvyn Kay, Tom Franks and Laurence Smith. Published in 1997 by E & FN Spon. ISBN 0 419 21840 8.

as an economic good", and presented information on the value of water in different end uses, and the supply and opportunity cost of water in different sectors and settings. This paper focuses on just two (but two major) water-using sectors -- urban water supply and irrigation.

1.1 Urban water supply

Urban water supply is a low-volume, high-value (typically between 10 and 100 US cents per cubic meter [1]) use. The supply costs (incurred in financing and operating the abstraction, transmission, treatment and distribution systems) are relatively high, while the opportunity costs (imposed on others as a result of use of the water) are quite low. Accordingly, the priority issue for the economic management of urban water supplies relates primarily to the supply cost axis (as illustrated in Figure 1).

Conventional economic wisdom suggests that users should be charged full marginal costs -- level IV in Figure 1. In most developing country situations, however, aiming for economic perfection is neither practical nor helpful. Instead, it is imperative that tariffs be set in a way that is understandable, transparent and legitimate and that forces suppliers to be accountable (and thus produce services efficiently). In the urban water supply sector, this "common-sense" pricing approach will therefore mean:
- focusing on supply costs and
- aiming to increase user charges first up to level II and then up to level III (in Figure 1).

1.2 Irrigation

In developing countries most irrigation -- 90% in World Bank-financed irrigation projects -- is for foodgrains. This type of irrigation is a high-volume, low-value user of water (with values generally less than 1 US cent per cubic meter). But there is an important and growing sector of high-value irrigation (often fruits and vegetables), with typical values between 5 and 15 cents per cubic meter of water [1].

The supply cost of irrigation water is usually modest, but when there is competition with either urban uses or high-value irrigation, the opportunity cost is high. In the Limari Basin in Chile, for example, the supply cost (which is partially subsidized) is about 0.5 US cents per cubic meter, whereas water trades at about 5 US cents per cubic meter [2]. In California, a typical financial charge to an irrigator is about $5 per acre foot (0.4 cents per cubic meter), whereas water trades at the equivalent of about $150 per acre foot (12 cents per cubic meter) [3].

From the perspective of treating water as an economic good, the great challenge in irrigated agriculture is how to ensure that farmers take into account the opportunity costs, which are often an order of magnitude higher than current charges. This is the essence of the appeal of the approach of water markets -- as described in the Chilean case [2] "the genius of the approach is that it ensures that the user will face the appropriate economic incentives, but de-links these incentives from the tariff, which is set on "common-sense" grounds". Or, in the case of the California Water Bank (which is a formal mechanism for pooling surplus water rights for rental to other users) the key is " the ability to increase the value of water without an increase in the cost to the farmer (that) is a politically acceptable way of sending the signal to users of the true value of water" [4]

There are many vital issues relating to charges for supply costs in irrigation. Experience [5] has shown that cost recovery in and of itself achieves nothing, unless the money collected is used efficiently to improve the quality of services provided. Recent experience in Mexico confirms this -- since management of irrigation systems has been transferred to users' associations, recovery of operation and maintenance costs has increased from 30% to about 80% .

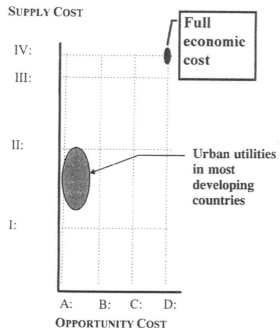

SUPPLY COST

IV:

III:

II:

I:

Full economic cost

Urban utilities in most developing countries

A: B: C: D:

OPPORTUNITY COST

Legend:

Supply Costs:
I: Operations and maintenance costs only
II: Average financial (capital + O&M) cost, with capital valued in terms of historical costs
III: Average financial (capital + O&M) cost, with capital costs computed in replacement terms
IV: Long run marginal cost of additional supplies

Opportunity Costs
A: Water can be used only by an individual user
B: Water can be leased or sold to neighbors
C: Water can be leased or sold within a limited district
D: Water can be leased or sold to any urban or agricultural user

Fig. 1. Supply cost, opportunity cost and full economic cost for urban water supply

In terms of economic signals to irrigators, however, as illustrated in Figure 2, the "vertical axis" (supply costs) is relatively short, and the "horizontal axis" (opportunity costs) is often long. That is, from the perspective of the economic allocation of water, the key challenge is to ensure that farmers are aware of the opportunity cost of the resource, and that there are institutional arrangements for ensuring that water moves to higher-valued uses. This paper will, accordingly, focus primarily on the opportunity cost issues and not the supply cost issues in irrigation.

The ICID review [1] concluded that it was inappropriate, on a number of counts, to think of rolling opportunity costs into water tariffs (as has been suggested in

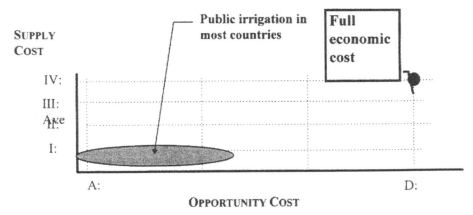

Fig. 2. Supply cost, opportunity cost and full economic cost for public irrigation

several countries, notably Chile [2] and South Africa). This is so for three main reasons:

- because the information requirements are very onerous (opportunity costs vary dramatically by place and season, and even sophisticated research studies cannot estimate them in a way that is universally accepted);
- because the levying of such charges would (usually correctly) be perceived as expropriation by those who currently use the water;
- because it would defy common sense -- using the numbers cited earlier in this paragraph it would mean that farmers in Chile, Australia and California would be asked to pay more than 10 times the cost of providing the services they receive!

Emerging international experience is clear -- from a conceptual, practical and political perspective, the appropriate approach for ensuring that the scarcity value of water is transmitted to users is to clarify property rights and to facilitate the leasing and trading of these rights.

2   Expanding the framework to take account of externalities

Discussions stimulated by the ICID paper have confirmed [6] that this is a useful and practical approach, but that the basic framework needs to be expanded to take account of return flows and the positive and negative externalities they generate.

2.1   Negative externalities

Use of water by one user commonly has negative impacts (externalities) on other users. For example, pollution from a town can mean that downstream users have to incur additional treatment costs. Similarly, drainage water from irrigation fields often carries high levels of salts, nutrients and pesticides, leading to losses of aquatic habitats [7]. These externalities are easily incorporated into the conceptual framework of the ICID paper, by simply increasing the supply costs to include the cost of mitigating the negative externalities (Figure 3).

The implications of Figure 3 are evident in practice in the real world. For example, in many countries cities are required to meet specified wastewater (or receiving water) quality standards. The utility does this by treating its wastewater, and passing on the costs of this treatment to its customers. Similar practices are now starting to emerge for non-point sources of pollution. In the Murray-Darling Basin in

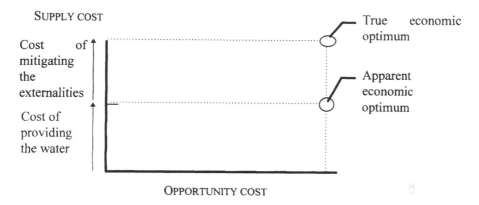

Fig. 3. Incorporating negative externalities

Australia, for instance, the major water quality problem is the high levels of irrigation-induced salinity [8]. The Murray-Darling Basin Commission has now specified maximum salinity fluxes from its different member states. Salinity control measures are required to stay within these limits, with the costs of these control measures being passed on to irrigators in their water bills. (These costs are considerable. In the state of Victoria, for instance, irrigators pay roughly equivalent amounts -- about 1 US cent per cubic meter for water and a similar amount as a salinity levy [9]. In the Colorado Basin it costs, on average, about $70 to remove a ton of salt [10]. For a program which aims to reduce salinity levels by 200 ppm, this equates to a cost of about 1.5 cents per cubic meter. This is a very substantial amount -- it compares to typical water levies of about 0.4 cents per cubic meter [3]).

2.2 Positive externalities
Return flows constitute a vital element of many hydrologic systems. For example, farmers in the Gangetic Plain who apply more surface irrigation water than is required for evapotranspiration are, in effect, recharging the aquifer which underlies their fields, thus performing an aquifer recharge service for farmers who use groundwater. As Seckler [11] and Fredericksen [12] point out, analyses which concentrate on (apparently low) farm-level irrigation efficiency do not take into account the fact that basin-level efficiency may be high. This observation is important because it means that the benefits of apparent efficiency improvements at the farm level may be illusory or even negative at the system level.

Consider the case where surface water irrigation is both meeting the evapotranspiration needs of crops, and recharging a groundwater aquifer which is subsequently used for irrigation. It is apparent that the surface water users are providing a "recharge service" to the groundwater users. What should the groundwater users pay the surface water irrigators for this service? An upper limit on this charge would be the cost of a formal recharge system (with which the groundwater users could recharge their aquifer). In such a case, the relevant charges to surface and groundwater users would be as shown in Figure 4. Once again, this is a "common sense" approach, with surface water users effectively getting a credit from groundwater users for the recharge service they are providing.

In many circumstances, of course, surface and groundwater users are the same population, in which case the notion of a monetary transfer from a user (to himself!) would not be sensible. Furthermore, in many areas where there is conjunctive use, there are countervailing "distortions". In Indian irrigation systems, for instance, headwater and tailwater users are charged the same (very low!) price for water. On the one hand, since tailwater users make greater use of groundwater, they "get an implicit subsidy" from the headwater users. On the other hand, the quantity and reliability of the service to the headwater users is far superior, probably more than offsetting the implicit subsidy.

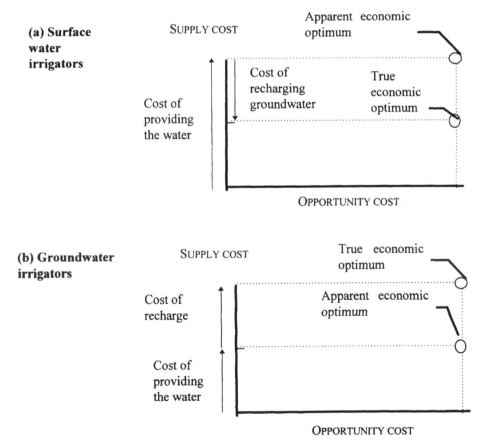

Fig. 4. Supply costs when recharge is important

Real-world experience shows that where there is a will there is a way -- there are practical solutions to the non-consumptive use issue [13]. In the Western United

States there are a variety of methods for taking account of return flows from irrigators. Under the laws of most Western states, return flows "belong to the stream". The practice of the Northern Colorado Water Conservancy District is consistent with this. While each water user has the full right to purchase, sell, trade or rent rights to the primary, consumptive flows, the District retains all rights to the return flows of water.

In this, as in all other aspects of water management, it is necessary to find a balance between the ideal and the practical. When water rights are sold in the Western US, the quantity of water that may be transferred to a new use will be limited to the amount of water which was deemed to be consumed historically. This is typically done by a State Water Engineer who makes an estimate based on the factors such as the type of crops cultivated, soil type and climate [13].

Finally, while the return flow issue is generally raised in the context of the pricing of irrigation services [11,12], the distinction between consumptive and non-consumptive use is not relevant only to irrigation. Indeed, taking the US as an example, consumptive use as a percentage of withdrawals was 56% for irrigation, compared with 17% for urban water supplies, 16% for industry and just 3% for thermoelectric power [14].

2.3   Negligible externalities

There are many situations in which return flows have neither a positive nor a negative value for other users in a basin. For example, where return flows accrue to low-value "salt sinks", they have no value. Or where return flows discharge into the ocean, they are also effectively "lost" from the system. In such cases there is no need for modification of the use-opportunity cost analysis presented on Figures 1 and 2. The 1989 agreement between the Imperial Irrigation District (IID) in California and the Metropolitan Water District (MWD)which serves Los Angeles, is a good illustration of such a case. Under this agreement, the MWD pays $120 million for conservation activities (mostly canal lining, but also operations modernization) in the IID. In turn, the MWD obtains the 130,000 ML of water which are conserved. This agreement was possible in part because there are few externalities -- the IID is at the lower end of the river system, and there are no opportunities to reuse return flows, which are therefore considered "wasted" [15].

3   Rules for implementing a reform program

There is an emerging global consensus on both the necessity for more effective management of water resources, and on the principles of effective management. The Dublin Statement of the pre-UNCED Conference on Water and the Environment [16] remains the clearest of such statements, articulating three principles. The three "Dublin principles" are:

- The "ecological principle", requiring the holistic management of water;
- The "institutional principle", requiring that management be participatory, with responsibility "at the lowest appropriate level", and with greater involvement of NGOs and the private sector and women; and

- The "instrument principle", requiring that water be managed as an economic resource.

Where the notion of "water as an economic good" was once an issue of interest primarily to theoreticians, in recent years the principle has been translated into practice in many settings, with varying degrees of success. What are the lessons to be gleaned from this growing body of experience?

3.1   There must be a demand for reform

The first requirement for reform is that there must be a demand for reform. Unless the shoe pinches, reform is unlikely to take place. This is an obvious and yet often-neglected fact, especially by water professionals who see basin-wide management, for example, as self-evidently necessary in all places. A couple of examples suggest where the impetus for managing water as an economic good might arise.

The most obvious type of water stress is scarcity. Accordingly, wherever there is scarcity, there has been an endogenous move by local citizens to develop some form of informal water market. That these markets have been invisible and usually illegal does not mean they have not existed, nor that they have not become very sophisticated, as illustrated by the water markets of Gujarat [17]. In such settings, formalization of water markets greatly reduces transactions costs, and is thus welcomed by those who have been trading "in the black market". The Australian case is informative -- "the major impetus for the development of water markets came from users rather than the government,... all the government did was remove the legislative obstacles to transferability" [9]. Similarly, in urban areas of most cities of the developing world there are sophisticated water vending systems which fill the void left by poorly-performing utilities [18]. Because the unserved (usually the poor) typically pay 10 times as much for a liter of water than do the served, there is a tremendous implicit demand for reform which can be tapped. (As in all black markets, however, there are those who thrive on the distortions. Jakarta, where vending is very widespread and very lucrative, proves a general point to which we will return later, namely that there will be losers in reform processes [19], and that their interests have to be identified and dealt with.)

Scarcity is a great impetus for change, and so is pollution. In 1857 the slogan of the day in London was "India is revolting and the Thames stinks". The result was Chadwick and the sanitary revolution in the United Kingdom. Similarly, in the early part of this century, the pollution of the Ruhr River in Germany, and its threat to the operation of the industrial heartland, mobilized industry and society to do something. The upshot was the Ruhrverband, a revolutionary approach to water quality management which was built on the twin principles of management by stakeholders and using economic instruments to provide incentives for efficient use and for waste reduction [20]. And when France started recognizing the importance of its river pollution problems, it adapted, in 1964, the Ruhr approach on a national level, with the result being the famed "River Basin Financing Agencies". And now, when the water quality situation in the Paraiba do Sul river (which is the sewer of Sao Paulo and the drinking water supply for Rio de Janeiro) becomes intolerable, the states involved (Rio, Sao Paulo and Minas Gerais) are forming a Ruhr-French type basin agency, again grounded on participation and the use of economic instruments, to deal with the problem [21].

All stresses and challenges do not have to be the "natural challenges" of scarcity and pollution. Stresses in the economic and institutional machinery can also act as an impetus to change. Of particular importance here is the overall view which a society takes of its economic development process. Thus, in settings as diverse as Chile, Australia and Peru, it has been the articulation of an open, export-oriented, growth-driven economic development strategy which has been critical in providing the impetus to improving the economic performance of water (and other factor) markets.

The possibilities of reform are, therefore, greatest when there is a confluence of "natural" challenge (scarcity or pollution) and institutional reform-mindedness, as exemplified by Chile in the 1980s and Australia today. The basic point here is simple -- creative responses in dealing with water as an economic resource will only happen when there is a problem to be addressed, when that problem is perceived to be important, and when there is a political climate conducive to reform.

RULE 1: Initiate reforms only when there is a powerful need, and demonstrated demand, for change.

3.2   Water is special -- dealing with the "exceptionalism syndrome"
Water is a good with special properties -- it is the basis of life itself, it is not produced, it is unitary, it is fugitive [22]. These particular attributes have long made water "special", in symbolic, religious, and legal terms. It is no wonder, therefore, that there is much skepticism and concern about the effects of reforms which purport to treat water as an economic good.

3.2.1. Concerns about "the end of irrigated agriculture"
A common concern is that treating water as an economic resource will mean "the end of irrigated agriculture". While this will certainly be true in some local areas, the overall effect on the quantity of water used in irrigated agriculture will be small. There are two reasons for this -- first, because irrigation is a dominant withdrawer and even more dominant consumer of water in all countries where irrigation is important. The changes in the United States illustrate the general point: between 1960 and 1990, while withdrawals for municipal purposes almost doubled, irrigation's share of total consumption fell by only 4%. As concluded by the US Department of Agriculture: "Growth in non-agricultural water needs, particularly in areas with limited supply-enhancement opportunities, may be met by relatively small shifts in national irrigation water use. However, small national shifts may mean large adjustments in local irrigated activity."[14] The implications of this are obvious -- it is essential that the reform debate be informed by solid analysis which can distinguish the legitimate (and in this case local and limited) concerns from extravagant "doomsday" claims.

3.2.2. Concerns about the thinness of the market
Another set of skeptics have the opposite concern. They claim that water markets are, in practice, very "thin", so thin that the notion of a functioning market allocating resources is an illusion.

First some facts. In Chile the number of transactions varies very widely, in a way that is systematically related to water stress and structural changes [2]. In most basins stresses are not yet great and so transactions are rare [23]. But in highly-stressed basins, there are many transactions. For example, in Santiago County, over a one-year period, 3% of total water rights were transferred (with 94% of transferred water moving from one farmer to another) [24]. In Australia (where there are currently both area-of-origin restrictions and restrictions on trades out of agriculture) about 2.5% of all water is leased each year and about 0.3% sold. The Northern Colorado Water Conservation District illustrates how important the cumulative effects of transfers can be -- in 1957, 98% of deliveries from the Colorado-Big Thompson Project were used in irrigation; in 1989 this figure was down to 73% [25].

3.2.3. Concerns about the impact on the poor
There is a common and reasonable concern that treating water as an economic good will inevitably be damaging to the poor, especially in terms of supplying "basic human needs". This concern acquires particular validity because most interventions to "keep tariffs low" are defended in the name of the poor.

In the urban water sector the evidence is compelling and consistent throughout the developing world. Giving politics a central role in determining tariffs has meant three things -- costs are too high because utilities are not accountable to users; coverage is low, with the poor always the last to get services; the underserved have to resort to buying from water vendors, typically at prices 10 times those which they would have to pay an efficient utility [18]. This has aptly been labeled "the hydraulic law of subsidies" [26]-- the subsidies follow the water, and the water flows to power and influence and away from the poor.

Recent experience in the urban sector has shown that with commitment and imagination, the poor can be much better off when water is managed as an economic resource. Three examples illustrate this.

In Santiago, Chile, the government realized that it was inherently contradictory to require that an urban water utility (EMOS) function as a commercial entity and provide subsidized services to the poor, since each subsidized person served would represent a loss of revenue to the utility. Accordingly, the government decided to institute a targeted, means-tested, government-administered "water stamps" program, whereby poor people would get "stamps" which would cover part of their water bill. The utility then not only strengthened its focus (getting out of the welfare business and focusing on becoming the most efficient utility it could), but it now had a clear incentive to serve the poor, who became revenue-generating customers like all others. The system works very well [27].

In Conakry, Guinea, the performance of the water utility in the late 1980s was catastrophic -- water for only a few hours a day, with the poor, as always "at the end of the line". The familiar "low-level equilibrium" prevailed -- service was poor, people were not willing to pay, revenues were inadequate, service got worse and so on. The government made creative use of a World Bank credit to get to a "high-level equilibrium". The assets were leased to a private operator who was paid a fee which reflected the full cost of the service. Users initially paid only about a quarter of this fee, with three quarters of the operator's fee covered by the World Bank credit. Users were informed that service would improve, and that as it did tariffs would be increased to cover costs over a five-year period. Although problems remain [28], this innovative approach worked well for the poor -- coverage increased from 15% in the 1980s to 52% in 1994.

In Buenos Aires, Argentina, the public water company performed poorly for years -- coverage was low, water was rationed every summer, and prices were high. The Government gave a concession contract to a private operator in 1993. At the end of 1995 the water tariff was 27% lower than it was when the utility was publicly run, 650,000 new water connections and 340,000 new sewerage connections had been made.

The experience of the urban water sector is clear and well-documented -- the poor are much better off when water is managed as an economic good. What of irrigation?

The inequities of existing command-and-control mechanisms for water allocation in irrigated agriculture have been widely documented (for instance by Wade [29] in South Asia). Because water has rarely been formally managed as an economic good in developing countries, however, there is little information on the equity effects of a market-oriented management system.

In the case of Chile, there are differing positions on the equity implications of water markets. As always, the counterfactual is of central importance and debatable! Proponents of the Chilean water markets [24] argue that the counterfactual is of subsidized infrastructure which inevitably (just as in the case of urban water) differentially favors the well-connected. They further argue that it is the poor that suffer disproportionately from the fiscal deficits and inflation to which such subsidies contribute. Critics of the Chilean markets [30], focus on the fact that no effort was made to address the specific problems which faced the poor when the market system was introduced: "In the 1980s the government undertook no campaign of public information or education about the Code's new features, not offered legal or technical advice about how to apply for new rights or regularize old ones". What is striking in Chile today is that post-Pinochet social democratic governments (who have a strong commitment to equity) have remained firmly committed to the use of water markets, while being equally committed to addressing the informational deficiencies which disproportionately affect the poor [23].

A rare empirical assessment of the equity impacts of water markets was done for Spanish and Western US irrigation in the 1970s. The authors [31] concluded: "although it is a doctrine of many welfare economists that procedures that rank high in efficiency will do poorly in distributing income equally among beneficiaries, while procedures that do well in distributive terms will be inefficient... this conventional wisdom does not apply to a wide variety of conditions in irrigated agriculture".

3.2.4. Concerns that the environment will be neglected or damaged
In the past there was a widespread perception that there is some inherent contradiction between "the capitalist economy" and the environment, and, therefore, concern with the environmental effects of treating water as an economic good. After the devastating effects of command-and-control policies on the environment became clear in the Soviet Union and elsewhere, perceptions have changed dramatically. It is now widely understood that it market mechanisms induce efficient resource use, and that

inefficiency is the enemy of the environment. And it is equally widely understood [18,32] that this is particularly true of water.

Many sophisticated environmental groups have, accordingly, become vigorous advocates of the concept of water as an economic good. In the case of some of these groups, such as the World Resources Institute [32] and the Environmental Defense Fund [33], the principal issue is that treatment of water as an economic good per se will mean much more efficiency, and far fewer environmentally-destructive investments. For others, the purchase of water rights becomes a cost-effective and practical method through which environmental requirements can be handled without expropriation. The Nature Conservancy, for instance, has spent $1.5 million to purchase water from farmers to leave instream in the Carson River in Nevada [34]. The US EPA is also starting to follow the same path -- Federal Clean Water Act funds now being used to buy water rights from irrigators in the Truckee River area (near Reno, Nevada), to increase river flow in dry summer months [35].

3.2.5. The "exceptionalism syndrome" -- "markets may work for selling cars, but they don't work for water; water markets may work in California, but they can't work in India,....."

Institutional change has always been resisted, precisely because it involves change. A universal argument against change is that "water is different" or "India is different". It is, however, becoming increasingly clear that there is a remarkable degree of commonality between the ingredients of successful reforms in quite different contexts, ranging from Taco Bell, to General Electric, to the New York Police Department, to the NGO Aravind Eye Hospital in Tamil Nadu [36], to the state government of Ceara in Brazil [37]. The "uniqueness" idea is not standing the test of time!

In the water sector there is a remarkable degree of similarity in the nature of "the problem" throughout the world. The political economy of contemporary public irrigation systems in, say, India [29], is remarkably similar in many respects to the political economy of public irrigation systems in the Western United States [38]. And there is much in common between the groundwater markets of Gujarat [17] and those of New Mexico [39]. And there is little distinction between the reform recipes for urban water supply in Adelaide and Abidjan. It is this very convergence which underlies the consensus around the Dublin principles!

This does not imply that the instruments used in different cases will not be strikingly different. The institutional arrangements are not the same in short, unregulated river basins in Chile as they are in continent-wide rivers like the Murray-Darling, but the underlying principles are similar.

What is striking in the contemporary water management arena is how a few hegemonic ideas are becoming universalized very rapidly. The Ruhr-French model (of participatory management, with the use of user and pollution charges) is proving to be well adapted to quality problems in the Paraiba do Sul river in Southeast Brazil [21] and for the management of coastal wastewaters in the United States [40]. And the water market/water bank model is finding wide application where there is scarcity -- in Chile, Argentina, Mexico and South Africa, just as in Australia and the Western United States.

3.2.6.   Discussion, debate and development of a consensus for reform

There are, essentially, two reform paths. Either a dictatorial government can simply declare that the resource will, henceforth, be managed differently. Or there is a process of open debate, with different stakeholders expressing their concerns and views. There are important examples of water reform which have taken place via the former mode (notably Chile), and there are certainly some countries which could follow that route today. But, happily, in the vast majority of countries of the world today, changes require the consent of the people. In these circumstances a vigorous, open debate is necessary for reform to take place. Three contemporary examples are illustrative of "good practice" in this regard.

The first example is Australia. As in every other situation of scarcity, informal water markets existed for many years. Starting in 1983, the government formalized and legalized this practice, simply by removing legal obstacles to transferability. These markets have worked well in many respects, and have become a fixture of the institutional landscape, supported by both sides of the political spectrum [9]. The markets have, however, been limited because there have been restrictions on inter-sectoral and inter-state trade. In recent years the Coalition of Australian Governments (COAG, comprising the Prime Ministers of the Federal and State Governments) has looked carefully at Australia's overall economic and trade policies, and concluded that economic growth requires that water (and other factor) markets operate efficiently. Accordingly, COAG has decreed that "the major goal of water resources management is to achieve the highest and best value of the limited resource... (and that)... the move towards property right regimes is intended to link the responsibilities and accountabilities for decisions on water use with the incentives and sanctions for achieving highest value use" [41]. A central element in achieving this goal is to ensure that the geographic and sectoral reach of water markets is much broader, with inter-state, inter-sectoral trade a prime objective. What was clearly understood was that high-level commitment was a necessary but not sufficient condition -- it was vital that there be a broad-based discussion of the why and how. And in this discussion it was vital to identify different stakeholders and their interests and concerns and address these specifically and systematically in a consultation process. The work of the "Murray Darling Basin Commission's Water Market Reform Working Group" [42] is a model of how this stakeholder identification and information/consultation process should be done.

The other two examples are developing countries which are now embarking on similar reforms. When Peru initiated discussions of "following the Chilean model", there was concern that the Pinochet dictatorship was a necessary condition for such radical reforms. The (as yet incomplete) experience in Peru showed that, on the contrary, broad-based discussions with interested parties strengthened rather than weakened the support for market-based reforms in water management [2]. South Africa is also undergoing a major reform in its water management practices, with the management of water as an economic resource a major objective. The debate on this policy issue is very broad and very open and is making use of a variety of traditional (public meetings) and modern (internet-based) instruments [43].

Finally, it is imperative that this debate be conducted in the appropriate language. As noted elsewhere: "Above all else, policy-makers need to demystify the academic literature, to strip away the jargon and explain to politicians, public servants and private stakeholders the advantages and disadvantages of market solutions to natural resource allocation" [9].

RULE 2: There are many concerns about the possible effects of treating water as an economic good: Many of these fears are misconceived, and some are fanned for nefarious purposes! All concerns must be recognized and addressed openly and clearly.

3.3   Tailor the reforms to the reality of the problem
While there are clear and universal principles on what constitutes effective water management, the details of what can and should be done are enormously variable. It is obvious that context -- historical, cultural, legal, institutional, political, economic and hydrologic -- matter a great deal, and that the particulars of appropriate solutions require careful and ongoing adaptation to particular circumstances.   A couple of examples illustrate this general point.

Within the Ruhr basin, pollution charges are the major economic instrument for managing point sources.   But as non-point source pollution has become a more important and recognized problem, different and creative approaches (such as subsidies for changed land-use practices) have been invented, and used successfully [44].

Within most water markets, a mix of different instruments are used.   Short-term leases are effective for addressing the needs for higher reliability, but long-term sales are more appropriate when there are structural shifts (in the location of agriculture, or from agriculture to industry) [4].   And as water markets mature, other niches are emerging -- with options and futures contracts now coming onto the scene in the Western United States, Australia and Chile.

In urban areas, there is a similar need for inventiveness and adaptation.   Thus there is innovation in how to direct subsidies to the poor (see the Santiago example discussed earlier), and innovation [26] in how to charge for water from public standpipes.

RULE 3: Context matters (a lot!) Basic principles apply, but details require adaptation of these principles to the institutional, political, economic and hydrological context

3.4   Keep expectations reasonable
Treating water as an economic resource is desirable for a wide variety of economic, equity and environmental reasons.   And the benefits of this approach are substantial. But precisely because context matters so much, there are no ready solutions which can simply be plucked off the shelf and no "final solutions".

Reform requires a complex mixture of impatience and patience.   Impatience is required to make paradigm shifts, but then it must be realized that implementation is a very long-term process, which requires persistence, patience and adjustment.   This is well illustrated by the contemporary processes in Australia and Chile.   In both places formal water markets have existed since the early 1980s and in both places these have been relatively successful. But both countries are now in the process of major adjustments -- in Australia to extend and deepen what were initially relatively timid reforms [9]; in Chile to adjust the framework to account for some distortions (hoarding of rights by hydroelectric companies) and to embed the markets in a more effective river basin management framework [2,23].

There is a further, very clear lesson from all efforts to deal with water as an economic resource, whether it be through pricing or marketing mechanisms. In all cases where economic instruments for water management work well -- such as the Ruhr, the French River Basin Financing Agencies, the Northern Colorado Water Conservancy District, the New Mexico water markets, the Murray-Darling Basin, the Elqui Basin in Chile -- this happens in the framework of an effective overall river basin management system. The issues of governance and technical management of the resources are at least as important as, and essential complements to, the use of economic instruments [45].

Experience shows that blind advocacy of, say, water markets as "the silver bullet which will solve all problems" is not only misguided but actually counterproductive (as has been apparent in the debate over water markets in Chile). Acknowledging broader issues and keeping expectations realistic is not a recipe for inaction, but essential if there is to be effective reform.

RULE 4: There are no silver bullets: Economic instruments work well only when they are part of an effective overall water resource management system.

3.5 Nothing succeeds like success

Reforming water management systems is never easy. Early successes are vital in demonstrating that change is possible and in building a broader constituency for reform. The strategy being followed in introducing inter-state water trading in Australia is a good example.

The background to this stage of the Australian reform process is recognition of the gains to be had from broadening and deepening water markets. As has been pointed out in the Western United States "since localized markets such as those within water conservancy districts and small river basins have been active for many years, some of the greatest opportunities for increased efficiency lie in interdistrict and interstate markets"[46]. In the lower Sevier River in Utah, for example, the actual gains in efficiency were measured following a relaxation of exchange restrictions -- exchanges were allowed between four irrigation districts (rather than just within a particular district). The average real rental price in the period after the free inter-district exchanges was more than three times that in the exchange-restricted period [47].

The practice of water trading in Australia is now well established and supported by all major stakeholders, and all political parties. However, the water markets in Australia are relatively restricted (to sales within agriculture, and to sales within specific states). High-level commitment to inter-state trading has been made; the question is where to start. The Murray Darling Water Market Reform Working Group [42] is clear and strategic:

"Initial community consultation on permanent interstate trade focused on the Mallee region for three reasons:

- Mallee irrigation enterprises in each state (New South Wales, South Australia and Victoria) are similar enough for a high degree of commonality in understanding of water entitlements;
- Since Mildura (the main town in the Mallee region) acts as a regional transport and processing hub for integrated produce from each state, many traditional interstate economic rivalries are relatively subdued in the Mallee;
- The Mallee is a likely net importer of water entitlements under each of the existing separate intrastate water markets. Consequently it is possible for a variety of community interests to discuss interstate trade as a potential win-win exercise."

The message is clear -- start with the relatively easy problems, get success there and then move on with the momentum of success to address the more difficult problems.

Similar strategies are evident in other parts of the world. In Brazil, for example, the establishment of tradable water rights in being undertaken on a pilot basis on a single-state basin, in a state (Ceara) where water is scarce and where there is a recent track record of effective, modernizing government [37]. In India the establishment of the first formal water market has been proposed as a method for effecting the voluntary transfer of water from low-value agriculture to high-value urban uses in the Madras Metropolitan area. The chances of success are increased by the fact that it is a classic "win-win" situation (in which the city could buy water at a fraction of the cost of alternatives, and farmers could get paid much more than the value of the water in irrigation) [48].

RULE 5: Pick the low-hanging fruits first – nothing succeeds like success.

3.6 Don't let the best become the enemy of the good

There is no such thing as the perfect water management system. Insisting on perfection is a recipe for inaction -- the best can become the enemy of the good.

All water management systems have to face many difficult issues -- systems which manage water as an economic resource are no exception. Consider, for instance, the thorny issue of allocation of initial rights in publicly-financed irrigation systems. In virtually every country in the world, these systems have been heavily subsidized, with (as always) the privileged getting disproportionate benefit from these subsidies. One legitimate perspective on the allocation of permanent rights is thus: "these people have had a privileged position for long enough; now is the time to allocate the rights on a more equitable basis." However, those who enjoy those rights never see it like this. Their generic position is most persuasively argued by a recent purchaser of land in the irrigation district: "When I bought my land, I implicitly paid for the right to water at the historic (subsidized) price. To take this right away from me now -- either by pricing or by re-allocation -- is expropriation, which I will resist fiercely and honorably."

There is reason (and often commitment) on both sides of this debate, and thus no elegant "perfect" solution which will be just, efficient and politically acceptable to all. A practical outcome to the solution will be different in different places. In New South Wales, for example, there has been considerable controversy regarding one-off capital gains some water-entitlement holders have been able to make on the sale at market prices of water rights acquired under government-financed and subsidized projects.

After much debate, the government approach been to accept a one-off imperfection in return for continuing future improvements in efficiency [9].

A similar debate is under way in South Africa, complicated greatly by the fact that it was whites who were the beneficiaries of apartheid-era pork barrel politics. Sanctification of these past inequities is out of the question, yet a policy of confiscation, as in other parts of the world, would meet with serious political opposition. While this debate is far from over, the task is clear -- try to "start playing the right ball game" (perhaps by buying out initial rights at the opportunity cost of water in irrigation, and then auctioning off all rights). This would probably mean, as in Northern Colorado [25], that high-value urban users would buy the rights, and then lease them back to farmers until such time as they were needed by the cities.

Finally, it should be recognized that these difficulties are not created by a water market system -- it is simply that a market system brings these issues to the surface. The bottom line is that there is no perfect solution -- the challenge is to find a reasonable, practical second-best solution which will start the vital process of treating water as an economic resource.

RULE 6: Keep your eye on the ball, and don't let the best become the enemy of the good.

3.7   Ensure that legitimate third party interests are addressed and transactions costs minimized

3.7.1   Concerns with economic and employment effects in "areas of origin"

The rationale for tradable water rights has been clearly articulated by the Water Science and Technology Board of the US National Academy of Sciences: "The classic rationale for all economic activity -- gains from trade - motivates most water transfers. Buyers perceive that the cost of purchasing existing water rights and transferring water to new locations, seasons, or purposes of use is less than the cost of alternative means of securing needed supplies. Conversely, sellers -- generally farmers -- sell when the price offered is greater than the economic value of the crops or livestock they produce. The net result is that the new use generates higher economic returns than the old use" [15].

Where they have been established, water markets have performed this re-allocation function well. In the State of Victoria in Australia, for example "water is tending to move away from badly-salinized mixed-farming land to dairying and horticultural areas where the returns are higher" [49]. In California the Water Bank has meant the transfer of water from low-value fodder and foodgrain crops to high-value fruit, vegetable and nut agriculture, and to municipal uses [4]. A study of trades in the Arkansas River Valley illustrates the typical imbalance between costs and benefits well -- net income losses in the area of origin were about $53 per acre foot (4.4 cents per cubic meter), and the market value of water in the urban areas $1,000 per acre foot (80 cents per cubic meter) [50].

Precisely because of these imbalances and shifts, the economic benefits from trades is substantial -- an estimated $104 million in 1991 alone from the California Water Bank [4]; about $5 million per year in the (presently-restricted) market in New South Wales [9] and an estimated $100 million per year in additional agricultural production in the southern part of the Murray-Darling Basin in Australia once inter-state trading becomes a reality [42]. In Chile, the gains-from-trade in the (small) Limari Valley are about $2 million per year [51]. The impacts on employment follow a similar pattern -- in the case of the 1991 California Water Bank about 1600 jobs lost in the area of origin, offset by 5,400 jobs gained in the importing regions [4].

These large net benefits notwithstanding, there are losers in water trades (as there are whenever there is an adjustment in any market). Naturally these third parties would like protection, usually by way of legislation on "area-of-origin restrictions". As has become evident in the case of Australia "advocates of restrictions on transfers ... argue that unrestricted transfers impoverish less-productive regions. Yet this should, in fact, be viewed as a vindication on the transferability of water -- it shows the resource moving to higher-valued use"[9]. Whatever the conceptual shortcomings of the area-of-origin claims, they remain a fact of life [45] and a potentially serious impediment to managing water as an economic resource because of the way in which they can increase transactions costs.

3.7.2 Minimizing transactions costs

While externalities are (as the above discussion suggests) too often the "first refuge of scoundrels", the very nature of water means that real externalities are, in fact, pervasive. Accordingly, "the main administrative problem in water markets is the existence of 'third-party' effects that take the forms of changed return flows, changed groundwater levels and water quality changes, and the main issue in making markets work more efficiently is to identify and quantify these effects accurately and quickly and to get agreement on their magnitudes so that compensation and/or adjustments to the original property rights can be carried out without excessive transactions costs" [46].

Because transactions costs are so important, the choice of institutional arrangements for dealing with these is critical. Detailed empirical investigations of transactions costs in the Western US are revealing. Policy-induced transactions costs (include attorneys' fees, engineering and hydrological studies, court costs and costs paid to state agencies) range from about $50 per acre foot (4 cents per cubic meter) in New Mexico (where the State engineer adjudicates these issues) to nearly $200 per acre ft in Colorado (where they are adjudicated by the courts). The delays follow a similar pattern -- an average of 4 months in New Mexico, and 30 months in Colorado [52].

The issue of transactions costs is particularly vital where short-term markets are dominant (as with the California Water Bank), since the problem is a rapid response to drought conditions [4]. Accordingly, much attention is now given to streamlining the systems that impose superfluous restrictions, costs and delays on the transfer process, and, at the same time, to devise new ways to account for the important interests that are now left out. For example, in California, "one mechanism being considered would decouple transactions and compensation by establishing a compensation fund from which third party claims would be paid. The fund would be kept solvent by a standard charge on all inter-basin water sales. While the system is open to aggregation inefficiencies and moral hazard costs, it is hypothesized that the reduction in transaction risk will more than compensate for these costs" [4].

A related, important issue for third parties is whether trades deal with consumptive use or withdrawals. One way of reducing transactions costs is to tie transfers to consumptive use [15]. While there is no perfect solution to the problem of determining consumptive use, the problem can be addressed effectively where there is a will. In New Mexico, for instance, all rights are consumptive, with a typical description of the water right as follows: "This dedication is for 6.82 acres of irrigated land having a diversion right of 20.46 acre feet of water per annum and having a consumptive use of 1.5 acre feet per irrigated acre for a total of 10.23 acre feet per annum of consumptive use" [34]. The consumptive use is estimated by the State Engineer, who considers the type of crops cultivated, soil type, climate and other factors that affect water consumption [15]. Protests are rare. In California, where farmers are paid <u>not</u> to use water, "(aerial) photography used by the US soil conservation service to verify and monitor crop areas for subsidy payments were used to establish the past cropping pattern of an area, and to verify that irrigation was discontinued once the fallowing agreement was in force. Using this data base, the Water Bank was monitored at low cost ($4 per acre ft, or 03. cents per cubic meter)" [4].

RULE 7: Transactions cost matter (a lot) – institutions should be designed to minimize transactions costs while providing protection for legitimate third-party claims.

4   Conclusions

This paper extends a previously-developed framework [1] for thinking about management of water as an economic resource, by showing how positive and negative externalities can be taken into account. The main focus of the paper, however, is on assessing the lessons of experience which emerge from successful reforms.

The conclusions of this review are that strategy is vital in implementing a market-oriented water reform. More specifically, the following emerge as a tentative set of "rules for reformers":

- initiate change only when there is a powerful, articulated need for reform;
- have a clear strategy for involving all interested parties in the discussions of reform, and for addressing fears seriously, with effective, understandable information;
- pay attention to general principles, but be sensitive and innovative in adapting these in different institutional and environmental contexts;
- do not advertise water markets as a silver bullet or a panacea, but ensure that they are part of an effective water resource management system;
- start with the relatively easy problems to get experience and build momentum for reform;
- acknowledge that there are no perfect solutions, and don't let the best become the enemy of the good;

- pay close attention to prescribing institutional arrangements which will address legitimate third-party issues, but which will simultaneously minimize transactions costs.

Finally, it is important to acknowledge that the idea of "water as an economic good" is but one of a triad of related ideas which will increasingly shape the way in which societies are organized (and water managed) in the latter part of the twentieth century. These ideas are:

- broad-based participation by civil society in decisions (including those on water management) which were previously often treated as the province of technocrats alone;
- the hegemony of the market model of development, and the corresponding move to using market-like and market-friendly instruments for managing all elements of the economy (including water);
- the emergence of the environment as a major focus of concern.

5   References

1. Briscoe, J. (1996) Water as an economic good: The idea and what it means in practice, *Special Session R 11, International Commission on Irrigation and Drainage*, Sixteenth Session, Cairo, Egypt.
2. Briscoe, J. (1996) Water Resources Management in Chile: Lessons from a World Bank Study Tour, Washington DC.
3. Water Strategist. (1989) Evaporating water markets? New contingencies for urban water use, *Water Strategist*, Claremont CA, July 1989, Vol 3, No.2, p 11,12.
4. Howitt, R.E. (1996) Initiating Option and Spot Private Water Markets: Some examples from California, paper presented at the *IDB Seminar on Economic Instruments for Integrated Water Resources Management*, Inter American Development Bank, Washington DC., 15 pages.
5. World Bank (1995) *The World Bank and Irrigation*, A World Bank Operations Evaluation Study, Washington DC.
6. Conley, A. (1996) The future of irrigation under increased demand from competitive uses of water and greater needs for food supply. *International Commission on Irrigation and Drainage, Sixteenth Session*, Cairo, Egypt.
7. US National Research Council. (1989) *Irrigation-Induced Water Quality Problems. Washington DC.*, National Academy Press.
8. Murray Darling Basin Commission (1996) *The Murray Darling Initiative*, Canberra, Australia
9. Sturgess, G.L. and Wright, M. (1993) *Water Rights in Rural New South Wales: The Evolution of a Property Rights System*, The Centre for Independent Studies, Policy Monograph No.26, St. Leonards, Australia
10. US Dept of the Interior (1995). Quality of Water: Colorado River Basin, Progress Report No. 17, Washington DC
11. Seckler, D. (1993) *Designing Water Resources Strategies for the Twenty-First Century*, Center for Economic Policy Studies, Winrock International, Arlington,

Virginia.
12. Fredericksen, H. (1996) The water crisis in the developing world: Misconceptions about solutions. *Journal of Water Resources Planning and Management*, Vol 122. No 1.
13. US National Research Council. (1996) *The Future of Irrigation.* Washington DC., National Academy Press.
14. Gollehon, N., Aillery, M. and Quinby, W. (1995) Water use and pricing in agriculture, pp 45-56 in *Agricultural Resources and Environmental Indicators, Agricultural Handbook No. 705,* United States Department of Agriculture, Washington DC.
15. US National Research Council. (1992) *Water Transfers in the West.* Washington DC., National Academy Press.
16. International Conference on Water and the Environment (1992) *The Dublin Statement*, World Meteorological Organisation, Geneva
17. Shah, T. (1993) Groundwater Management and Irrigation Development: Political Economy and Practical Policy, Oxford University Press, New Delhi, India.
18. World Bank (1992) World Development Report: Development and the Environment, Oxford University Press, New York.
19. Porter, R.C. (1996) The Economics of Water and Waste: A case study of Jakarta, Indonesia, Avebury, Aldershot, UK
20. Ruhrverband (1992) *Tasks and Structure*, Essen, Germany
21. Kelman, J. (1996) Building a water resources management system in Brazil -- A status report, Federal University, Rio de Janeiro, Brazil.
22. Rogers, P. (1992) *Comprehensive water resources management: A concept paper*, World Bank Infrastructure and Urban Development Department, WPS 879, Washington DC.
23. Pena, H.T. (1996) Water markets in Chile: What they are, how they have worked and what needs to be done to strengthen them, Paper presented at the *Fourth Annual World Bank Conference on Environmentally Sustainable Development*, Washington DC.
24. Gazmuri Schleyer, R., and Rosegrant, M.W. (1996) Chilean water policy: The role of water rights, institutions and markets. *Water Resources Development,* Vol 12, No.1, pp. 33-48.
25. Water Strategist (1990) Trading Federal Project Water: The Colorado-Big Thompson Project, *Water Strategist*, Oct 1990, Vol 4, No 3. p 1-13.
26. Serageldin, I. (1994) Water supply, sanitation and environmental sustainability: The financing challenge. Directions in Development, World Bank, Washington DC.
27. Empresa Metropolitana de Obras Sanitarias S.A. (1995) *Memoria 95*, Santiago.
28. Brook Cowen, P. The Guinea water lease -- five years on, *Viewpoint Note No 78*, World Bank, Washington DC.
29. Wade, R. (1986) Corruption, delivery systems and property rights. *World Development,* XIV(1), pp 127-132.
30. Bauer, C. (1996) Bringing water markets down to earth: The political economy

of water rights in Chile, 1976-1995, University of California at Berkeley, California.

31. Maass, A., and Anderson, R. (1978) *And the Desert Shall Rejoice,* MIT Press, Cambridge, USA.

32. Repetto, R. (1986) Skimming the Water: Rent-seeking and the Performance of Public Irrigation Systems, World Resources Institute, Washington DC.

33. Moore, D. and Willey, Z. (1991) Water in the American West: Institutional Evolution and Environmental Restoration in the 20th Century. *University of Colorado Law Review,* 62,4, pp 775-825.

34. Anderson, T.L. and Snyder, P.S. (1995) *Water Crisis: Ending the Policy Drought* (Revised Edition, draft), Political Economy Research Center, Mozeman, Montana, US.

35. New York Times (1996) Novel use of clean-water loans brightens outlook for a river, *New York Times,* Oct 31, 1996.

36. Harvard University School of Business Administration (1997) Teaching materials for World Bank Executive Development Program, Boston Massachusetts.

37. Tendler, J. (1997) *Good Government in the Tropics.* The Johns Hopkins University Press, Baltimore, Maryland.

38. Worster, D. (1992) Rivers of Empire: Water, Aridity and Growth of the American West. Oxford University Press, New York.

39. Dumars, C. (1992) *Water marketing in the Western United States: A description and commentary.* University of New Mexico School of Law, Albuquerque, New Mexico.

40. US National Research Council. (1993) *Managing Wastewater in Coastal Urban Areas.* Washington DC., National Academy Press.

41. Coalition of Australian Governments (1995) Water Allocation and Entitlements: A National Framework for the Implementation of Property Rights in Water, Canberra.

42. Murray Darling Basin Commission (1996) *Report of the Water Market Reform Working Group,* Canberra.

43. Department of Water Affairs (1997) *Home page on the World Wide Web* (http://www.gov.za/dwaf/). Pretoria, South Africa.

44. Briscoe, J. (1995) The German water and sewerage sector: How well it works and what this means for developing countries, World Bank ,Report TWU 21, Washington DC.

45. Simpson, L. (1992) Water resources marketing: The Northern Colorado experience and its applicability to other locations, Northern Colorado Water Conservancy District, Fort Collins, Colorado.

46. Howe, C.W., Shurmeier, D.R. and Shaw, W.D. (1986) Innovative approaches to water allocation: The potential for water markets, *Water Resources Research,* Vol 22, No 4, pages 439-445, 1986.

47. Gardner, B.D. and Fullerton, H.H. (1968) Transfer restrictions and misallocation of irrigation water, *American Journal of Agricultural Economics,* 50(3):556-571.

48. World Bank (1996) Water resources supply and management issues in Chennai (formerly Madras), Internal Working Document, Washington DC.

49. Stringer, D. (1995) Water markets and trading developments in Victoria. *Water*, March 1995, Melbourne, Australia.

50. Howe, C.W., Lazo, J.K., and Weber, K.R. (1990) The economic impacts of agriculture to urban water transfers on the area of origin: A case study of the Arkansas River Valley in Colorado. *American Journal of Agricultural Economics* 72(5):1200-1204.

51. Rios Brehm, M. and J. Quiroz (1995) *The market for water rights in Chile.* World Bank Technical Paper No.285, Washington DC.

52. Colby, B.G. (1990). Transactions costs and efficiency in Western water allocation, *American Journal of Agricultural Economics*, 1184-92.

INTERSECTORAL WATER MARKETS: A SOLUTION FOR THE WATER CRISIS IN ARID AREAS?

Theory and a case study of Jordan

M. SCHIFFLER,
German Development Institute, Berlin, Germany

Abstract

Water markets are sometimes hailed as the most effective and efficient way to alleviate water scarcity in arid areas. In irrigated agriculture, which accounts for more than 80% of water use in most arid areas, the economic value of water is much lower as compared to the value in increasing municipal and industrial (M & I) uses. Traditional legal instruments - such as non-tradable licenses for water use - lock up water in these inefficient uses. It suffices to allow trade in water rights, so the argument goes, and water will gravitate to the highest value uses ("water flows upstream to money"), while agricultural water users are compensated through the market. As all water users would face prices which reflect the opportunity cost of water, unlike the often ridiculously low subsidized prices charged in many cases at present, they would face a powerful incentive to conserve water instead to waste it, which would free up high amounts of water for other uses. Because of market imperfections, an appropriate regulatory framework is necessary to allow such markets to function.[1]

The basic line of this argument shall not be disputed in this paper. The trouble, however, often is with the numerous difficulties which arise if one tries to implement the recommendations resulting from a basically sound economic analysis in practice. Therefore, I will first discuss the theoretical concept of water markets, its preconditions and economic rationale, second empirical studies on water markets will be summarized, third the results of empirical research in Jordan will be presented, and finally some conclusions will be drawn.

Keywords: Water markets, regulation, economics, Jordan, Arizona, Chile

1   What are Intersectoral Water Markets?

Intersectoral water markets are institutions which establish the rules for the

Water: Economics, Management and Demand. Edited by Melvyn Kay, Tom Franks and Laurence Smith. Published in 1997 by E & FN Spon. ISBN 0 419 21840 8.

decentralized transfer of water rights between the agricultural and the urban (municipal and industrial) sector. Such markets are yet rare in the real world. They exist only in few areas in the world, such as the Southwestern United States and Chile (see below).

Intersectoral water markets, as they are discussed here, should not be confounded with other types of water markets. First, they are different from relatively small-scale intrasectoral water markets in agriculture. These agricultural water markets are frequent in irrigation systems worldwide. In systems based on surface water they consist of shifts in irrigation turns between irrigators; compensation can be in terms of other water turns, in kind, or - in a few cases - in money. In systems mainly based on groundwater, those farmers who own wells and pumps often sell surplus water to smaller farmers who are unable to afford such an investment.[2] Neither of these water markets within agriculture are the subject of this paper. Nor are markets for the one-time-sale of water contained in tankers - as opposed to water *rights* - the subject of this paper. The sale of water through tankers - on motorized trucks or small carts pulled by animals - is a frequent sight in many developing countries, where often the majority of the urban poor depend on this source of water supply.

Although in many cases former agricultural wells are the source of water for these tankers, the amount of water transferred and the distance covered are limited, and the costs of the transfer is high.

The types of water markets discussed in this paper involve the sale of water rights - property rights over the flow of surface water or over a certain amount of groundwater allowed to be abstracted - by agricultural water users to urban water users, be they private or public urban water utilities, industries, or any other type of urban users.

2   Preconditions for Successful Intersectoral Water Markets

An essential precondition for the functioning of an intersectoral water market is the existence of fully-specified property rights to water. Property rights are fully specified, if they are well defined, exclusive, secure, indefinite, enforceable and transferable.[3] The definition of property rights to water is particularly difficult because of the characteristics of the resource. First, water is mobile and its availability varies through time. It is therefore more appropriate to define surface water rights as a share of the total flow in a river. As groundwater availability is to a lesser degree subject to seasonal fluctuations, groundwater rights may be defined in absolute terms, but fast depletion of groundwater stocks has to be avoided by setting a limit for total abstraction. Second, water is usually not fully "consumed". A part of the water returns to the natural environment with degraded quality, causing externalities to downstream users. Fully-specified property rights to water would thus necessitate the specification of the return-flow coefficient and the quality of the return flows.

Specification of property rights can be through customary law and traditional institutions, or through formal law and state institutions.[4] Customary water law in developing countries usually defines water rights on a local scale. Formal water law, in many cases only introduced relatively recently in developing countries, often enshrines the public ownership of all water resources, notwithstanding that traditional private rights to use water are de facto respected. In many cases, neither customary water law, nor modern water law are well suited to provide a basis for intersectoral water markets. Large-scale water transfers are unknown to traditional water law. Modern water law, however, is often based on centralized public decision making, which leaves little scope for private property rights to water. Only few countries have water laws which stipulate both public ownership of water and a simultaneous private right to use water subject to certain conditions set by a public body. Such a system would be the best basis for intersectoral water markets: A public body defines and enforces the limits for private water use (e.g. limits to groundwater depletion, preservation of minimum flows for aquatic ecosystems, protection of water quality), while private water users decide within these limits at what time they use how much water for which purposes, and whether they wish to lease or sell their water rights to other water users.

The initial allocation of tradable property rights to water is a sensitive issue. The allocation can be done on the basis of existing use ("grandfathering"), but such a procedure would reward inefficient water users and penalize those who have undertaken efforts to save water. It is also difficult to implement in a situation where total water use is higher than the desirable water use on environmental grounds. To avoid these problems, the initial allocation can be done on the basis of coefficients which take into account the best-practice water use for certain production processes, and all water rights can be furthermore proportionately reduced in the case of overuse.

Intersectoral water markets can only function if property rights are well-defined and enforced. Otherwise the introduction of markets can have a negative impact on the environment and on equity. In a clientelistic system, water rights are attributed to the wealthy and influential, who can violate conditions of their licenses unharmed, and who may reap windfall profits in a water market. A condition sine qua non for a successful, intersectoral water market thus is a strong and independent regulatory body, whose decisions are transparent and whose staff is accountable to water users.

3   Economic Rationale for Intersectoral Water Markets

The basic economic rationale of intersectoral water markets has been expressed by Young in two simple inequations.[5] The first inequation states that water should be transferred whenever the benefits in M & I uses exceed the development cost (DC) and transaction costs (TC) of the transfer, as well as foregone direct and indirect benefits (FDB and FIB) in agricultural uses. The benefits of M & I uses are reduced by the share of unaccounted-for water (UFW) lost in the municipal distribution network.

$$B * (1 - UFW) \quad > \quad DC + TC + FDB + FIB \tag{1}$$

The second inequation states that the costs of a transfer should be lower than the cost of an alternative project for supply management (AC), which usually is a long-distance pipeline combined with either a dam, a wellfield or a seawater desalination plant.

$$DC + TC + FDB + FIB \quad < \quad AC \tag{2}$$

Some items of these equations are fairly simple to quantify. Development costs (DC) include transport and pre-treatment costs. The investment, operation and maintenance costs per unit of water over the lifetime of the investment can be calculated through standard dynamic cost discounting. The same applies for the costs of an alternative water supply (AC).

The economic value of water itself at the source (its opportunity cost; used to quantify benefits and foregone direct benefits) is more difficult to determine in the absence of market prices for water. There are, however, a number of techniques to establish shadow prices for water.[6] Concerning water in productive uses, such as irrigated agriculture and industry, its value can be assessed through two methods. The *residual imputation method* subtracts the value of all other production factors (capital, labor and land) from the value added in an economic activity, to arrive at the "water profit" or "water rent". If only the value added per unit of water ("water productivity") is calculated, which can be justified for comparative purposes, the economic value of water in absolute terms is overstated. The *land price differential method* compares the value of otherwise similar plots of land with and without water resources. The price difference per unit of water yields the shadow price of water.

Agricultural and industrial policies in the form of protectionism (import tariffs or quantitative restrictions), of subsidies or of distorted exchange rates can push the financial value of water (the value from the viewpoint of the farmer) above or below its economic value (from the viewpoint of the country).[7] The appropriate measure to calculate foregone direct benefits in agriculture or the direct benefits in industrial uses is the economic rather than the financial value of water.

For municipal water uses, water tariffs are a poor indicator of the economic value of water, as they are usually heavily subsidized in developing countries. In other cases, they may be distorted upwards through the monopoly power of a water utility. The economic value of water can, however, be assessed through observing the prices paid to water vendors or through surveys using the contingent valuation method.

Two items remain however in the two inequations quoted above which are difficult to quantify: transaction costs and foregone indirect benefits. Transaction costs, which are a central concept of neoinstitutional economics, are the costs incurred while searching for information, bargaining, making contracts, monitoring contractual partners, enforcing contracts and protecting property rights against third-party encroachment.[8] The level of transaction costs strongly depends on the degree of trust between market participants: when trust is low, transactions costs tend to be high. Government regulation can help in reducing transaction costs by establishing and enforcing a clear legal framework. Empirical studies suggest that transaction costs are far from negligible. The second item, which is difficult to quantify, is foregone indirect benefits. In intersectoral water markets these are the benefits associated with maintaining irrigated agriculture: prevention of migration to the cities, preservation of rural livelihoods, a higher degree of food self-sufficiency, forward and backward linkages to suppliers of agricultural inputs and to the food-processing industry, and the prevention of desertification. How important these benefits actually are differs from case to case. It often depends on the values in a society: For example, the preservation of rural livelihoods may be less valuable to urban inhabitants in the United States than to the urban inhabitants in a developing country, who migrated only recently from rural areas and maintain close familiar and emotional links to it.

The two inequations mentioned above may thus be transformed so that all the items which can be quantified in monetary terms are found on the left side:

$$B(1 - UFW) - DC - FDB > TC + FIB \qquad (3)$$

$$AC - DC - FDB > TC + FIB \qquad (4)$$

The basic precondition for a successful intersectoral water transfer is that the left hand of the two equations yields a positive value. Otherwise, urban water demand management or an alternative supply option are preferable to a water transfer. If this precondition is fulfilled, policy-makers have to compare the resulting net benefits with their subjective estimate of the transaction costs and the foregone indirect benefits. If policy-makers believe the net benefits to be higher than the costs on the right hand of the equation, further steps toward an intersectoral water transfer should be undertaken.

4   Experience with Intersectoral Water Markets in Chile and the USA

Intersectoral water markets have been successfully introduced in Chile (mainly for surface water) and in the Southwest of the United States (mainly for groundwater).

Chile's water resources consist of a large number of relatively short rivers running from the Andean mountains to the Pacific ocean. In the semi-arid middle-northern part of Chile, irrigated agriculture is the main water user and water was mainly used for low-value crops. When a system of tradable water rights was introduced in 1981, both water transfers among farmers and between farmers and urban water users occurred.[9] The transfers were, however, geographically limited inside the relatively small river basins, so that almost no infrastructure for water transport was necessary. Transaction costs were low, because farmers were grouped in well-functioning water user associations even before the introduction of the water market. The quantities traded were limited. Water prices remained low (under 1 US Cent/m³), reflecting the low marginal value of these small amounts of water in agriculture. The water purchased at low prices was, however, sufficient to prevent the building of expensive new dams for urban water supply. Farmers used the receipts from the sale of their water rights to invest in more efficient irrigation techniques. At the same time, many farmers shifted to high value crops for export. Farmer thus managed to increase their income while using less water. Under the favorable conditions found in Chile, intersectoral water markets apparently were a "win-win"-option for all participants.

Intersectoral water markets exist in many Southwestern States of the USA.[10] A particular interesting example is the case of Arizona. The M & I water demand in Arizona is rising rapidly because of population growth, while agriculture is by far the largest water user. The first documented intersectoral water transfer occurred in 1968, when the city of Tucson purchased agricultural groundwater rights in its urban surroundings. The quantities traded are relatively large. Until today Phoenix, the largest city of Arizona, has bought over 40,000 hectares of farmland, which corresponds to 10% of the irrigated area of the state, with water rights amounting to more than 370 MCM/year. Unit water prices were between 3 and 12 US Cent/m³, but these prices were still below the cost of the next supply option, the transfer of water from the Colorado River over a distance of more than 500 km. Rural areas of origin were located less than 80 km away from the purchasing cities. In some cases, existing pipelines were used, and in others new pipelines were built to transport the water. Farmers responded by planting less low-value crops, which yielded a lower value per unit of water than the price offered by urban water utilities. The problem of groundwater depletion has not been solved in Arizona, and targets for the reduction of groundwater abstraction have not been met.

The experience of both Arizona and Chile shows that intersectoral water markets can be beneficial to both farmers and urban water users. In both cases, however, favorable conditions eased the functioning of these transfers. Among these conditions are the use of large amounts of water for low-value crops, relatively low transport costs, high urban water tariffs, the existence of water user associations (in the case of Chile), the institutional separation of operating and regulatory functions, and a generally market-friendly attitude in society. These conditions are not often to be found together in other cases.

5   Prospects for and Limits to Intersectoral Water Markets in Jordan[11]

Jordan is one of the countries with the scarcest water resources in the world. Its four million inhabitants have about 200 m³/capita and year of renewable water at their disposal, far less than an arid country needs to satisfy all its water needs, including self-sufficient food production from irrigated agriculture. The country uses about 70% of its water resources for irrigation, almost half of which takes place in the Highland based on groundwater, and the remainder is located in the Jordan Valley based on surface water. Jordan has, however, still enough water to satisfy its M & I water needs alone, even if the population doubles. Already by today 78% of Jordan's population lives in urban areas, and agriculture's contribution to Gross Domestic Product is only 7%.

Because of population growth, many arid and semi-arid areas in the Middle East and in other regions around the world, which still have sufficient water resources per capita today, will face increasing water scarcity such as is currently the case in Jordan. Thus, the water management experience of Jordan today may give insights into the challenges which many other countries will face tomorrow.

The Government of Jordan adopted the Water Policy Framework in 1994, which fixes the general principles of water resources management. The policy framework contains, among other things, a commitment to the public ownership of water, and it sets allocation policies as well as priorities for planning and investment. Public ownership of water, nevertheless, is said to be consistent with rights to use water by private entities, which may even be transferred under certain conditions. The water allocation policy gives first priority to "reasonable needs" for domestic consumption, second priority to industry and tourism, and only third priority to agriculture. In the field of planning and investment, priority is given to projects to cover M & I demand.

Additional water supplies are said to be needed, "even if full advantage is taken of conservation and reuse programs."

The economic value of water in industrial uses is 60 times higher, and in municipal uses it is about 6 times higher than in irrigated agriculture (based on water productivity in agriculture and industry, as well as on prices paid to water vendors for municipal uses). This difference by itself, however, does not yet justify a water transfer. Water losses are in the order of 50%, municipal use (rather than industrial use) is the marginal urban water use and transport costs arise. Therefore the benefits available to compensate farmers are limited, although the left-hand side of equation 3 is positive. Nevertheless, intersectoral water transfers could result in the transfer of about 40 MCM/year from agricultural to M & I uses in the Highlands at a bulk cost (DC + FDB) of about 0.36 JD/m³ (0.54 US $/m³) on a largely renewable basis. This cost is lower than the cost of expanding water supply (AC) either through a pipeline from the Disi aquifer in the South of Jordan or from the Yarmouk River in the North of Jordan (the left-hand side of equation 4 is positive). Compared to a bulk cost of about 0.65 JD/m³ (0.98 US $/m³) for the exploitation of fossil groundwater in the Disi aquifer, the application of such a policy to groundwater in the Highlands could save about JD 12 million per year (US $ 18 million). Additional savings are possible, if transfers of surface water from irrigation in the Jordan Valley are considered. Here the quantities are larger, but the transport costs are higher because of the high difference in altitude between the Valley and Amman.

Policy-makers have to assess whether the net benefits of intersectoral water transfers, in their view, exceed transaction costs and foregone indirect benefits, which are difficult to quantify in monetary terms (equation 3). Any reduction of agricultural water use is likely to face opposition by the farming lobby, which emphasizes the indirect benefits of agriculture to the country. The rural population of Jordan is a backbone of support to the King and his government. This has resulted in political and economic privileges for rural areas, including relatively open access in the past to licenses for drilling wells for irrigation in the Highlands of Jordan. The analysis suggests, however, that employment losses in agriculture in the Highland are only in the order of 3,500 jobs, half of which are immigrants from Egypt and other countries. Employment per unit of water is more than 13 times higher in industry compared to agriculture.

Although intersectoral reallocation of water from agriculture to M & I uses apparently is beneficial to Jordan, policy-makers have expressed strong reservations regarding market decisions on intersectoral water transfers.[12] Furthermore, reallocation will not free up large amounts of water for M & I uses in a short time. Water markets thus are no panacea for the Jordanian water crisis, and they need to be complemented by an expansion in water supply.

Great hopes have been attached to a solution of the water crisis through regional cooperation coupled with supply management.[13] Jordan has obtained water rights in the October 1994 Peace Treaty with Israel, but the amount of water immediately available as a result of the Treaty is limited (30 MCM/year), and the rights of both Syria and the Palestinian Authority in the Jordan waters have not been mentioned in the treaty. Regional supply options remain politically risky in a yet uncompleted peace process, and past experience suggests that cooperation on water issues is difficult to achieve, as long as basic political problems are not settled.[14] Some of the regional supply options, such as a canal either from the Mediterranean or from the Red Sea to the Dead Sea coupled with seawater desalination, are also much more costly than intersectoral water allocation and other options for supply and demand management.

In sum, intersectoral water markets can be beneficial to Jordan, but the analysis suggests that any enthusiastic support would be misplaced. If, on the other hand, intersectoral water transfers are ruled out because of assumed high foregone indirect benefits and transaction costs, policy-makers have to face the high cash costs of the expensive supply management options.

6   References

1   The arguments in favor of water markets are summarized by ROSEGRANT, M. / H. BINSWANGER: Markets in Tradable Water Rights: Potential for Efficiency Gains in Developing Country Irrigation, World Development 22, No. 11, 1994; THOBANI, M.: Tradable Property Rights to Water. How to improve water use and resolve water conflicts, in: Private Sector, 1995, pp. 9-12; SIMPSON, L.D.: Conditions for Successful Water Marketing, in: LE MOIGNE et al. (eds.): Water Policy and Water Markets, World Bank Technical Paper 249, Washington, D.C. 1994.

2   SHAH, T.: Groundwater markets and irrigation development, Bombay 1993.

3   PANAYOTOU, TH.: Green Markets, San Francisco 1993, pp. 35-37.

4   cf. CAPONERA, D.A.: Principles of water law and administration, Rotterdam, Brookfield 1992.

5   YOUNG, R.A.: Why Are There So Few Transactions among Water Users?, in: American Journal of Agricultural Economics, December 1986, pp. 1143-1151.

6   YOUNG, R.A.: Measuring Economic Benefits for Water Investments and Policies, World Bank Technical Paper No. 338, Washington D.C. 1996; YOUNG, R.A. / R.H. HAVEMAN: Economics of Water Resources, in: A.V. KNEESE / J.L. SWEENEY (eds), Handbook of Natural Resource and Energy Economics, Vol. II, Amsterdam 1985.

7   cf. GITTINGER, J.P.: Economic Analysis of Agricultural Projects, Baltimore, London 1982, pp. 3 - 8.

8   EGGERTSON, T.: Economic behavior and institutions, Cambridge 1990, p. 15. Other definitions include transport costs as well, but they are excluded in our definition.

9   cf. ROSEGRANT, M.W. / R. GAZMURI SCHLEYER: Tradable Water Rights. Experiences in reforming water allocation policy, USAID, Washington D.C. 1994; HEARNE, R.R. / K.W. EASTER: Water Allocation and Water Markets: An Analysis of Gains-From-Trade in Chile, Washington D.C. 1995; WERNER, F.: Die Bedeutung von Wassermärkten für ein nachhaltiges Wassermanagement in ariden Regionen. Eine geographische Analyse am Beispiel der Wassermärkte

Chiles unter besonderer Berücksichtigung des Elqui-Tals, Heidelberg 1996 (unpublished manuscript).

10   SALIBA, B.C. / D.B. BUSH: Water markets in theory and practice, Boulder 1987; REISNER, M. / S. BATES: Overtapped Oasis. Reform or Revolution for Western Water, Washington D.C. 1990.

11   this section is based on SCHIFFLER, M.: The Economics of Groundwater Management in Arid Areas. Theory, International Experience and a Case Study of Jordan, German Development Institute, Frank Cass Publishers, London (forthcoming).

12   cf. SHATANAWI, M.R. / O. AL-JAYOUSI / G.A. ORABI: Water Rights in Jordan, in: M.W. ROSEGRANT / R. GAZMURI SCHLEYER (eds.), op. cit. 1994.

13   cf. WOLF, A.T.: Hydropolitics along the Jordan River: Scarce water and its impact on the Arab-Israeli Conflict, Tokyo 1995, pp. 162-169.

14   cf. LOWI, M.R.: Water and power. The politics of a scarce resource in the Jordan River Basin, Cambridge 1993, pp. 192-204.

ANALYSING THE LINK BETWEEN IRRIGATION WATER SUPPLY AND AGRICULTURAL PRODUCTION: PAKISTAN

Economic issues in irrigation systems in Pakistan

P. STROSSER
International Irrigation Management Institute, Lahore, Pakistan
T. RIEU
Irrigation Division, Cemagref, Montpellier, France

Abstract

With increasing water scarcity, a move from supply to demand oriented interventions is often proposed for irrigation sector policies, with predominant role given to prices and markets. This shift increases the need to better understand the economic dimension of decisions taken by water users. The present paper uses examples from one irrigation system in Pakistan to investigate selected issues related to the economics of irrigation. First, the paper examines the relationship between canal water supply and farm decisions regarding crop choice and groundwater extraction, using economic models developed for individual farms. Aggregated at the level of a tertiary unit, these economic models support the analysis of the functioning of existing (tubewell) water markets. Issues related to market perfection, monopoly and value of water are discussed. The examples are used as starting point to emphasise the complexity of economic issues as a result of the high heterogeneity and spatial variability in farming systems and water constraints within an irrigation system, along with the existing structure of the agrarian society in Pakistan.
Keywords: economics, groundwater market, irrigation, Pakistan

1   Introduction

The discrepancy between stagnant agricultural production and increasing demand for agricultural products has been well documented and is a major issue in the agenda of policy makers and funding agencies in Pakistan [1]. With the importance of irrigated agriculture that covers 16 million hectares and current low agricultural productivity per unit of land (yields of the major crops in irrigation systems in Pakistan are among

Water: Economics, Management and Demand. Edited by Melvyn Kay, Tom Franks and Laurence Smith. Published in 1997 by E & FN Spon. ISBN 0 419 21840 8.

the lowest in the world), most of the attention is given to potential improvements in agricultural productivity coming from the irrigation sector.

Low crop yields are often related to poor canal water supply performance in terms of adequacy, variability and equity. Also, major problems of waterlogging, salinity and sodicity are encountered and constrain agricultural production [1]. To improve water supply performance and mitigate environmental problems, various interventions have been proposed for irrigation systems in Pakistan. Those include: technical interventions such as lining of secondary canals or distributaries to reduce seepage losses and increase canal water supplies, building new storage capacity to transfer water from the *kharif* (summer) season to the *rabi* (winter) season, or building drainage systems to reduce waterlogging and salinity; institutional changes promoting decentralisation of the provincial irrigation departments and increasing involvement of water users into the management of the irrigation system to improve the quality of irrigation services and reduce the budgetary deficit of the irrigation sector; or, encouraging the development of water markets to improve allocative efficiency and increase agricultural production and productivity.

Surprisingly, very little is known about the potential impact of these interventions on crop yields and agricultural production. Several reasons may explain this situation: absence of methodology to estimate potential impact of changes, lack of appropriate information or difficult access to this information, lack of interest from actors involved in the irrigation sector as a result of the political process followed for policy decisions. To better understand the magnitude of potential impacts would require detailed understanding of the relationship between water and agricultural production and a good analysis of decisions taken by water users to achieve objectives under given constraints.

With increasing water scarcity and the recognition of the potential of demand-based interventions and economic instruments to manage the irrigation sector in Pakistan, the need to better understand the relation between water and agricultural production and the economics of irrigation is increasingly required to support policy decisions. This would enhance the evaluation of potential benefits (and opportunity costs) of intervention in the irrigation sector, provide estimates of the value of water as indicator of productivity of water (of use for investigating the potential for market mechanisms), or strengthen the development of appropriate water pricing mechanisms [2].

The present paper investigates some of the issues related to the economics of irrigation in Pakistan, with emphasis on micro-level analysis of the relationship between irrigation water and agricultural production.

2   Linking water supply and agricultural production

Surface water and groundwater resources are used conjunctively by farmers in large parts of the Indus Basin Irrigation System in Pakistan. Surface water is distributed through a huge network of primary canals and distributaries to tertiary units or watercourses. Operation and maintenance of primary canals and distributaries is the task of the provincial irrigation departments that aim at distributing constant and equitable supplies to distributaries and watercourses. Below the watercourse head, water is shared between individual farmers following a weekly schedule of water turns or *warabandi*. As a response to inadequate and unreliable canal water supplies, farmers have installed a large number of private tubewells within the watercourse command areas (more than 400,000 in the Indus Basin Irrigation System). Farmers have increased flexibility in irrigation water supplies through the development of canal and tubewell water markets within watercourse command areas [3]. In some areas, public tubewells installed and operated by the irrigation departments complement surface water supplies.

Linking water supply and agricultural production is a complex research issue, as it integrates different dimensions of water supply and several decisions taken by farmers at different periods of time (planning of farming activities, water scheduling, water use, etc). Farm level decisions are central to the analysis of the link between water supply and agricultural production. These decisions use field level information on the soil-water-plant relationship, and watercourse level information regarding potential access to (or sharing of) water resources. Irrigation management activities are usually individual interventions but often include interactions with other water users. Decisions taken by farmers include:

Crop choice based on anticipated water supply at the farm gate: decisions consider canal water supply quantity and variability, potential access to (and price of) groundwater resources (from farmer's own tubewell or through participation in groundwater markets), problems of sustainability related to salinisation and use of poor quality groundwater. Farm objectives and other constraints (labour, land, credit, etc,) are considered at this planning stage.

Adjusting canal water supplies to farm needs: canal water supplied to the farm may not be sufficient (the majority of the cases) or may be in excess. Farmers' decisions relate to tubewell operation, participation in water markets, or direct (and illegal) interference in the management of the irrigation system to increase canal water supplies.

Allocation of irrigation water supplies to individual fields: farmers base their allocation of canal water supplies to different fields (complemented or not with tubewell water) on past irrigation events, crop types and physical characteristics of the fields, expected production and farm objectives and constraints.

Field level application: Crop status and physical characteristics of soil, along with past irrigation of the field and expected output at any given point of time are considered to define quantities of irrigation water that are applied to specific fields.

Economic related research issues include analysing water use efficiency (comparing opportunity and marginal costs) whether at the field, farm or other scale of the irrigation system, estimating the value of water and comparing values between users to investigate allocative efficiency, analysing factors that constrain the functioning of water markets and limit their efficiency. In the context of Pakistan, the understanding of farm level decisions is to be considered in the framework of *peasant* economics [4]. This implies considering objectives other than maximisation of output (risk-minimisation or auto-consumption), recognising that input and output markets are incomplete and imperfect markets, looking at inter-household relationships, and recognising the social dimension of farmers' decisions. Also, the heterogeneity of situations within an irrigation system is an important issues that is to be investigated.

The following sections present two examples investigating economic issues in irrigation systems in Pakistan. The first example concentrates on the relationship between canal water supplies and agricultural production to assess the impact of changes in canal water supply performance on cropping pattern and gross income at the farm level. The second example focuses on the functioning of tubewell water markets and on the seller-buyer relationships within these localised markets. The information used in the two examples has been collected in the Chishtian Sub-division, an irrigation system located in the cotton-wheat agro-ecological zone of the Punjab. Primary information on farm characteristics, agricultural production and irrigation water supply has been collected for the *kharif* 1992 and *rabi* 1992-93 seasons in the context of several research studies conducted by the International Irrigation Management Institute (IIMI) in this irrigation system.

3   Impact of canal water supply performance on agricultural production

3.1   Characterising canal water supply performance

Canal water supply performance has been analysed by several authors in the Chishtian Sub-division [5], [6]. The main emphasis of this analysis has been on the gap between actual and official canal water supplies and the identification of factors that explain this gap. Various performance indicators have been computed at different levels of the irrigation system (farm, watercourse, distributary) and for different time periods (week, month, season), using daily discharge information collected by IIMI field staff. Three important performance related issues have been stressed by the analysis:

The *significant gap between official and actual discharges* and volumes delivered. Poor implementation of operational rules, physical constraints and users' interference into the management of the irrigation system explain this gap for distributaries. Inadequate supplies to watercourses are caused by low supplies at the distributary head, inadequate maintenance along the distributary and changes in outlet dimensions of upstream watercourses.

The high spatial variability in this gap between farms, watercourses and/or distributaries. This variability leads to a *high inequitability in canal water supplies*, with some farmers, watercourses and distributaries receiving more than their due share while others receive significantly lower volumes as compared to design. Local preferences given by gate keepers to distributaries (for technical or non-technical reasons), differences in socio-political power of water users, and position along the irrigation system (i.e. head versus tail) are factors explaining the high inequity observed in the Chishtian Sub-division.

The *high temporal variability and unreliability* of canal water supplies at all levels of the irrigation system. In fact, the variability in discharges received at the head of the Chishtian Sub-division is amplified throughout the irrigation system as a result of gate keepers' operations, lack of appropriate information system and users' interference.

3.2   Impact of water supply adequacy and dependability on cropping pattern

Analysis of the diversity of farming systems is performed for sample areas of the Chishtian Sub-division to identify farm groups with specific strategies and constraints [7]. Linear programming models are developed for farms representative of each group to analyse the link between canal water supply and agricultural production (cropping pattern and gross income). The models maximise farm gross income under water, land and credit constraints and include home-consumption objectives for wheat (rabi season). Farmer's choices that are considered in the models include selecting the area under different crops (cotton, wheat, sugarcane, rice, fodder), selecting between different quantities of irrigation water applied to wheat and cotton, using groundwater (at a given cost or price per cubic meter) for cultivation through own tubewell operation or purchase of tubewell water, and selling groundwater (for tubewell owners only) at a given price.

Risk related to crop yield and gross margin has been specified for the two main crops, wheat and cotton (kharif season) using the Minimisation Of Total Absolute Deviation (MOTAD) model [8]. Canal water supply performance is included in the models using a generalised form of the MOTAD applied to water constraints [9], using both the *total monthly volume of canal water received at the farm* (i.e. sum of daily volumes computed for each month), and the *monthly temporal variability of daily canal water supplies* (expressed by the standard deviation of daily volumes computed for each month).

To illustrate the impact of changes in canal water supplies on agricultural production, a linear programming model built for a relatively large farm (10 hectares of cultivated area) is used. The farm has good access to credit facilities and to input and output markets, and has its own private tubewell. The farm is located in a watercourse with high canal water supply (i.e. + 20 percent on average per year as compared to design), but also a high temporal variability of this supply (average monthly coefficient of variation of daily volumes equal to 0.3). Several scenarios of canal water supply (similar changes being applied to each monthly canal water supply) have been tested using this economic model:

Scenario 1 (reference scenario): actual volume and temporal variability of daily volumes;

Scenario 2:   actual volume but no temporal variability (i.e. monthly standard deviation of daily volumes equal to 0);

Scenario 3: design volume and actual temporal variability (i.e. using the design monthly volumes and the actual monthly coefficient of variation of daily volumes to compute monthly standard deviations of daily volumes);

Scenario 4: design volume and no temporal variability.

Table 1 presents some of the output of the economic model obtained for each scenario.

Table 1. Impact of changes in canal water supply on gross income and tubewell water use (1 Rs equivalent to 0.025 US cents)

Indicator	Scenario 1	Scenario 2	Scenario 3	Scenario 4
Gross income per unit area (in Rs/ha)	3500	5700	3200	5000
Gross income per unit of water (in Rs/m^3)	400	500	410	540
Tubewell water use (in m^3)	57,000	45,000	57,000	47,000
Tubewell water in total irrigation water use (in %)	65	40	70	50

Results presented in Table 1 shows that canal water supplies effectively influence farm decisions and agricultural production. Table 1 stresses the large impact of a reduction in canal water variability (from Scenario 1 to Scenario 2, for example) on farm gross income per unit area. The reduction in variability is accompanied by a reduction in tubewell water use of 20%. This stresses the dual role of private tubewells in a context of highly variable canal water supplies: (i) to increase the quantity of irrigation water available at the farm; (ii) to reduce risk (and uncertainty) related to canal water supplies.

The impact of canal water supply dependability on gross income opens a way for restoring equity in canal water supplies. Reallocating water from high supply to low supply areas may be more easily accepted by farmers that would lose canal water supplies *if* accompanied by a significant reduction in canal water supply variability that would allow them to keep identical gross income. As a result, tubewell water use and pressure on groundwater resources will decrease, leading to less problems related to the use of poor quality groundwater and increasing tubewell water supplies potentially available for tubewell water purchasers.

The illustrative objective of this example is to be stressed again. The variability in water constraints [6] and farming systems [7] within the Chishtian Sub-division clearly emphasises the need to repeat simulations with economic models built for other farm types and with different canal water supplies.

4   Functioning of tubewell water markets

4.1   Description of tubewell water markets
Water markets have developed spontaneously in irrigation systems in Pakistan. Water markets are localised water markets, involving most of the time a limited number of water users and always developed within the boundaries of the watercourse command area. Water transactions include exchange of partial and full canal water turns, sale and purchase of canal water turns, sale and purchase of tubewell water, or exchange of tubewell water for canal water. In areas with good groundwater quality, tubewell water

transactions are the most important activity in terms of volumes of water transacted. Exchange of canal water turn are predominant in irrigation systems with poor groundwater quality and limited number of private tubewells [3].

In the Chishtian Sub-division, more than 70 percent of farmers participate in tubewell water purchases while tubewell owners represent 40 percent of the farm population. High variability in tubewell water sales in terms of volumes or percentage of tubewell water sold are observed from watercourse to watercourse. Differences in water supply conditions and spatial variability in farming systems within the irrigation system explain part of this difference in tubewell water market development. Most of the transactions are linked to cash payments paid whether at the time of use or at the end of the season. Tubewell water prices vary from tubewell to tubewell as a result of different operation and maintenance costs, but also between purchasers of a given tubewell. Tenancy contracts and family links are the main reasons explaining these price differences.

Research undertaken so far on tubewell water markets in Pakistan have concentrated on the impact of these markets on agricultural production and on the characteristics of tubewell water sellers and purchasers [3]. The link between sellers and purchasers, however, has rarely been investigated. Also, limited insight has been gained on the trade-off faced by tubewell owners between using tubewell water on their own farm and selling water to other farmers. The following section analyses these issues based on a case study developed for one watercourse (labeled here FD 62-R) of the Chishtian Sub-division.

4.2   Developing tubewell water supply and demand curves

A watercourse-based economic model is built for the watercourse FD 62-R, using linear programming models developed for each farm group and aggregated at the watercourse level according to the relative importance of each farm group in this watercourse. Individual linear programming models are run for different tubewell water (purchase or sale) prices to develop demand and supply curves for tubewell water for each farm type (i.e. relationships between total yearly quantity of tubewell water purchased and sold for different tubewell water prices, respectively). Individual demand and supply curves are then aggregated at the level of the whole watercourse command area where demand and supply of tubewell water are confronted.

FD 62-R watercourse covers a total cultivable command area of 130 hectares and has 4 major types of farms within its command area [10]. Three farm types with different strategies and constraints, i.e. (i) low input use and cash constraints, (ii) owner-cum-tenant with cotton-wheat specialisation, and, (iii) very small holdings with highly intensive input use, are non-tubewell owners relying upon groundwater resources through tubewell water markets. The forth group of larger and more prosperous farms is composed of tubewell owners (the seller group). The analysis of individual demand curves shows a large heterogeneity of elasticity of demand with respect to price along these curves (and for a realistic range of tubewell water prices), with significant differences being recorded between farm groups: arc elasticity of demand ranges from - 0.4 to -3.6 at current average tubewell water price (0.4 Rs/m^3). Figure 1 presents aggregated demand and supply curves for the watercourse FD 62-R. The computed equilibrium price between both curves is very close to the actual average tubewell water price for the watercourse.

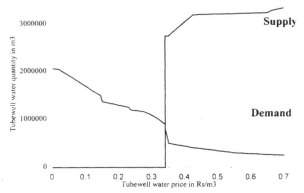

Fig 1. Demand and supply curve of tubewell water in the FD 62-R watercourse

The shape of the supply curve in the range 0.35-0.7 Rs/m^3 expresses the competition between using tubewell water for agricultural production or selling it on the market. As long as tubewell water prices are lower than tubewell operation and maintenance costs, tubewell owners do not sell water. As soon as prices are set higher than these costs, tubewell owners have an interest to sell any water that is not used on their own farm. As prices further increase, the competition between own use and water sale takes place with a progressive reduction in own use. At the extreme, tubewell owners become solely tubewell water sellers.

4.3   Analysing supply and demand curves: strategy of tubewell water sellers, interlocked markets and the value of tubewell water

The analysis of the individual demand curves show that tubewell water quantity estimated from the demand curve for the group of farms that include sharecroppers is lower than actual tubewell water use at current tubewell water price. Several reasons may explain this difference:

Tubewell water may not be available whenever required by tubewell water purchasers (For example, in periods of higher water demand with tubewell owners giving priority to irrigating their own fields against selling tubewell water)

The price of tubewell water may not be the value of water effectively faced and considered by purchasers in their decisions. Transaction costs for the purchaser (including seepage losses taking place between the tubewell and the place of use) may explain this difference. Also, payments in kind that are not recorded in tubewell water price but that are part of the sharecropping arrangement linking tubewell water seller and purchaser are to be considered.

The localised nature of tubewell water markets, with one tubewell owner selling water to a few purchasers, offers ideal conditions for monopolistic conditions. In fact, as farm types do not have clear spatial distribution within the watercourse command area, each tubewell water seller is expected to adjust his selling strategy according to his own operation and maintenance costs, potential competing water sellers (a rare situation) and the price elasticity of demand of potential purchasers. According to neoclassical economic theory, the elasticities of demand obtained for the different farm groups in FD 62-R would induce sellers to sell low quantities of tubewell water for high prices to maximise their profit. In reality, however, expected high prices that would be imposed by water sellers do not occur as tubewell water prices paid by purchasers are very close (and often equal) to operation and maintenance costs of tubewells [11]. Several reasons may explain the absence of high tubewell water prices [10]:

Social pressure or social cohesion within the watercourse command area;

The use of interlocked markets for setting high prices: although tubewell water price remains low, purchasers may have to provide other services (for example, labour) to tubewell water sellers, whether explicitly specified in a sharecropping arrangement (see above) or informally agreed between the seller and the purchaser for each transaction;

The quantity dimension of tubewell water only has been taken into account in the development of the economic models. However, the quality of irrigation water is an important dimension that influences water-users' decisions. Thus, poor groundwater quality may explain the relatively low tubewell water prices.

In FD 62-R, the existing equilibrium between tubewell water demand and supply does not lead to an apparent competition (although this situation may appear temporally during periods with high water demand) for tubewell water sellers between use of tubewell water on their farm and tubewell water sales. This situation would occur if the number of tubewell owners (potential sellers) would be reduced as compared to the number of potential purchasers. Changes in the composition of potential purchasers would lead to shifts in the demand curve. And changes in canal water supply at the watercourse head would produce different supply and demand curves. The economic models developed offer the possibility to estimate different supply and demand curves based on different canal water supplies and relative importance of different farm types. This would facilitate the analysis of factors that explain differences in the importance of tubewell water markets from one watercourse to the other.

5   Conclusion

The examples presented in this paper have provided insight into several important issues related to the relationship between water and agricultural production and economics of irrigation in Pakistan :

The use of micro-economic (linear programming) models has led to initial quantification of the impact of changes in canal water supply (in terms of volumes supplied and temporal variability) on cropping pattern and farm gross income;

The output of model simulations illustrates the role of private tubewells, whether as an insurance against canal water supply variability or to increase adequacy of total irrigation water supplies;

Estimates of price elasticity of tubewell water demand stresses the variability of these elasticities along the tubewell water demand curve and between different types of farm;

The analysis of tubewell water prices leads to the postulate that prices do not represent the value of tubewell water (at least for the farm groups considered).

Transaction costs for tubewell water purchasers, or other costs "hidden" in tenancy contracts or in-kind arrangements must be added to estimate the value of tubewell water.

Groundwater quality is also to be considered in the analysis of the value of tubewell water. Several dimensions such as quantity, quality, variability or reliability characterise tubewell water as an economic good and must be considered simultaneously.

As a follow-up, similar analysis will be undertaken looking at both canal water and tubewell water, and for the variability of situations existing within the Chishtian Sub-division in terms of farming systems, canal water supply and tubewell water prices, whether within watercourse command areas or between watercourses.

6   References

1.   World Bank. (1994) *Pakistan - Irrigation and Drainage Issues and Options*, The World Bank, Washington D.C.

2.   Winpenny, J.T. (1996) The Value of Water Valuation, in *Water Policy: Allocation and Management in Practice* (ed. P. Howsam and R. Carter), E & FN Spon, London, pp. 197-204.

3.   Strosser, P. and Meinzen-Dick R. (1994) Groundwater Markets in Pakistan: an Analysis of Selected Issues, in *Selling Water: Conceptual and Policy Debates over Groundwater Markets in India* (ed. M. Moench), Viksat, Pacific Institute, Natural Heritage Institute, pp.73-91.

4.   Ellis, F. (1988) *Peasant Economics - Farm Households and Agrarian Development*, Cambridge University Press, Cambridge.

5.   Kuper, M. and Kijne, J.W. (1993) Irrigation Management for the FD Branch Command Area, South East Punjab, Pakistan, in *Advancements in IIMI's Research - 1992*, International Irrigation Management Institute, Colombo.

6.   Kuper, M. and Strosser, P. (1996) *Current Water Management in Pakistan: Lessons from Current Practices for Proposed Policy and Management Changes*, draft repor//t, International Irrigation Management Institute, Lahore.

7.   Rinaudo, J.D. (1994) *Development of a Tool to Assess the Impact of Water Markets on Agricultural Production in Pakistan*, Rapport de DEA, Université de Montpellier I, Montpellier.

8.   Hazell, P.B.R. and Norton, R.D. (1986) *Mathematical Programming for Economic Analysis in Agriculture*, Macmillan Publishing Company, New York.

9.   Bouzit, A.M., Rieu T. And Rio P. 1994. Modélisation du Comportement des Exploitants Agricoles Tenant Compte du Risque: Application du MOTAD Généralisé. *Economie Rurale*, No. 220-221. pp. 69-73.

10.   Richard, S.E. (1996) *Water Markets in Pakistan: Quantitative and Qualitative Analysis Using an Economic Modeling Tool*, Rapport de Fin d'Etudes, Ecole Nationale du Génie Rural, des Eaux et des Forêts, Paris.

11.   Strosser, P. and Kuper, M. 1994. *Water Markets in the FD/Eastern Sadiqia Area: An Answer To Perceived Deficiencies in Canal Water Supplies?* IIMI Working Paper No.30, International Irrigation Management Institute, Colombo, Sri Lanka.

GROUNDWATER MANAGEMENT IN SOUTH ASIA: WHAT ROLE FOR THE MARKET?

R.W. PALMER-JONES
School of Development Studies, University of East Anglia.

Abstract

Until recently public policy towards groundwater exploitation in South Asia involved state control of technology and institutions, siting and abstraction. Public or co-operative ownership and management of Deep Tubewells (DTW) was favoured; when DTW under-performed technical and institutional fixes were promoted, but without success. Privately owned shallow tubewells (STW) came to play the major role in groundwater exploitation., and groundwater came to be distributed in what are called water markets. Recent policy prescriptions advocate privatisation of groundwater partly to facilitate these markets. However water per se is not being traded in these markets, which are better understood as irrigation services markets. In many areas these markets are associated with severe environmental problems. Three broad agro-hydrological regions pose different groundwater management and policy problems. North-West South Asia (both Punjabs) has groundwater quality problems related to salinity which will require careful regulation and manipulation of private groundwater abstraction. Peninsular South Asia and Western India face extreme groundwater scarcity, and pervasive over-pumping, with saline intrusion and pollution problems in some locations, which again will need strong regulation. Eastern India has abundant groundwater and relatively isolated quality problems, where exploitation of groundwater should largely be left to the market.

1   Introduction

Recent initiatives to improve the management of water resources in the developed and developing world have focused increasingly on the use of economic methods [1][2]. The origins of this 'New Water Resources Agenda' (NWRA) lies in the diagnosis of the widely agreed deficiencies of public water development and management institutions. The approach is usefully summarised as 'commercial management, competition,   and   stakeholder

Water: Economics, Management and Demand. Edited by Melvyn Kay, Tom Franks and Laurence Smith. Published in 1997 by E & FN Spon. ISBN 0 419 21840 8.

involvement' [3]. This paper reviews the diagnosis and the potential for these policy instruments for management of groundwater in South Asia.

In the 1950s the state and international donors promoted groundwater development in South Asia by publicly or co-operatively owned deep tubewells (DTW). DTW have not proved efficient, equitable or sustainable despite the introduction of many technical and institutional innovations. Shallow tubewells (STW) under private ownership have been much more successful [4]. The discovery of what are called 'water markets' associated with private groundwater exploitation has lent support to an extension of the NWRA involving the establishment of private property rights in water and water markets [5]. However, South Asia is not without groundwater management problems, which vary by region. The South Asian experience of management of groundwater resources is of general interest because of the scale of groundwater in the region, its major role in the growth of agricultural production over the last three decades[1], and its importance for domestic and industrial water supply. The paper concludes that the 'the economic approach' puts too little emphasis on the regulation of private groundwater abstraction, and the associated need to improve the regulatory performance of the state.

2   Groundwater irrigation developments in South Asia

Well irrigation is extremely ancient and widespread in South Asia but in the colonial period development effort concentrated mainly on large scale canal irrigation. Well irrigation was appreciated as efficient and cost effective, even superior to canals, and as a protection against famine; but development was limited because '[if] the inhabitants of a tract of country have found well-sinking unremunerative,   ... Government would certainly not [find well sinking remunerative]' GOI, 1881:V:103). Until the advent of the green revolution canals often led to the abandonment of wells [6]. Wells were sometimes seen as being in competition with the Government's canal irrigation [6].

In the early decades of this century technical innovations - the diesel engine, the tubewell, and rural electrification, increased the potential for groundwater irrigation. In Gujarat there seems to have been a spontaneous adoption of these new technologies around the time of the First World War, apparently as a transfer of technology used in the urban and industrial sectors. In the Punjab the Department of Agriculture undertook some experiments with pumped groundwater, partly in response to a need to combat water-logging and salinisation through drainage [7][8]. There was some support for private development of groundwater by zamindars. A major groundwater irrigation project was implemented in those parts of Uttar Pradesh which could not be served by canal irrigation [9]; this project used mainly electric powered surface and deep set suction mode pumps of technologies but also seems to have put in a few force mode pumps. The water was managed by a state employed operator who 'sold' water to farmers based on the electricity consumed. Collection of these volumetric water charges seems to have been as problematic as it was in earlier experiments at volumetric water charging [10][11][12][2].

There were several lesser projects in the Punjab justified by perception of drainage and salinity needs, which evolved into the massive World Bank sponsored Salinity Control and Reclamation Projects (SCARP) in Pakistan [8]. The World Bank also supported State Tubewell Projects in Uttar Pradesh, Bihar, and West Bengal, and most Indian States supported DTW projects[3]. In the 1960s the SCARP technology was used in the North Bengal Tubewell Project at Thakurgaon in East Pakistan, and in the late 1960s the East Pakistan (later Bangladesh) Agricultural Development Corporation (EP(B)ADC) took over an evolution of this tubewell technology. Co-operative management of 2 cusec TWs was developed by the Academy for Rural Development in Comila under Akhter Ahmed Khan, but the 2 cusec force mode DTW was spread throughout Bangladesh, led by World Bank supported projects. The World Bank supported DTW projects in Nepal and Indonesia[4]. World Bank sponsorship of the technology was crucial to its legitimisation for other agencies.

The public sector DTW projects in South Asia received subsidies to capital, operation and maintenance costs. STW were promoted mainly through credit programmes, but at much lower rates of subsidy; however, STW have become much more significant in terms of numbers, area irrigated and productivity per unit area. Where STW are technologically inoperable (the dynamic water table is well below the suction limit), tubewell irrigation either has not survived, or only survives due to massive public subsidy in direct form (as in the Barind Tract in Bangladesh), or indirectly - e.g. through electricity, credit, or other agricultural subsidies - as in much of South and Western India).

Statistics on the areas irrigated by different technologies show that by 1990-1 there were some 63.5 thousand public DTW in India while there were 4.7 million private STW. After allowing for differences in capacity and its utilisation, and taking account of the likely higher productivity of privately supplied irrigation water, STW are far the more important source of irrigation (and drainage). In Pakistan, there were some eight and a half thousand Salinity Control and Drainage Project (SCARP) DTW by 1980, but more than 176 thousand STW which contributed the bulk of groundwater pumped, and had higher agricultural productivity because of the control exerted by their owners over the timing and quantity of irrigation [13][14][15]. In Bangladesh DTW were popularised from the late 1960s, but here too, while government and aid donor attention focused on DTW, STW have come to play the much more important role in agricultural production and productivity. In 1996 there were more than 500,000 STW compared to some 25,000 DTW. While DTW irrigate a greater area than STW per unit the overwhelming numbers of the latter mean that the bulk of the groundwater irrigated area obtains its water from STW.

3   The choice of technology and institution for groundwater development

Public sector support for DTW for the SCARPs was justified by the argument that Drainage is a public good benefits of which are long term, widely dispersed, hard to observe, etc., so would not be provided by the private sector. However, economic justification required irrigation benefits so most of the SCARP DTW have been installed in areas of fresh groundwater [16][5]. A private sector STW approach to groundwater development was put forward as an alternative to the 'massive assault' of the SCARPs [[17][18], but this was rejected as inappropriate to the problems of water-logging and salinity[6]. In India and Bangladesh the main justification fir DTW was the provision of irrigation in areas without access to surface irrigation where the agrarian structure obstructed private ownership of tubewells [19][20]. The scale of the more technically efficient DTW technology was though to be inconsistent with private ownership [20][21]. Engineers and agronomists were generally in favour of the force mode Deep tubewell (DTW) technology which had the lowest cost of water at the well-head [22]), and, with its pump set well below the static water table, was less likely to be affected by fluctuations in the water table. Both equity and efficiency reasons were given for choice of DTW under public or co-operative management [23][19]. The optimum technology was a compromise between the economies of scale and the managerial difficulties of posed by larger water flows[7]. In the face of continual poor performance of DTW the World Bank introduced canal lining, more control structures, buried pipe conveyance systems, automatic operation using header tanks and dedicated electricity feeder lines, rotational irrigation, 'day-area committees' (a sort of Water Users Association), and so on [23][24][19] which were put forward as more appropriate than STW and water (rental) markets [24].

In Bangladesh the World Bank emphasised DTW, adopting the agrarian-populist model of community co-operative irrigation formulated initially by the Comila Academy for Rural Development (as mentioned above). Despite providing some support for the adoption of STW in Bangladesh from the mid 1970s the World Bank and other donors continued to give prominence to DTW projects to be managed by informal farmer's groups, co-operatives, or NGO sponsored groups. Official documents produced with the Bank's participation continued to use a model of 'water lords' managing STW in an exploitive manner [20]. Academic and development research studies recognised that management of DTW was monopolised by rural elites, but rejected the possibility of water markets [25]; a number advocated 'appropriate' manually operated irrigation [26].

Both engineers and social scientists made mistaken diagnoses in their choice of technologies and institutions for groundwater exploitation; engineers preference for minimum cost of water at the well head should be rejected in favour of the cost of water in the field including conveyance and transaction costs [27]. The 'class' model of agrarian society used by social scientists predicted that market processes would be inefficient and inequitable; this model needs to be reconsidered in the light of reductions of poverty following the GR in South Asia and elsewhere.

4   Irrigation service markets

From the late 1970s a number of studies documented and then advocated the role of irrigation water markets in South Asia. In Pakistan a study reviewing the SCARP DTW in Pakistan, was one of the first to recognise the significance of irrigation water markets [28]. The most influential work on irrigation water markets stemmed from the development of DTW in Gujarat [29]. Water markets were independently recognised in Bangladesh [30][31][32]. Shah claimed that groundwater irrigation water markets were monopolistic even while they caused rapid growth of agricultural production. His policy prescriptions for flat rate electricity pricing had considerable influence on policy[8]. Other work also put forward unflattering perspectives of groundwater irrigation developments [33] but see [34]. By the end of the 1980s the World Bank came out strongly in favour of private sector development of groundwater, but with considerable qualifications based in large part on Shah's model [35][1].

However, water market evangelists draw quite different conclusions from Shah's work; e.g. details the development of highly competitive water markets in India. Competition among sellers contains arbitrary or collusive behaviour and ensures access to groundwater by land poor farmers who cannot afford to invest in their own wells. Groundwater prices are responsive to the availability to changes in energy costs and the availability of canal water. On equity as well as efficiency grounds, these highly developed groundwater markets have attractive and socially desirable properties' [5]. While Shah's work is seriously flawed and over generalised, it does not advocate and unregulated market in water. It employs a naive model of monopoly and neglects the ecological differences between the sites [27].

It is important to recognise that there has been as yet no market in groundwater per se in the sub-continent, and no property rights in groundwater; what can be observed are for the most part 'water services' markets in which free, water is raised, transferred, delivered and used, and it is the delivery rather than the water that is paid for. There is generally no rent or royalty for the water itself (see [36]). This distinction is an important insight which the water market evangelists do not adequately recognise. Furthermore, South Asia contains a huge range of agro-ecological, hydro-geological and socio-economic conditions that are relevant. We can identify three broadly different regional situations [37], which are of course far from uniform.

The first region is North-West India, including the Indus plains of Pakistan. Here there is groundwater related to surface water mainly as the result of seepage from large scale canal systems, often associated with water-logging and salinity. Fresh water is concentrated along natural river courses and perennial canals. Pumping of groundwater by privately owned STW has played the major role in managing the longer term problems of water-logging and salinity [14]; and an irrigation service market has developed [38], although it seems to be quite imperfect [15]. Secondary salinity is emerging due to the poor quality of groundwater pumped, particularly by private STW at the bottom ends of watercourses [39]. Since the groundwater originates in seepage from canal water both canal and surface water should be managed in a conjunctive manner. The interests of top-enders have a spatial monopoly of both surface and groundwater. Pumping itself (together perhaps with canal water distribution) could be franchised out to local entrepreneurs (who might also be farmers), subject to state supported regulation by stakeholders.

The second region covers most of peninsular India where groundwater and surface water are generally extremely scarce. Groundwater is spatially quite confined either in quite narrow tracts of alluvium, or in fissures in rock formations in strictly limited quantities (see various papers in [40]). There are significant negative externalities between pumpers and between ground and surface water users [37][40]. Farmers have in many States obtained electricity subsidies and flat rate electricity pricing regimes to alleviate escalating pumping costs, in Tamil Nadu electricity is free for irrigation. These subsidies constitute a major price distortion. The 'rent' of water is dissipated in a drilling and pumping races as wells are deepened and surface systems undermined by groundwater pumping. While the incentive to privatise water may well be strong, there are numerous problems of definition of property rights which would make their establishment and enforcement extremely difficult. Local collective management is no guarantee that water will not be appropriated by a minority at the expense of the majority [41], and is anyway limited compared to the extensive hydraulic system.

The alluvial tract in north-eastern Gujarat associated with the former course of the Sabarmati river has been highly exploited by private groundwater abstraction since early 1900s. By the 1970s groundwater was being mined at increasingly alarming rates [42][43]. In the some of these areas pumping has led to saline intrusion with consequent damage to the resource.

In much of Eastern India, in the Brahmaputra-Gangetic delta, groundwater is generally very abundant, close to the surface and largely free from salinity problems[9]. It is here that the circumstances are most auspicious for irrigation services markets, which have developed widely [30][31][32]. But it is exactly here that no market for water per se need, since water can be treated a free good, subject to some minor caveats. There is sufficient groundwater in this region to allow effectively unlimited exploitation with little risk of other than intra-year drawdown of the water table. The net social benefits of STW irrigation outweigh the social costs of installing deep set hand tubewells for potable water [44]. Appropriate policies aimed at regulating excessive groundwater use in the other regions could do great damage in the circumstances of Eastern India [45].

5   Conclusions

Groundwater exploitation by the private sector in South Asia, facilitated greatly by irrigation service markets, has resulted in major social and economic benefits. In Eastern India where groundwater is abundant and of good quality there is enormous potential for these markets and generally no need to regulate private groundwater abstraction, the location of pumps, or the terms and conditions of irrigation services contracts, since actual and potential competition results in pricing and allocation of resources on which it would be hard to improve. However, it is important to realise that these are not water markets per se, even though they are often treated as such. This 'market' solution will have equity implications, but whether these should be addressed by interventions in the 'irrigation service market' is a separate issue. Promoting and sustaining the rapid (and efficient) spread of irrigation service markets by appropriate electricity and other infrastructure provision, and agricultural price policy may do more for equity and poverty alleviation than more targeted policy alleviation polices [45]. There is no need for special incentives to technologies (DTW) or institutions (co-operatives or NGO supported groups) to pump and distribute groundwater in these areas. Potential competition in contestable irrigation services markets seems to be able to prevent the emergence of monopolistic practices in these groundwater abundant areas[10]. Elsewhere significant problems are associated with irrigation service markets, and, although there is undoubtedly an important role for these markets, they should be subject to strong and well informed regulation, which aid and other policy should also promote. It seems unlikely that there is a role for private property in groundwater and water markets per se. The history of groundwater exploitation in South Asia does not lend support to a generalised application of the prescriptions of 'the' economic approach to water management, although there are roles for some components of these approaches in most situations. The need to tailor policies to locally specific situations means that generalised policy prescriptions should carry appropriate health warnings.

6   References

1.   World Bank, 1993, Water Resources Management: a World Bank Policy Paper, Washington, World Bank.
2.   Winpenney, J., 1994, Managing Water as an Economic Resource, Routledge, London.
3.   World Bank, 1994, World Development Report, 1994: Infrastructure for Development, Oxford, Oxford University Press for the World Bank.
4.   Palmer-Jones, R.W., 1993, Deep Tubewells for Irrigation: the Life and Times of an Inappropriate Technology, mimeo, School of Development Studies, UEA, Norwich.
5.   Rosegrant, M. W., and Binswanger, H. P., 1994, Markets in Tradable Water Rights: Potential for Efficiency Gains in Developing Countries, World Development, 22, (11):1613-1626.
6.   Whitcombe, E., 1972, Agrarian Conditions in Northern India: Volume 1: The United Provinces under British Rule, 1860-1900, Berkeley, Univ. of California Press.
7.   Dhawan, B. D., 1969, Under-utilisation of Minor Irrigation Works: a Case Study of State Owned Tubewells in Uttar Pradesh, Delhi, Institute for Economic Growth.
8.   Michel, A. A., 1967, The Indus Rivers, New Haven, Yale University Press.
9.   Stampe, Sir. W., 1935, State Tubewell Irrigation Scheme, Lucknow, United Provinces Public Works Department, Irrigation Branch.
10.   Baird Smith, R., 1861, First Revenue Report of the Ganges Canal, Calcutta, Selections from the Records of the GoI (PWD), Report of the Revenue Returns of the Canals under the Superintending Engineer of Irrigation of the North West Provinces for 1859-60, PWD Press.
11.   GoI, 1881, East India (Report of the Famine Commission) Appendix V Irrigation as a Protection Against, London, HMSO, C-3086.-V.
12.   Bolding, A., Mollinga, P., and van Straaten, K., 1995, Modules for Modernisation: Colonial Irrigation in India and the Technological Dimension of Agrarian Change, Journal of Development Studies, 31, (6):805-44.
13.   Johnson, S. H., 1982, large-scale Irrigation and Drainage Schemes in Pakistan: a Study of Rigidities in Public Decision Making, Food Research Institute Studies, 18, (2):149-180.
14.   Johnson, R., 2989, Private Tubewell Development in Pakistan's Punjab: Theory, Quantification and Applications, Unpub. PhD. Cornell University, Ithaca, N.Y.

15. Meinzen-Dick, R., 1996, Groundwater Markets in Pakistan: Participation and Productivity, IFPRI, Research Report No. 105, Washington.

16. WSIP, 1990, Water Sector Investment Plan Volume 1: Main Report, Lahore, Federal Planning Cell, (co-ordinated by Sir M. MacDonald & Partners).

17. Mohammed, G., 1964, Waterlogging and Salinity in the Indus Plain: a Critical Analysis of some of the major conclusions of the Revelle Report, Pak. Dev. Rev., 4, (3)

18. Mohammed, G. et. al., 1967, Programme for the Development of Irrigation and Agriculture in West Pakistan: an Analysis of Public versus Private Groundwater Development Programme and the IBRD Draft Report, Karachi, Pakistan Institute of Development Studies.

19. World Bank, 1991, India: Uttar Pradesh Groundwater Developments. Issues and Options, Washington, World Bank, Agriculture Operations Division, Country Department V, Asia Region.

20. World Bank, 1982, Bangladesh Minor Irrigation: a Joint Review by the Government of Bangladesh and the World Bank, Dhaka, World Bank.

21. World Bank, 1981, Bangladesh: Northwest Tubewell Project (CR. 341-BD). Project Completion Project, World Bank, Washington.

22. Stoner, R. F., Milne, D. M., and Lund, P. J., 1979, Economic Design of Wells, Quarterly Journal of Engineering Geology, 12, :63-78.

23. World Bank, 1980, India: Uttar Pradesh Public Tubewells Project, Washington, Staff Appraisal Report, South Asia Department, World Bank.

24. World Bank, 1983, India: Second Uttar Pradesh Public Tubewells Project, Washington, Staff Appraisal Report, South Asia Projects Department, Washington.

25. Boyce, J. K., 1987, Agrarian Impasse in Bengal: Institutional Constraints to Technical Change, Oxford, Clarendon Press.

26. Biggs, S.D., and J. Griffiths, 1987, Irrigation in Bangladesh, in F. Stewart (ed), Macro-Policies for Appropriate Technology in Developing Countries, Westview Press, Boulder, Colorado.

27. Palmer-Jones, R. W., 1994, Water Markets in South Asia: a Discussion of Theory and Evidence, in Moench, M., ed., 1994.

28. WAPDA, 1979, Revised Action Plan for Irrigated Agriculture, Master Planning and Review Division, WAPDA, Lahore.

29. Shah, T., 1993, The Political Economy of Groundwater Markets in India, New Delhi, Oxford University Press.

30. BAU, 1985, Evaluating the Role of Institutions in Irrigation Management, Mymensingh, Department of Irrigation and Water Management, Bangladesh Agricultural University.

31. BAU, 1986, Water Market in Bangladesh: Inefficient and Inequitable?, Mymensingh, Department of Irrigation and Water Management, Bangladesh Agricultural University.

32. Wood, G. D., and Palmer-Jones, R. W., 1991, The Water Sellers: a Co-operative Venture by the Rural Poor, West Hartford, Kumarian.

33. Janakarajan, S., 1994, Trading in Groundwater: a Source of Power and

Accumulation, in Moench, M., ed., 1994.

34. Kolavalli, S., and Chicoine, D. L., 1989, Groundwater Markets in Gujarat, India, Water Resources Development, 5, (1):38-44.

35. Kahnert, F., 1989, Assisting Poor Rural Areas Through Groundwater Irrigation: Exploratory Proposals for East India, Bangladesh and Nepal, World Bank, Washington, Internal Discussion Paper, Asia Regional Series.

36. Saleth, R. M., 1994, Groundwater Markets in India: a Legal and Institutional Perspective, in Moench, M., ed., 1994.

37. Ballabh, V., and Shah, T., 1989, Efficiency and Equity in Groundwater Use and Management, Anand, IRMA.

38. Anson, R. e. al., 1984, Public and Private Tubewell Performance: Emerging Issues and Options, Washington, World Bank, South Asia Department.

39. Kijne, J., and van der Velde, E., IIMI, 1990, Salinity in Punjab Watercourse Commands and Irrigation System Operations: the Imperative Case for Improving Irrigation Management in Pakistan, IIMI, Colombo.

40. Moench, M., ed., 1995, Groundwater Availability and Pollution: the growing Debate over Resource Condition in India, Ahmedabad, VIKSAT.

41. Mosse, D., 1992, "Community Management" and Rehabilitation of Tank Irrigation Systems in Tamil Nadu: a Research Agenda, Paper presented at GAPP Conference on Participatory Development, July.

42. Bhatia, B., 1991, Lush Fields and Parched Throats: Political Economy of Groundwater in Gujarat, Economic and Political Weekly, 27, (51-2):A-143-65.

43. Moench, M., 1992, Chasing the Water Table: Equity and Sustainability in Groundwater Management, Econ. and Pol. Weekly, 27, (51-2):A-171-77.

44. Gisselquist, D., 1989, Groundwater Resources in Bangladesh: an Overview, mimeo, Agricultural Sector Team, Ministry of Agriculture, Dhaka.

45. Palmer-Jones, R. W., 1992, Sustaining Serendipity? Groundwater Irrigation, Growth of Agricultural Production, and Poverty in Bangladesh, Economic and Political Weekly, 28, (39):A-128-140.

1 Especially its crucial role in providing supplementary irrigation for the 'green revolution' (GR) technologies in the early years of the GR in North West India and more recently in Eastern India.

2 These earlier experiences are most instructive, but there is unfortunately not space to pursue this here.

3 These State sponsored DTW in Gujarat, under the Gujarat State Tubewell Corporation, are separate from the many DTW which have been installed by the private sector, whether by individuals or small partnerships (Shah, 1993).

4 The Philippines also had a number of DTW programmes initially supported by UNDP and the Japanese Government (see Palmer-Jones, 1993 for further details).

5 The logic of public sector provision of drainage in fresh groundwater areas would appear to completely contradict the main argument used for public sector provision.

6 Of course a range of complex technical and political economy reasons should be called on to explain the choice of SCARP technologies, and the emphasis on DTW in India and Bangladesh (Thomas, 1975).

7 The authors of the paper referred to here were senior members of the consulting engineering firm in volved in groundwater development projects in Pakistan, India, Bangladesh, Indonesia, and in a number of Middle East countries.

8 Although in fairness to Shah it should be said that the policies that have resulted do not follow all his prescriptions, and were in large part due to the political pressure of farmers' movements rather than directly following from his policy discussions.

9 Arsenic has recently been found in toxic quantities in groundwater for drinking. It is not clear how this is related to increased use of groundwater for irrigation.

10 In contestable market theory, if entry and exit costs are low and incumbents cannot set price above long run average cost (Armstrong et al, 1994).

MANAGING CHANGE IN ROMANIA'S IRRIGATION AND DRAINAGE SECTOR

Irrigation and drainage in Romania

D.B.STACEY
Binnie Black & Veatch, Redhill, UK
D.H.POTTEN
Hunting Technical Services Ltd, Hemel Hempstead, UK
MOGUR BOERU
Binnie Black & Veatch, Romania

Abstract
The irrigated area in Romania expanded rapidly between 1965 and 1989. There are now some three million hectares under command but in recent years only about twenty percent of this area has been irrigated. The drive behind the rapid development was more political than economic. Since 1989 economic realities have intruded. Low scheme utilisation, falling agricultural output, poorly maintained infrastructure, high energy costs and the impacts of institutional and land tenure changes signalled to the Government that a review of the irrigation and drainage sector was required. A study of existing schemes and sectoral strategy has recently been completed for the Romanian Ministry of Agriculture and Food.

The study concluded that about 1.7 million hectares of the currently commanded area could be rehabilitated and would be economically viable. A sector development programme has been proposed. This includes a programme of scheme rehabilitation accompanied by progressive withdrawal of central Government and parastatal involvement from the control and management of schemes, by attraction of private investment into ownership of irrigation facilities, by the promotion of irrigation under appropriate circumstances and by the restructuring of the subsidy and tariff regimes in order to promote rational use of irrigation water.
Keywords: Drainage, economics, irrigation, institutions, policy, strategy, tariffs

1 Introduction

Romania has about 15 million ha of agricultural land, of which nearly half could

Water: Economics, Management and Demand. Edited by Melvyn Kay, Tom Franks and Laurence Smith. Published in 1997 by E & FN Spon. ISBN 0 419 21840 8.

theoretically be irrigated. In 1965 there were 200,000 ha of irrigated land. By the end of the 1980s, there were more than one hundred schemes in operation commanding about three million hectares. These are principally large pressurised sprinkler systems, many of which serve more than 50,000 ha each, supplied by water pumped from the River Danube and its tributaries (principally the Olt, Siret, Arges and Prut). Of the area served, ninety-five percent is arable land; the remainder pasture, vineyards and orchards. The national distribution of irrigation schemes is illustrated in Figure 1.

The drive behind the rapid development of three million hectares of irrigation systems was political rather than economic. The Ceausescu Government aimed to maximise agricultural output for both domestic consumption and exports. Although "Feasibility studies" of individual schemes were carried out, financial prices only were used, discounting methods were not employed in cost benefit analyses and the assumptions on benefit accrual appear to have been rather optimistic. The main decision criterion used was the payback period. Energy costs were based on those prevailing in Central Europe before 1989. This may have been reasonable at the time but could well have biased investment even more towards the high-energy use schemes which are a striking feature of the present scene in Romania.

The majority of schemes have been developed to use medium to high pressure impact hand move sprinkler equipment for the application of water although side roll and centre pivot equipment has been employed in a number of instances. Supply networks are pressurised either from fixed electrically-powered pumping stations (each serving between 400 and 3,000 ha) or via mobile electrically or diesel driven pumps.

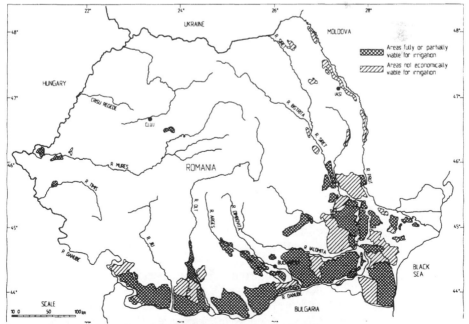

Fig. 1. Distribution of irrigation and drainage schemes in Romania

The majority of schemes have been developed to use medium to high pressure impact hand move sprinkler equipment for the application of water although side roll and centre pivot equipment has been employed in a number of instances. Supply networks are pressurised either from fixed electrically-powered pumping stations (each serving between 400 and 3,000 ha) or via mobile electrically or diesel driven pumps.

Much of the irrigable land is on terraces high above the water source. In some instances, schemes have been developed to irrigate land more than two hundred and fifty metres above the source of water. The principal crops grown are maize, wheat, sugar beet and sunflower. Surface and sub-surface drainage, associated with the irrigation schemes, has been widely established. Tile drains have been installed in about 76,000 ha.

The one hundred and eight schemes comprise extensive conveyance and distribution systems and more than 4,000 pumping stations. The annual energy consumption for irrigation pumping in the late 1980s averaged about 2.5 million MWh[1].

2   Impact of political changes

Political changes since 1989 and moves towards a market economy have had important effects on irrigation not only with regard to investment in the irrigation and drainage infrastructure and its operation and maintenance but particularly with respect to institutions, land tenure and farming organisations. The changes in land tenure have been far reaching in their effects. Formal and informal farmer associations as well as individual farmers are the successors to the old agricultural cooperatives. State farms have been transformed into commercial joint stock companies and commercial societies are beginning to emerge slowly in some areas where entrepreneurs are renting land from groups of landowners. Cropping strategies and productivity are now highly influenced by constraints on the farm management systems adopted.

Agricultural production from the irrigable areas has fallen since 1989. In 1993, only about 20 percent of the area commanded by the irrigation infrastructure was irrigated and the areas irrigated were fragmented and widely dispersed within each scheme. Much of the infrastructure is now in need of rehabilitation and upgrading and there is a lack of irrigation equipment.

The effect of the present underutilisation of the irrigation infrastructure has been to reduce the efficiency of operating the network to a very low level making the financial terms of operating schemes extremely unattractive to the Regie Autonome, the owner-operator currently responsible for the supply of water to the farmers. In such circumstances a clear strategy is required on which to base future planning and investment decisions.

A study on the viability of the existing irrigation systems on which a development strategy could be developed has recently been completed for the Ministry of Agriculture and Food (MAF)[2]. The study concluded that of the three million hectares served by irrigation and drainage infrastructure:

- complete schemes commanding more than 1.3 million hectares of the three million hectares commanded by the existing irrigation infrastructure could justifiably be included, on economic grounds, in a long term rehabilitation programme;
- some of the other existing schemes have part of their command area more than 100 metres above the source of water. Some of these schemes (up to 0.15 million hectares) could prove economically viable in the long term if the higher terraces were excluded;
- further areas covering about 0.2 million ha in the Danube flood plain, where good farm management organisations are established but where incremental yields to irrigation are low, are close to the water source and have well maintained infrastructure: these areas would be viable in the medium term at least.

3   Development strategy

A general conclusion from the studies is that investment in irrigation equipment (mobile pipework and sprinklers) and removal of operating constraints are economically sound immediate and medium term strategies to pursue for those schemes which are shown to be viable in the long term. In the present circumstances, the decision on when to invest in full rehabilitation of a scheme depends on the uptake of irrigation and the degree to which irrigation is re-established on contiguous areas of land within a scheme area. The poor economic performance of most of the existing schemes arises from the current scattered nature of the land being irrigated, the low utilisation of the existing infrastructure and the high cost of operating and maintaining the infrastructure per hectare actually irrigated. A key component of any development strategy must, therefore, include the encouragement of irrigation in proven viable areas.

There are a number of areas where work is required immediately. These include:

- study of priority schemes to identify constraints on irrigated agriculture (in particular to identify why farmers who currently have access to irrigation are not using the available resource) followed by the implementation of proposals: (i) for the removal of these; and (ii) to encourage the uptake of irrigation at these schemes;
- support to farmers and farmers groups for the provision of irrigation equipment (through credit, hire or other means) and promotion of ownership of irrigation equipment by as diverse a range of proprietors as possible;
- an information programme to signal to irrigators the Government's commitment to support irrigation in priority areas;
- advice to farmers on the use of irrigation equipment and initiation of a programme to support the formation and development of water users' associations;
- advice to operating companies on ways of adapting existing distribution networks for use by different sizes of farmers groups;

- implementation of an institutional development programme for operation and maintenance of schemes at system and regional levels;
- design for full rehabilitation of the most viable schemes and implementation at schemes where and when irrigation is substantially re-established;
- pilot project development at three or four priority schemes (of about 1,000 ha each) to demonstrate and support the concepts of a future long term development programme.

In addition to these, important policy decisions are required on tariffs and subsidies if future development is to be effective. These are principally related to: (i) the introduction of energy cost based (but subsidised) tariffs; (ii) restructuring the subsidies provided to the current operating companies (or their successors) to a uniform level based on irrigated area; and (iii) complementing the subsidy programme by ensuring agricultural incomes are not exempted from income tax.

The establishment of future irrigation fee levels in Romania raises a number of issues of wider significance. The returns to irrigation (per hectare and per m^3 of water applied) vary substantially by agro-climatic zone; by crop; and by farming system. The last of these raises very sensitive issues. When the irrigation systems were built they were expected to supply water to very large farming units, State Farms or Agricultural Cooperatives. Following the fall of President Ceausescu, there has been a major land reform programme and much of the land has been redistributed. There are at present a variety of farm management systems in the commanded areas ranging from large concerns which used to be State Farms (2,000 ha up to 70,000 ha) through Farmers' Associations, often in practice a number of privately owned farms which are operated by an entrepreneur as a single entity (of up to about 1,000 ha), to individually owned and operated farms which may be as small as one or two hectares.

The long term financial viability of economically sound irrigation systems depends on recovery of at least a significant part of the total costs and, therefore, on the willingness of the farmers to pay for irrigation.

Farmers growing high value crops, as well as the large farming enterprises based in areas where returns to irrigation are relatively high, are expected to be willing to pay irrigation tariffs at levels which could yield full cost recovery. If, however, tariffs are raised to these levels they would almost certainly be unacceptable to many of the individually owned and operated enterprises as: (i) these farms were only returned in the early 1990s to their past owners who are mainly fairly unsophisticated farmers who are obtaining poorer yields than the large enterprises; and (ii) these farmers are unfamiliar with irrigation and other relatively high-cost farming practices and are likely to require a high return to justify adopting a "high-cost, high return" farming system as opposed to the low cost, risk minimisation strategy that has characterised their farming since they obtained the land.

If tariffs are set at a level which deters these small and new farming enterprises, it is likely that only a small proportion of the commanded area will ever be irrigated. An opportunity to utilise a substantial past investment in infrastructure and to expand economically viable agricultural output will be lost. If, however, tariffs are set at an appropriately low level, small and new farmers could be encouraged to participate, provided that they are growing crops which give an acceptable financial return to irrigation. It would not be practicable to charge the larger or more efficient farming enterprises at higher rates (per m^3) and these, therefore, would earn significant economic rent. This rent could be partially recovered through taxes on corporate and individual earnings from agriculture. The Government, therefore, could potentially achieve cost recovery on the irrigation systems through a combination of direct revenues from tariffs and indirect incremental tax revenues.

Any development programme must be prepared in the context of the government's policy objectives if it is to be effective. In Romania, the relevant objectives in the agricultural sector are to:

- increase the production of irrigated food crops to meet national requirements;
- replace imports in the short term and recover, in the longer term, the country's traditional role as a food exporter;
- increase employment in the agricultural sector, reversing recent rises in unemployment;
- encourage crop production and irrigation by small and private farming units, including individuals and informal farming associations and maintain irrigated agriculture wherever it is viable;
- promote a market economy in the sector and reduce the direct role of Government in the provision of agricultural support services.

These agricultural sector policies have major implications for the irrigation sub-sector and these are reflected in the objectives of the strategy and investment programme being pursued by the Romanian Government. They include:

- bringing back under irrigation and ultimately rehabilitating and modernising the areas where irrigation is viable;
- supporting progressive withdrawal of central Government and parastatal involvement from the control and management of operational activities;
- promoting viable irrigation under a broad range of farm management types;
- discouraging and eventually terminating irrigation where it is not viable, thereby reducing subsidy costs.

Figure 2 illustrates the overall impact of the proposed development programme.

4   Institutional and policy implications

Meeting the objectives of the development programme will depend not only on the provision of adequate financial resources but also on the implementation of complementary policy and institutional changes. Prerequisites for the successful implementation of the investment programme include:

- progressive withdrawal of Government from direct control and management of operational activities;
- the establishment of commercially managed organisations responsible for the provision of an efficient irrigation service;
- implementation of a policy for progressive reduction in the irrigation subsidy, including a transparent mechanism which shows beneficiaries the

level of subsidy being received from Government and encourages economically efficient water use;

- liberalisation of water pricing as a means of balancing supply and demand and of directing investment to economically efficient locations;
- demonstration of the benefits to be obtained from irrigation to encourage farmers to irrigate viable areas;
- competitive tendering for construction and rehabilitation of physical infrastructure;
- attracting private investment into ownership of in-field irrigation networks and equipment;
- greater involvement of water users in the operational control of the irrigation and drainage networks;
- establishment of procedures to monitor and evaluate the impact of sub-sector policies and investment.

Development programmes have been prepared based on the assumption that these policies and institutional initiatives will be implemented by the Ministry of Agriculture and Food (MAF). Wherever possible, the programmes proposed include activities which will complement and reinforce the desired institutional changes.

5   Conclusions

The Romanian irrigation and drainage sector is undergoing a period of transition which has put considerable strain on both farmers and farmers' organisations and on the institutions involved in the provision of water for irrigation. A recent re-evaluation of the sector by the Ministry of Agriculture and Food, supported by the World Bank, has been undertaken. This applied economic principles to the future rehabilitation and development of one of Europe's major sprinkler irrigation networks and concluded that a programme of institutional change, encouragement of farmers groups, the introduction of water tariffs and restructuring of subsidies should go hand in hand with phased rehabilitation of economically viable schemes.

The study concluded that investment in rehabilitation and modernisation should be concentrated on about sixty percent of the existing infrastructure and decisions should be based on the long term viability of irrigable areas. This has led to the conclusion that some areas of the existing systems should be taken out of use and land returned to other uses. There are social and political implications in adopting this policy but the long term efficacy of applying appropriate economic criteria to the provision of water for irrigation and the introduction of cost recovery is now accepted as the only sustainable approach to the country's needs.

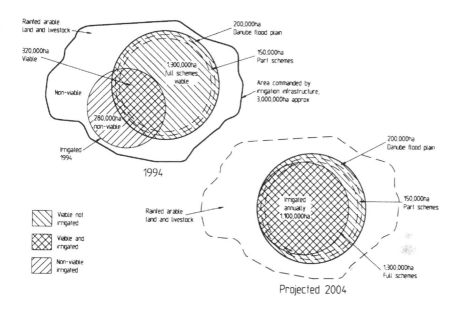

Fig. 2. Impact of development programme

6   Acknowledgements

The authors acknowledge, with thanks, the support of the Ministry of Agriculture and Food, Romania, the World Bank and Consultants staff in providing data and reports for the preparation of this paper.

7   References

1.   Berbeci, V. and Stacey, D.B.(1993) *Development of Irrigation and drainage in Romania,* ICID 17th European Congress, Varna.
2.   Binnie & Partners et al (1994) Study of Irrigation & Drainage in Romania, Ten Year Investment Programme, Ministry of Agriculture and Food, Bucharest.

WHICH POLICY INSTRUMENTS FOR MANAGING THE IRRIGATION SECTOR IN PAKISTAN?

Irrigation sector policy in Pakistan

M. MONTGINOUL
Irrigation Division, Cemagref, Montpellier, France
P. STROSSER
International Irrigation Management Institute, Lahore, Pakistan

Abstract

The prevalent recognition of poor irrigation system performance in Pakistan has recently generated passionate debates on potential solutions to improve this performance. Drastic changes, such as decentralization and development of water markets, have been proposed. And implementation of some of these changes is already under way. Surprisingly, many facets of the new policies are still unclear and need to be spelt out before further implementation of any change. The present paper investigates the potential for water pricing, quota and water markets to address current policy objectives and constraints of the irrigation system in Pakistan. The analysis highlights the need to combine policy interventions that consider issues related to both the demand for, and supply of, water.
Keywords: irrigation sector policy, Pakistan, quota, water markets, water pricing

1   Introduction

Agriculture plays a major role in the economy of Pakistan and provides job opportunities for more than 55 percent of the labour force, counts for 26 percent of the Gross Domestic Product, and contributes to 80 percent of the total export earnings of the country. With a ratio of irrigated area over non-irrigated area equal to 4, irrigation is central to most of the agricultural sector policies in Pakistan and is vital for meeting demands for food products, raw material for industrial processing and exports.

Although past achievements of the irrigation sector in terms of area irrigated and total agricultural production are recognized, current irrigation sector policies are increasingly questioned. Irrigation sector policies and irrigation management rules have not been updated since the inception of large scale irrigation in Pakistan more than a century ago, and have not been able to address successfully issues of low

Water: Economics, Management and Demand. Edited by Melvyn Kay, Tom Franks and Laurence Smith. Published in 1997 by E & FN Spon. ISBN 0 419 21840 8.

agricultural productivity, poor water supply performance and poor financial and resource-base sustainability [1]. And current macro-economic and sectoral policies promoted by international lending institutions in the context of structural adjustment programs have also added to the increasing pressure for change in irrigation sector policies.

In this context, radical changes in the institutional framework and policy instruments for managing the irrigation sector are proposed by government agencies and donors. The following sections of this paper investigates the potential for water pricing, quota and water markets to management surface water resources in the irrigation sector in Pakistan.

2   Policy instruments for irrigation sector management: some theoretical issues

2.1   Generalities

Policy instruments as used in the present paper are broadly defined as interventions devised by the State/government agencies/local communities to achieve well defined policy objectives. Policy instruments modify or influence (whether directly or indirectly) decisions taken by water users, which in turn has an impact on water use, agricultural production, environmental protection, etc. Specific policy objectives include to develop   new areas, to increase agricultural production for autoconsumption or for industrial processing, to ensure equitable water supply, or to promote sustainable use of water resources. Objectives vary over time as a result of changes in the economic development of the country, along with changes in water use and water scarcity. To develop new areas or increase the number of users are often initial objectives of irrigation sector policies, while increasing water scarcity leads to emphasizing the economic nature of water and promoting economic efficiency and optimal allocation of water resources.

Examples of policy instruments or interventions are summarized in Table 1. Interventions may target the supply of, or the demand for, irrigation water. They may modify physical infrastructure, management activities or address issues related to the institutional framework and the environment within which irrigation related activities are performed by different stakeholders institutions involved in the provision of water for irrigation. A recent re-evaluation of the sector by the Ministry of Agriculture and Food, supported by the World Bank, has been undertaken. This applied economic principles to the future rehabilitation and development of one of Europe's major sprinkler irrigation networks and concluded that a programme of institutional change, encouragement of farmers groups, the introduction of water tariffs and restructuring of subsidies should go hand in hand with phased rehabilitation of economically viable schemes.

Table 1. Policy instruments in the irrigation sector

Intervention	Modification of physical system	Changes in management activities	Changes in enabling environment
Supply-based	Construction Rehabilitation Changes in infrastructure	Operation Maintenance Information management	Institutional framework Legal system Budgetary policy
Demand-based	Development of new irrigation technology	Farm scheduling Information management	Water pricing, quotas water markets Legal system Macro-economic policies (input, output, etc)

Of increasing interest for irrigation sector policies are economic instruments that influence the demand for water. In situations with scarce water resources, competing use between and within sectors of the economy, environmental concerns and differences between private and social costs, supply-based interventions may not be adapted anymore and/or too costly. Economic instruments offer an alternative by modifying the economic environment under which private users operate and influencing users' decisions so their actions are in accordance with the objectives and overall benefits of the society. Three economic instruments are further investigated: water pricing, quota and water markets.

2.2 Water pricing

From the theoretical point of view, water pricing is the basic instrument that help to distribute limited water resources to users and to determine allocation of these resources by providing appropriate signals/incentives. At the same time, water pricing is used for cost recovery purposes. Several conditions are required for water pricing to achieve allocative efficiency: the demand for water is to be sensitive to water prices; water pricing mechanisms are to be easily understood by water users; and, these mechanisms must be stable enough over time to be in accordance with time horizons considered by farmers/water users to take their decisions.

Other aspects to be considered for the establishment of water pricing include to enforce payments to users, the need to consider marginal cost issues and externalities, and financial issues such as the cost of implementing water pricing mechanisms and the financial autonomy of the managing agency.

Although water pricing is aimed at affecting farmers' decisions and the demand for water, demand and/or supply parameters may be considered while developing water pricing mechanisms. Objectives of demand-based pricing may be to appropriate part of the benefits made by water users, to maximize their total utility by putting users in competition, or to take into account more socio-political aspects (for example equity considerations). Supply-based water pricing considers the costs of delivering water to the plant (operation and maintenance costs, investment costs) and other (social) costs related to negative external effects and support by the society.

The parameters used to develop and compute water prices, along with the water pricing structures, differ from country to country, and even from irrigation project to irrigation project [2]. Extreme cases of water price structure (or water charge schedule) are flat-rate or fixed charge (whether area-based or user-based) and volume-based charge.

2.3   Quota

The main objective of the quota is to limit water use. Contrary to water pricing, quotas do not provide incentive to water users but simply constraint their demand. A quota has an effect on the demand only if, for a given water price, the estimated demand for water would be higher than the imposed quota. Different types of quota may be proposed: (i) quota in time, with the possibility to use water during a limited period of time per year or periodically every week; (ii) quota in volume defined (for a given period of time) per user or per hectare cultivated; (iii) quota related to crops and defined as a percentage of the maximum evapo-transpiration. Quotas may be fixed over time or may vary according to the availability of water resources (different quotas for low and high water scarcity periods, for example).

If the elasticity of water demand compared to price is equal or close to zero, quota are more efficient than water pricing to constrain water users to save water. If the demand is elastic with regard to prices, then the economic theory shows that quota and water pricing lead to the same results in terms of water use when the demand for water is stationary. However, if increases in the demand for water are expected as a result of changes in other sectors of the economy, for example, then quota are the preferred option as they constraint the demand at the same level while changes in water demand would be expected under water pricing.

2.4   Water markets

Water markets are an allocation mechanism based on an initial allocation of water rights. Based on the confrontation between water supply and water demand, water is (re-) allocated between users at an equilibrium price established on the market. Unlike for most of the markets, water sellers have also a direct utility in using water on their own farm. Thus, they compare the marginal value product of water on their farm with the (expected) equilibrium price to decide whether they participate or not in water transfers.

Requirements for well functioning water markets include water scarcity, well defined and transferable water rights, a large number of purchasers and sellers, no or limited transaction costs and the existence of an appropriate information system. In practice, however, the definition of water rights is often the central issue [3]. Water rights can be defined as a volume of water (with different levels of priority) or as a proportion of available water resources. A specific infrastructure is also required to transfer water between participants in water markets.

Water markets include permanent transfers of water rights, or lease of these rights and transfers of volumes of water. Existing water markets are localized or between regions or water basins, involve changes in water use (from irrigation to municipal or industrial use), are developed for groundwater or surface water, and involve individual users or user groups, private companies and public institutions [4]. Markets can be centralized (one organisation collects water rights and sell them to individuals) or decentralized when transfers are the results of direct negotiations between individuals.

Water markets often function at the margin, i.e. involving only a limited portion of existing water resources. High transaction costs (as compared to the value of water and expected benefits), inadequate physical infrastructure and legal framework, or socio-political resistance are factors that explain this situation [5]. Many difficulties arise as a result of potential externalities and third-party effect of water transactions.

3   Irrigation sector policy in Pakistan: an historical perspective

3.1   A bit of history

Up to the end of the colonial period in 1947, the British administration promoted the construction of large irrigation schemes that resulted in the existing Indus Basin Irrigation System. At that time, the main objectives of irrigation related policies were to mitigate famines by spreading water resources thinly and equitably over large areas of land. Also, the establishment of a comprehensive hydraulic network and its related administration strengthened the control of the British over large areas and populations.

In irrigated areas, the British established a system of quota (still in use nowadays) to share limited canal water supplies (it is not clear whether the quota played its role initially as the pressure on water resources was lower than today). The quota, expressed as the combination of a duration of use at the farm level (i.e. the water turn of the *warabandi* schedule that shares canal water between users within the tertiary unit or *watercourse*) and a share of water flows (specified by the dimensions of the watercourse outlet) was enforced through social control and physical infrastructures. A limited number of control structures along main canals were provided to minimize operation requirements. With imposed scarcity, efficient use of canal water was also expected. The British established a system of area/crop-based water charges to cover operation and maintenance costs.

Since the Independence in 1947, there has been no major change in irrigation sector policies. These policies have focused on water supplies to increase agricultural production, improve self-sufficiency in major food crops, and promote equity in water supplies. Major efforts were targeted towards replacing the supplies "lost" to India under the 1959 Water Treaty by building link canals to bring supplies from the Western to the Eastern part of the country, and constructing dams to increase storage capacity [6]. However, cropping intensities more than irrigated areas increased during this period. At the same time, further emphasis was given to mitigating adverse effects of irrigation such as waterlogging and salinity. Large Salinity Control And Reclamation Projects have been implemented since the 1960s, initially with the installation of large public tubewells that provided also extra irrigation water supplies in areas with good groundwater quality.

3.2   The irrigation system to date

The Indus Basin Irrigation System today is the result of one century of supply-based policies. Surface water is supplied to more than 14.5 million hectares or 89,000 watercourses through an intensive network of main canals and secondary canals or *distributaries*. The system of quota and area/crop-based water charges is still in use, although discrepancies exist between official rules and rules in practice, with for example the development of localized canal water markets [7]. The 1980s, however, have brought major changes that have stressed the inadequacy of the supply-oriented engineering-driven policies implemented so far. In this context, three issues are seen as particularly important.

First, the availability of financial resources has drastically decreased. Budgetary constraints (as a result of pressures from international lending agencies), along with the increasing competition from other sectors of the economy, have led to a reduction in funds available for the construction of increasingly expensive (and with often limited tangible benefits) irrigation and drainage projects.

Second, the 1980s have recorded an important increase in the pressure on water resources. Farmers have installed a large number of private tubewells to complement canal water supplies and cultivate cropping intensities double than design figures. However, in large areas with good groundwater quality, mining of the aquifer has been already taken place [8]. In areas with poorer groundwater quality, secondary salinity has become a major issue. At the same time, increasing financial deficits of the irrigation sector have led to poor maintenance of the system with negative impact on water supply performance [1].

Third, and much less documented, is the increasing interference of water users into the management of the irrigation system [7]. The operation of the irrigation system is becoming more and more a *political process* [9], not only because it involves politicians but also because it is dominated by economic and institutional power relationships of the different stakeholders. It is translated into problems of assignment of responsibilities, information sharing, enforcement, and ultimately irrigation system performance in terms of water supply and financial sustainability. As changes in macro-economic environment and water scarcity have not been accompanied by changes in irrigation sector policies for the last 100 years, water users have adapted to changing circumstances with a resulting (increasing) gap between official and existing rules. Table 2 summarizes changes in the irrigation sector that have taken place since the inception of the irrigation system.

During the 1980s, the recognition of the inadequacy of official policies to tackle existing problems and constraints led to a shift in interventions away from purely supply-oriented and engineering approaches. Examples of interventions include developing water user association to improve water management below the watercourse outlet, increasing water charges or transferring public tubewells to private users to reduce financial deficits of the irrigation sector and improve water use efficiency. In most of the cases, however, the engineering component only of these projects has been successfully implemented.

Table 2. Irrigation sector policy in Pakistan: an historical perspective

	Initial	Actual
Objectives (by order of priority)	Settlement/occupy space Mitigate famines Equity Efficiency in use via imposed scarcity Recovery of O&M costs	Increased agricultural productivity & sustainability Financial autonomy Equity
Environment & constraints	Surface water resources available Need to consider existing irrigation schemes (rules and rights, social setup, etc) Resettlement of populations	Limited water resources Limited financial resources Political interference in irrigation system management Existing irrigation system in place
Strategy	Construction of irrigation schemes Increased water supplies for existing schemes	Decentralization and privatization Financial autonomy of line agencies Mix of supply and demand-based policies
Economic instruments	Quota in time (watercourse) and quantity (main, secondary canals) linked with infrastructure Area and crop related water charges	Future of existing economic instruments not specified although their limitations recognized. Water markets existing at local level and further development proposed as a potential alternative.
Main problems encountered	Poor irrigation system performance Insufficient cost recovery Increasing interference by users in management of the irrigation system	------

Recently, however, as a result of the recognition of the current problems of the irrigation sector and under pressure from international lending agencies, drastic changes in irrigation sector policies have been proposed. Those include decentralization with increasing involvement of water users, and development of financially autonomous irrigation authorities (Provincial level) and area water boards (canal command area level). Also, water markets have been proposed to solve problems of economic efficiency and increase agricultural productivity [1]. However, policy instruments that would lead to foreseen improvements in irrigation system performance have not been specified yet.

4   Appropriateness of water pricing, quota and water markets for the irrigation sector in Pakistan

The present section discusses the potential for water pricing, quota and water markets for the irrigation sector in Pakistan, using theoretical considerations and the historical analysis of objectives, constraints and irrigation sector policies presented in the previous sections.

4.1 Water pricing

That the area-based structure and low level of current water charges do not promote efficiency in Pakistan is recognized [10]. However, volumetric water pricing would not impact on water use efficiency as water users do not control water supplies with the existing infrastructure. Thus, implementing volumetric water pricing would require significant (and costly) changes in infrastructure for control and measurement of volumes.

While the costs of changes required for volumetric pricing for individual water users would out weight expected benefits (due to the existing too low value of water in irrigation in Pakistan), it may be possible to consider intermediary approaches where volumetric pricing is implemented for groups of users (whether for watercourse command areas or specific sections of distributaries). However, to ensure an impact of pricing on water use efficiency, effective water user organisations are required with possibilities to request irrigation department staff for changes in water flows or with direct control of water flows. Also, as changes in flows would affect simultaneously a large population of farmers, the expected benefits in productivity are expected to be minimal.

Regarding cost-recovery, a water charge per hectare without differentiating between crops would be the preferred option: it reduces the potential for users' interference during assessment of water charges, and has limited information requirements. To provide an incentive to the irrigation department to supply water adequately, a fixed charge per user could be added to take into account the successful delivery of canal water to individual farmers or watercourses (farmers would pay this charge whenever they have received canal water more than x days per season).

4.2 Quota

The design discharge linked to outlet dimensions is similar to a quota for the watercourse. However, modification in outlet dimensions that are regularly performed by influential farmers lead to changes in this quota with negative impact on downstream water users [6]. A quota defined at the head of the distributary (with specific design of the infrastructure so no discharge higher than the design discharge can be delivered to the distributary) would be more easily enforced. But this would limit the buffer capacity of the irrigation system that is required to take care of unpredicted events such as rain or breaches.

Because of the unreliability of water supplies, quota in time as already existing within the watercourse command area are easier to be enforced than quotas in volume. The implementation of quotas in volume for individual water users would lead to similar measurements difficulties than for volumetric water pricing, and would require control structures to cut off water supplies as soon as the quota has been reached. At higher levels of the irrigation system, quota in time would be equivalent to a well-designed and implemented rotation between secondary canals or primary canals. However, due to problems in operation, hydraulic behaviour, siltation, technical skills are required for proper implementation. Also, as users' interferences would be possible, the enforcement of these quotas would be more problematic.

Quotas in volumes could be implemented at higher levels of the irrigation system by specifying, for example, a maximum discharge at specific control points. In such cases, quotas should be physically enforced by limiting the size of control structures so that no higher discharge than the quota divided by the period of time for which the quota is defined could pass through the control structure. With the present socio-political system, it would be difficult to imagine quotas legally enforced. A quota per crop could also be envisaged to promote productivity objectives. However, this would require an accurate assessment of cropping pattern, a process that is likely to be influenced by water users to increase their quota. Different quotas may also be desirable between periods (seasons) with distinct levels of water scarcity, but it would require complex infrastructures.

4.3   Water markets

A prerequisite for water market development is the initial definition of water rights [3]. The question: how to allocate water rights to water users? relates to the need to define the users (whether water users or landowners, individuals or user groups), the basis for the initial allocation (official or existing water supplies, or any other allocation criteria), and the variables considered for the definition of the water right itself. Water rights can be defined as a volume of water received during a given period, as a probability distribution of a discharge (to take into account variability in supplies) at a given point, or as a percentage of the water flow at a given point. The duration of use (similar to the water turn of the warabandi schedule) can also enter into the definition of the water right. Using the existing set-up of the irrigation system, individual canal water rights may be defined by the duration of the individual water turn, a percentage of the flow along the distributary (linked to the dimensions of the watercourse outlet), and a daily volume with probability distribution at the head of the secondary canal. The last element of the water right, however, is unknown and often unpredictable, thus making the definition of the water rights without the establishment of an information system.

According to the level at which water transfers are envisaged, costs of implementation and benefits are expected to vary greatly. Exchanges could be envisaged between primary canals without major changes in the existing infrastructure. And significant productivity gains are expected from water transfer developed between canal command areas from different agro-ecological zones. However, this would require well functioning water user organisations at that level (something that is difficult to foresee at present), or independent organisations that would be in a position to sell and purchase water rights defined at the canal command area level. How sales and purchases would be agreed upon by numerous users, along with financial and water related implications of water transfers for individual (groups of) users are issues for discussion.

Between secondary canals, transfers are possible with the present infrastructure (as illustrated by the large range of discharges currently observed at the head of most of the secondary canals). Water sales, however, would be limited to 30 percent of the distributary design discharge, as problems in distribution between watercourse outlets arise above this limit [6]. As secondary canals have often similar average farm populations, gains in productivity are expected to be limited and water markets between distributary would function at the margin. Organisational requirements and costs (to collect information and confront potential sellers and purchasers) would probably be too high for these expected gains in productivity.

Between tertiary units, gains in productivity are expected to be high with minimal organisational requirements. However, the existing infrastructure does not allow any transfer from one watercourse to the other and changes in infrastructure that would make such transfers possible (for example, installing gates at each watercourse outlet) would lead to complex system operations with potential users' interference.

Within the watercourse command area, lease or transfer of water rights are possible (and already exist) with the existing infrastructure and organisation. Also, productivity gains are expected to be relatively high due to the relative heterogeneity of farming systems within watercourse command areas. To identify factors (unpredictable canal water supplies, water quality, costs) that explain the current low level of canal water transfers would help identifying interventions to remove existing constraints to water transfers.

5   Discussion and conclusion

Several lessons can be drawn from the above discussion on the potential for water pricing, quota and water markets for irrigation sector policy related to surface water resources in Pakistan. First, it is important to use the three economic instruments in combination to meet the multiple objectives of irrigation sector policies. Each instrument is defined in relation to a simple (*marginal*) objective, taking into considerations objectives of other instruments, along with infrastructure and organisational requirements. For example: (i) area-based water charges (land tax) complemented by an independant delivery-based part (water tax) to recover costs; (ii) specific quotas based on parameters similar to the ones used for the original design of the canal irrigation system to share water scarcity among users. And these quotas may be the basis for the definition of water rights; and, (iii) water markets for an optimal allocation of water, within the watercourse command area to increase seasonal canal water supply and/or reduce canal water supply variability (lease of water rights), and within the watercourse (decentralized) and between canal commands (centralized) for long term changes in water allocation (transfers of water rights).

Second, and although demand-based policies are seen as more suitable to address water scarcity and sustainability issues, those are to be accompanied by supply-based interventions and changes in the legal framework to achieve the appropriate match between infrastructure, irrigation system management and economic instruments. In the case of Pakistan, the existing infrastructure is clearly a major factor that limit the range (or increase the costs) of possible options for economic instruments. Problems of enforcement are also seen as very important, with conflicts between decreasing flexibility to reduce enforcement needs, and increasing flexibility for higher (long-term) efficiency but with potentially higher levels of interference by users in irrigation system management.'

Third, to improve the management of scarce water resources increases information needs. To compute water prices, to develop a quota system, or to define water rights and develop water markets requires information on past, actual and potential water supply and water demand. In the case of Pakistan, no information system is currently available to support irrigation sector policies, existing information being sparce and not combined.

Finally, with the importance of socio-political processes that influence irrigation system management, it is clear that issues related to enforcement and users' rights will be central to any new irrigation sector policy in Pakistan. Political willingness will be a prerequisite for successful implementation of new policies. In this context, efficient water user organisations may also play an important role as lobby groups to ensure that the rights of the organisations are respected, but also to enforce water rights within the organisations.

6   References

1.   World Bank. (1994) *Pakistan - Irrigation and Drainage Issues and Options*, The World Bank, Washington D.C.

2.   United Nations. (1980) *Efficiency and Distributional Equity in the Use and Treatment of Water: Guidelines for Pricing and Regulations*. Natural Resources/Water Series No.8, United Nations, New York.

3.   Rosegrant, M.W. and Binswanger, H.P. (1994) Markets in Tradable Water Rights: Potential for Efficiency Gains in Developing Countries Water Resource Allocation. *World Development*, Bol. 22, No. 1. pp. 1613-1625.

4.   Strosser, P. And Rieu, T. (1993) *A Research Methodology to Analyze the Impact of Water Markets on the Quality of Irrigation Services and Agricultural Production*, presented at Internal Program Review of the International Irrigation Management Institute, Colombo.

5.   Young, R.A. (1986) Why are There so Few Transactions among Water Users? *American Journal of Agricultural Economics*, Vol. 68, No. 5. pp. 1143-1151.

6.   Nazir, A. and Chaudhry, G.R. (1988) *Irrigated Agriculture of Pakistan*, Shahzad Nazir, Lahore.

7.   Kuper, M. and Strosser, P. (1996) *Current Water Management in Pakistan: Lessons from Current Practices for Proposed Policy and Management Changes*, draft report, International Irrigation Management Institute, Lahore.

8.   National Engineering Services Pakistan (1991). *Contribution of Private Tubewells in the Development of Water Potential*, Ministry of Planning and Development, Islamabad.

9.   Mollinga, P.P. and Van Straaten, C.J.M. (1996) The politics of Water Distribution - Negotiating Resource Use in a South Indian Canal Irrigation System, in *Water Policy: Allocation and Management in Practice* (ed. P. Howsam and R. Carter), E & FN Spon, London, pp. 243-250.

10.   Chaudhry, M.G., Majid, A. and Chaudhry, G.M. (1993) *The Policy of Irrigation Water Pricing in Pakistan: Aims, Assessment and Needed Redirections*, presented at the Ninth Annual General Meeting of the Pakistan Society of Development Economists, Islamabad.

SECOND GENERATION PROBLEMS OF PRIVATIZED IRRIGATION SYSTEMS
Second generation problems

M. SVENDSEN
Price Creek Consultants, Kings Valley, Oregon, USA

Abstract
This paper identifies second generation problems arising from the successful transfer of irrigation management responsibility from government agencies to user-based groups based on the experience of five developing countries. It discusses these problems from the perspective of irrigation associations, government irrigation agencies, farmers, and the government. Support services are singled out as key but neglected elements of a solution set for these problems.
Keywords: Financial autonomy, irrigation associations, irrigation districts, irrigation financing, irrigation management transfer, privatization, water rights

1   Introduction

In the last five years, there has been a major shift in thinking about the character of water. Where in the past, the social nature of water has been dominant, today it is the economic side of water which receives the greatest emphasis. To be sure, both aspects remain essential characteristics of the resource, but changing conditions (and changing ideologies of development) have today changed the focus of attention and debate.

Policy prescriptions arising from this new intellectual perspective on water have ranged from simplistic calls to make water and water rights entirely private goods, to more selective approaches aimed at particular aspects and uses of water and at particular functions of water control. Such approaches often focus on "getting the incentives right", by using the power of markets or market-like institutions and incentives to improve particular aspects of water allocation and use. We can identify three basic problems which can serve as the object of such efforts.

1. Achieving economically sound decisions on the initial allocation and exploitation of water resources

Water: Economics, Management and Demand. Edited by Melvyn Kay, Tom Franks and Laurence Smith. Published in 1997 by E & FN Spon. ISBN 0 419 21840 8.

2. Achieving efficient and sustainable use of water within existing patterns of allocation
3. Reallocating already exploited water resources to further national objectives (increased economic efficiency, species preservation, etc.)

The focus in this paper is on the second of these problems. One of the most widely advocated measures to achieve greater efficiency in the irrigation sector is to "privatize" or reduce the degree of government involvement in the provision of irrigation services to farmers. This usually involves increasing the level of beneficiary involvement in managing service provision, a process commonly termed Irrigation Management Transfer (IMT). Recent transfer programs have created locally-controlled Irrigation Districts in Mexico and Turkey, and longer-standing programs have created such institutions in other developing countries. This paper grows out of a workshop sponsored by the Economic Development Institute of the World Bank (EDI) and the International Irrigation Management Institute (IIMI) which looked at such programs. It was held in Cali, Colombia in February 1997 and attended by about 35 practitioners and other professionals experienced with these processes. The workshop aimed to identify "second-generation" problems which were likely to affect irrigation management institutions following successful transfer to locally-based management. The purpose of the paper is to draw lessons from five of the case studies presented at the workshop (Mexico [1], Turkey [2], Colombia [3], Argentina [4], and the Philippines [5]) and from ideas expressed during workshop discussions.

The term "second generation" requires some explanation. Transferring substantial management authority to a locally-based organisation is a complicated undertaking and, in addition to solving problems, will almost certainly create new problems which did not exist before or were not previously evident. An example might be inadequate technical capability of new irrigation field personnel. In addition, there may situations, such as low agricultural productivity, which were present prior to the transfer, but which were not acute problem when irrigation fees were low or non-existent. Problems of both types are termed second generation problems. Some second generation problems may be a result of faulty processes used to introduce the new management system. Some may be a result of conscious choices made during implementation to defer consideration of certain potential problems in the interests of advancing program coverage. Others may be practically unavoidable.

Problems are typically problems for some and not for all. Traffic congestion in an area may be an acute problem for those who live in or pass through that area, but may not inconvenience at all people living on the opposite side of town. Other situations may represent two-sided coins, where a solution for some represents a problem for others. Routing traffic off of clogged major highways and through residential neighborhoods to ease congestion is an example. The discussion below addresses problems from the perspectives of (a) the irrigation association, (b) the irrigation agency, (c) farmers, and (d) the government. Note that a change such an increase in irrigation service fees may be a problem for water users but a benefit for the irrigation association. Solving such problems can involve a variety of measures including revision of laws or implementing regulations, changes in organisational structure, organisational rules and processes, and revised funding mechanisms.

In order to solve second generation problems and for sustainable future operations, associations will generally require supporting services. Support services are services which come from outside the institution itself but which are necessary for it to carry out its mission. They include such things as financing for equipment purchases, legal advice, computer programming assistance, and financial auditing services. Such services may be either impossible for an institution to generate for itself, such as heavy equipment financing or external audits, or be used only infrequently and hence be too expensive to maintain on a full-time basis.

In the past, it has usually been assumed that any such services must come from government agencies. Today it is recognized that higher quality and less expensive services may be obtained from other sources and that the government should generally serve as a "provider of last resort." This represents another important way in which the water resource sector can be "privatized."

2   Irrigation associations

2.1   Insecurity of water rights

Water rights which are absent, poorly-defined, or insecure, can (a) inhibit investment in new system facilities or rehabilitation, (b) encourage short-term thinking and behavior on the part of association managers and farmers, (c) result in heavy expenditures on legal costs to defend a poorly-defined water right, and, ultimately, (d) a reduction in water supply and system collapse.

An effective water right should provide security, such that the association can depend on it over time, but must also be adaptable so that water can be diverted to other more productive or higher priority uses as economic and demographic conditions change. In this event, there must be provision for appropriate compensation to those who are giving up the water right by those who gain it. In Mexico, for example, the right of an association to water for irrigation use is always subservient to present **and future** municipal demands. This creates considerable insecurity for associations which share water sources with growing municipalities and violates both the principle of security and of just compensation.

An effective water right must also be specified in both quantitative and qualitative terms. Water quality degradation by upstream effluent discharges, as from a factory or an inadequate municipal sewage treatment plant, can render the water unusable by downstream agricultural right holders. It can also make the water suitable only for grain or fodder crops, since biologically or chemically contaminated water may not be useable for production of higher value fruits and vegetables. This will become an increasingly serious problem as water reuse increases in response to growing demand from all sectors.

Establishing a water rights system where it is lacking, as in Turkey, or clarifying water rights where they are weak, ineffective, or inequitable, as in Mexico, will usually require action from the national legislative body or from top level executive leadership, or both. It is thus extremely important for associations to have adequate representation of their interests when these issues are taken up.

Two different types of support services are identified as being crucial for associations in attempting to establish or firm up their water rights. The first of these is legal advice and representation when the association faces challenges to its rights. Such representation is best secured from private law firms, if available. Such services secured from government sources may be of lower quality, and may be subject to pressures which would compromise their objectivity. Such representation is also important during the formative stage of an association, at the stage when negotiations with the government irrigation agency establish the contract or concession which will control the relationship of the association and the government. Unfortunately, associations which are just forming may be unaware of the importance of high quality legal advice at this stage or may be unable to afford it. A national federation of associations could be valuable as a source of legal advice and assistance to newly forming associations.

The other type of support service required by associations is lobbying on their behalf in government policy making councils. Since other interests, such as municipalities and industrial water users, are usually larger than individual associations and likely to be more powerful politically, it is important for associations to establish regional or national federations representing many associations and a large number of farmers. This will give them political influence with which to counter the power of competing interests.

Another potential role for a federation would be to represent irrigation associations on the board of directors of the national irrigation agency. In Colombia this is currently the case.

2.2   Financial shortfalls

A central feature of the IMT programs undertaken in all of the primary case study countries is financial autonomy. Financial autonomy is the condition where an organisation generates all of the revenue it needs to support itself and to perform it's primary functions. It implies that the association is not directly subsidized by the government, or that if it is subsidized, that the subsidy is a fixed amount which does not vary according to the condition of the association's balance sheet. The principal source of revenue for most associations is irrigation service fee (ISF) collections. Financial shortfalls affect a number of associations in Colombia and are one of the major factors in limiting the continuing spread of the transfer program in Mexico.

Financial shortfalls are a function of several factors, including ISF rates, ISF collection percentages, the contribution of other sources of revenue, and expenditure patterns. One important parameter is the structure of the ISF. Fees can be levied on flat rate or volumetric bases, each with advantages and disadvantages. A recommended structure for fees is a two-part one comprising both fixed and volumetric components. The flat portion would constitute a "connection charge", a charge for simply being within the boundaries of the system's service area whether or not water was actually taken from the system. This would reimburse the association for expenses incurred in maintaining the physical and administrative capacity to deliver water to the farm. The absence of this component in the fee structure of

Mexican associations has created severe problems during years when drought greatly reduced the available water supply to the system's water users. The other portion of the fee would be based on the volume of water actually delivered during a cropping season, or some proxy for this amount, such as area irrigated and number of irrigations given. This would cover the costs incurred by the association which are related to the amount of water given, and would serve to limit excessive demand for water.

Revenue from ISF is also dependent on the percentage of the fees assessed which are actually collected, though associations in most countries do a reasonably good job in this regard. The Philippines is an exception to this rule, and this low collection percentage has been a persistent problem for NIA, the national irrigation agency.

Solving problems of revenue shortfalls that relate to fee levels and collection efficiency is largely an internal association responsibility. Outside assistance may be useful, in some cases, in estimating farmers ability to pay particular ISF levels, and in analyzing management systems set up for collecting revenue. This is discussed further in a following section.

Underlying difficulties in generating sufficient ISF revenue to sustain system operations, in many cases, is the low productivity of irrigated agriculture in system command areas. Low productivity can result from many factors, but is often associated with small farm size, a subsistence orientation, production of low value crops such as grains, inappropriate agricultural policies, a poor natural resource base, and inadequate agricultural support services. In such cases, a solution to the association's financial problems may only be possible if the underlying problems in the agricultural sector are addressed. If these problems cannot be solved, then the options are for the government to (a) take back responsibility for system management and financing, (b) provide the association with an explicit subsidy, or (c) to close the system. Irrigation service fees typically constitute only 3 to 10 percent of total production costs, so reducing them will generally not solve an underlying problem of high production costs and low productivity.

2.3   Rehabilitation

All irrigation systems require periodic rehabilitation and modernization. While usually less expensive, in real terms, than the original construction, rehabilitation is a costly undertaking, and is usually beyond the financial and technical means of an association to undertake. A number of problems are associated with system rehabilitation. One is the usual absence of a clear and consistent government policy on responsibility for rehabilitation[1]. In the absence of such a policy, the tendency is for associations to defer needed maintenance in the hope that the government will step in and take responsibility for the rehabilitation, and to under invest in system improvements between rehabilitations. This tendency is reinforced by the fact that, in all five primary cases, the government has retained ownership of system physical facilities, while transferring to associations the "use rights" of the facilities. Associations may thus regard the responsibility to rehabilitate those facilities as belonging to the government unless the actual policy is clearly spelled out in the agreement between the government and the association.

Another problem is the often unclear nature of the cost sharing formula which is to be applied to rehabilitation. Because irrigated agriculture benefits people beyond the ranks of system irrigators, and because full coverage of rehabilitation costs is usually beyond the means of the irrigators themselves, a sharing of costs is appropriate. Assuming responsibility for even a share of the costs involved will tend to counteract the tendency of an association to defer maintenance, as noted above. A cost sharing formula should thus also be a part of the agreement between government and associations.

To cover its share of future rehabilitation costs, associations will usually need to accumulate a capital replacement fund over a number of years. There must thus be a legal basis for establishing such a fund and retaining funds in it from year to year. At the same time there should be incentives for establishing and contributing to such a fund. Unfortunately fiscal and monetary policies in many countries, such as Turkey and Mexico, have led to high rates of inflation and low or negative real interest rates on savings, which acts as a powerful deterrent to fund accumulation. Governments must either adjust these policies or create special investment opportunities for associations which allow them to earn reasonable rates of return on accumulated funds.

Likewise, there should be incentives for making selective improvements in physical infrastructure between rehabilitations. One way to do this is to establish a trust fund, perhaps with donor financing, from which associations could request funds to complement their own investment funds. The matching ratio for such a funding facility should be established and made known in advance.

A number of support services are required to support system rehabilitation. One is a regular assessments of the condition of system facilities. This can be done jointly by the association and the government agency, as in Turkey. It can also be contracted out to an engineering consulting firm acceptable to both the association and the agency. Such an assessment can be used as a basis for annual maintenance planning, to suggest the need for selective improvements in system facilities, and for planning whole-system rehabilitation.

If an association is not able to accumulate its share of rehabilitation cost prior to rehabilitation, it will need a source of credit. Credit can come from private banks, government banks or other lending facilities, or from insurance pools in the case of rehabilitation induced by natural disasters such as floods, typhoons, or volcanic eruptions. Such a credit facility could also be used as a source of financing for capital equipment needed for system maintenance.

Rehabilitation will generally require external technical services for design and construction. Because of the sharing of costs, both the association and the government should be involved in decisionmaking relating to the selection of consultants and contractors and monitoring their performance. Advice and guidance to the association on handing these tasks might usefully be given by a federation of associations, since rehabilitation occurs only infrequently in any one association.

2.4   Lack of financial and administrative management expertise
There are several possible responses to this problem. One would be skill enhancement through staff training programs. Skills can also be enhanced by

replacing less skilled people with more capable ones. Contracting out for specialized services is another important way of addressing management deficiencies in associations.

One extremely important step in improving the quality of association management is to increase the transparency of management processes. This has a number of positive effects. It can (a) reduce the potential for misappropriation of funds, (b) help insure that salary levels and benefits are realistic, (c) help insure that maintenance allocations are appropriately targeted, (d) reduce favoritism in making personnel appointments, and (e) improve responsiveness of association staff to users.

A number of steps can be taken to increase transparency in association management. These include:

- regular external audits of financial accounts
- use of standardized budgeting and accounting frameworks
- wide dissemination of simplified budgets, plans, and financial statements to association members and clients
- active involvement of the board of directors in forward planning, budgeting, and auditing
- broad representation from among users on the association board of directors

There is broad scope for employing external support services in this problem area. Services that may be required include:

- advice on establishing and revising management systems and procedures
- advice on establishing financial budgeting and accounting systems, including software
- establishment of standard budgeting and accounting formats
- mandatory standards and requirements for regular external audits
- regular external audits
- management training

These services can be obtained by the association from a variety of sources, including private firms, a national or regional federation of associations, NGOs, government agencies, and universities and training institutes. Some services must generally come from a particular type of source, while others could be obtained from alternative sources. For example, the government is the logical party to establish mandatory standards and requirements for external audits. The audits themselves, however, could come from a government agency or from a private firm of chartered accountants. The later is probably preferable, in many cases, in terms of speed, quality, and objectivity. Other services, such as management and accounting system advice could come from a variety of sources, with private sources being generally preferred.

One argument in favor of the provision of these services by government agencies will often be that they can be obtained at no or low cost. What makes this lower cost provision possible, however, will generally be implicit government subsidies to the service providers. A preferred alternative would be to provide the funds supporting these subsidies instead to the associations as grants or cost-sharing subsidies to be used for obtaining management support services. This would allow the associations to shop for these services among alternative providers and select the sources which were most efficient and best met their needs. The demand-driven competition thus induced would be a very healthy force acting to hold down service prices and improve quality of services delivered. Provision of such grants during the transition phase from government agency to association management could be a very useful institution-strengthening activity.

3   Irrigation agencies

3.1   Dislocation of staff

This is the most prominent problem experienced by agencies following IMTand is typically dealt with is several ways. First, O&M staff levels are reduced by attrition — vacant positions are left vacant or filled by internal transfers rather than new hiring. Second, financial incentives are often provided for early retirement of older staff. Third, existing staff are transferred to other positions which become vacant rather than filling them from outside the agency. In some countries such as China, where it is difficult to lay off staff, sideline enterprises are created which can generate income for the irrigation district and cover, at least the salary costs of the involved personnel. In some cases, redundant agency staff may be reemployed by the associations which take up management responsibility for the schemes. Such employment must be at the complete discretion of the association, however.

3.2   Loss of technical capacity

This is a result of the downsizing occasioned by the transfer of operational responsibilities. To address this problem, agencies can
• obtain specialized expertise from outside consulting firms as needed
• improve the salary schedule to attract and retain high-quality staff
• provide in-service training opportunities for staff
• revise job descriptions to bring in new staff with the desired qualifications
Support services helping to address this problem include
• technical consultancies to provide specialized skills
• contracts for in-service training of personnel

3.3   Defining a new role for the agency

Operational responsibilities will often have been a major part of an agency's mandate. With these functions transferred to associations, scope exists for designing a new role to address emerging problems. Doing this requires that discussion takes place both among staff within the agency and at higher levels of the government, with broad participation by all involved parties. The aim should be to build broad consensus and political commitment for the new role. In some cases, changes in legislation may be required to enable the assumption of new responsibilities.

The new mandate should contain a clear definition of roles and responsibilities and should define skill requirements to carry out the new responsibilities. It should also contain a timetable for accomplishing the shift to the new mandate.

Support services which could be useful in this process include

- comprehensive diagnosis of the agency/association relationship and the association support needs
- professional assistance with the agency's strategic planning process
- consulting services to design new management information systems for the agency
- research to solve problems experienced by associations and assess possible solutions

In Colombia, the national agency, INAT, is using professional consultants to help them define a new role for themselves under an Inter-American Development Bank credit.

4   Farmers

Second generation problems experience by individual farmers relate mainly to the need to increase farm productivity to pay higher irrigation fees and to take advantage of possible improvements in irrigation service quality. Support services needed in this regard comprise the traditional roster of agricultural support needs, but with emphasis on a possible shift to higher value crops.

Although government agencies are the traditional source of many of the needed services, in many countries, private or other organisations can play an expanding role in supplying some or all of the services listed above. There is also the question of the potential role of the association itself in providing other agricultural services, in addition to irrigation service. In general, the association should have demonstrated competence in its core activity before considering such ancillary activities.

5   Government

The principal second-generation problem for government, beyond those already identified for the irrigation agency, is the reduced control which it will have over irrigation activities, and a diminished ability to use control of irrigation as a tool to implement other national policies and priorities. An example might be the government's wish to promote cultivation of upland crops rather than rice during a particular season. In the past it might have worked through the national irrigation agency to adjust water delivery schedules and volumes to try to achieve this end. Following transfer, this would be more difficult. There are other tools, such as support prices and subsidies, to achieve the same ends, however, and this should not pose a significant problem for agricultural policymakers.

6   Summary and conclusions

Experience is now available from a number of countries which have recently implemented programs aimed at privatizing irrigation management functions. Accompanying initial successes in devolving management responsibility are a number of emergent problems which must be addressed for the new managing entities to achieve long-term operational viability.

The nature of second generation problems experienced differs according to the perspective of the involved party. For irrigation associations, important second-generation problems include insecure water rights, inadequate revenues, uncertainty about responsibility for future rehabilitation, and weak financial and administrative skills. For irrigation agencies, important second-generation problems include the difficulty of reducing staff numbers, the erosion of agency technical capacity, and the need to develop a new role for the agency. Second generation problems for farmers center around the need to make agriculture more productive in order to pay higher irrigation fees and possibly rehabilitation costs. For government, the major problem is reduced direct control over irrigated agriculture.

Solutions to these problems involve a mix of measures — policy changes through amended laws and implementing regulations, changes in organisational structure and operating procedures for associations and agencies, changes in funding mechanisms, and improved support services. The importance of an adequate set of supporting services, particularly for irrigation associations, is often underappreciated. Important supporting services for associations include

- legal advice and representation
- collective political representation
- maintenance assessments
- credit
- technical training
- design and construction services
- management consultants
- financial consultants
- audits
- management training

These supporting services can be provided by either public or private sectors. The presumption should be with the private sector unless a strong comparative advantage can be shown for government. Any subsidies to be provided for supporting services should be given through the managing associations rather than to the service providers.

7   References

1. Palacios, E. Benefits and Second Generation Problems, the Case of Mexico. Paper presented to the International Workshop on Participatory Irrigation Management: Benefits and Second Generation Problems, Cali, Colombia, 9-16 February.

2. Svendsen, M. and Nott, G. (1997) Irrigation Management Transfer in Turkey: Process and Outcomes. Paper presented to the International Workshop on Participatory Irrigation Management: Benefits and Second Generation Problems, Cali, Colombia, 9-16 February.

3. Quintero-Pinto, L. E. (1997) Participatory management of irrigation: benefits and second generation problems, the case of Colombia. Paper presented to the

International Workshop on Participatory Irrigation Management: Benefits and Second Generation Problems, Cali, Colombia, 9-16 February.

4.   Chambouleyron, J., Herrera, J. E., and Mathus, M. (1997) Second Generation Benefits and Problems of Participatory Irrigation Management: Argentina Case Study.   Paper presented to the International Workshop on Participatory Irrigation Management: Benefits and Second Generation Problems, Cali, Colombia, 9-16 February.

5.   Raby, N. (1997) Participatory Irrigation Management in the Philippines: The Learning Process Approach in the National Irrigation Systems.

1. This policy may be left deliberately vague during a transfer program, while implying that support would be forthcoming, to increase farmer acceptance of the transfer.

SUBJECT INDEX